Trees, Forested Landscapes and Grazing Animals

In this comprehensive book, the critical components of the European landscape – forest, parkland, and other grazed landscapes with trees are addressed. The book considers the history of grazed treed landscapes, of large grazing herbivores in Europe, and the implications of the past in shaping our environment today and in the future. Debates on the types of anciently grazed landscapes in Europe, and what they tell us about past and present ecology, have been especially topical and controversial recently. This treatment brings the current discussions and the latest research to a much wider audience.

The book breaks new ground in broadening the scope of wood-pasture and woodland research to address sites and ecologies that have previously been overlooked but that hold potential keys to understanding landscape dynamics. Eminent contributors, including Oliver Rackham and Frans Vera, present a text that addresses the importance of history in understanding the past landscape, and the relevance of historical ecology and landscape studies in providing a future vision.

Ian D. Rotherham is Professor of Environmental Geography, Reader in Tourism and Environmental Change, and International Research Coordinator Professor in the Faculty of Development and Society, Sheffield Hallam University, UK.

Trees, Forested Landscapes and Grazing Animals
A European Perspective on Woodlands and Grazed Treescapes

Edited by Ian D. Rotherham

LONDON AND NEW YORK

This first edition published 2013
by Routledge
2 Park Square, Milton Park, Abingdon, Oxon, OX14 4RN

Simultaneously published in the USA and Canada
by Routledge
711 Third Avenue, New York, NY 10017

First issued in paperback 2017

Routledge is an imprint of the Taylor & Francis Group, an informa business

© 2013 Selection and editorial material, Ian D. Rotherham; individual chapters, the contributors

The right of Ian D. Rotherham to be identified as author of this part of the work has been asserted by him in accordance with sections 77 and 78 of the Copyright, Designs and Patents Act 1988.

All rights reserved. No part of this book may be reprinted or reproduced or utilised in any form or by any electronic, mechanical, or other means, now known or hereafter invented, including photocopying and recording, or in any information storage or retrieval system, without permission in writing from the publishers.

Trademark notice: Product or corporate names may be trademarks or registered trademarks, and are used only for identification and explanation without intent to infringe.

British Library Cataloguing in Publication Data
A catalogue record for this book is available from the British Library

Library of Congress Cataloging-in-Publication Data
Trees, forested landscapes, and grazing animals : a European perspective on woodlands and grazed treescapes / edited by Ian D. Rotherham. – First edition.
pages cm
Includes bibliographical references and index.
ISBN 978-0-415-62611-8 (hbk) – ISBN 978-0-203-10290-9 (ebk)
1. Grazing–Environmental aspects–Europe–History. 2. Browsing (Animal behavior)–Environmental aspects–Europe–History.
3. Forests and forestry–Europe–History. 4. Landscapes–Europe–History.
I. Rotherham, Ian D.
SD427.G8T74 2013
636.08'45–dc23
2012035583

ISBN 13: 978-1-138-30448-2 (pbk)
ISBN 13: 978-0-415-62611-8 (hbk)

Typeset in Baskerville by
Fish Books Ltd., Enfield

Contents

Notes on contributors ix
List of illustrations xi

PART I
Grazed treed landscapes   1

1 Grazed treed landscapes: Overview and introduction   2
    IAN D. ROTHERHAM

2 Woodland and wood-pasture   11
    OLIVER RACKHAM

PART II
The lessons of history   23

3 Woods, trees and animals:
    A perspective from South Yorkshire, England   24
    MELVYN JONES

4 Re-wilding the landscape: Some observations on
    landscape history   35
    DELLA HOOKE

5 Rethinking pannage: Historical interactions between
    oak and swine   51
    PÉTER SZABÓ

6 The post-glacial history of grazing animals in Europe   62
    DEREK YALDEN

PART III
Landscape dynamics   71

7   Reinterpreting wooded landscapes, shadow woods and the impacts of grazing   72
IAN D. ROTHERHAM

8   The dynamics of pre-Neolithic European landscapes and their relevance to modern conservation   87
KEITH J. KIRBY AND AMBROISE BAKER

9   Can't see the trees for the forest   99
FRANS VERA

10   Ancient trees and wood-pastures: Observations on recent progress   127
TED GREEN

11   Grazed wood-pasture versus browsed high forests: Impact of ungulates on forest landscapes from the perspective of the Białowieża Primeval Forest   143
TOMASZ SAMOJLIK AND DRIES KUIJPER

12   The influence of grazing animals on tree regeneration and woodland dynamics in the New Forest, England   163
ADRIAN C. NEWTON, ELENA CANTARELLO, ALEXANDER LOVEGROVE, DINA APPIAH AND LORRETTA PERRELLA

13   Forest and land management options to prevent unwanted forest fires   180
CAROLINE BOSTRÖM, ANA SEBASTIÁN, CARMEN HERNANDO LARA, ROSA PLANELLES, ARMANDO BUFFONI, ROSARIO ALVES, MARIELLE JAPPIOT AND JESÚS SAN MIGUEL AYANZ

PART IV
Case studies   189

14   Grazing Refuge Habitats and their importance for woody plants in the west of Scotland   190
RICHARD GULLIVER

15 Legacies of livestock grazing in the forest structure
of Valonia oak landscapes in the Eastern Mediterranean 208
TOBIAS PLIENINGER, HARALD SCHAICH AND THANASIS KIZOS

16 Palaeoecological records of woodland history during recent
centuries of grazing and management examples from
Glen Affric, Scotland and Ribblesdale, North Yorkshire 223
HELEN SHAW AND IAN WHYTE

17 Integrated conservation of a park and its associated cattle herd:
Chillingham Park, northern England 242
STEPHEN J.G. HALL

PART V
Conservation, management and wildscapes **255**

18 The impacts of the reintroduction of wild boar in the
Forest of Dean, Great Britain 256
MARTIN GOULDING

19 Wild cattle and the 'wilder valley' experiences: The introduction
of extensive grazing with Galloway cattle in the Ennerdale Valley,
England 269
GARETH BROWNING AND JOHN GORST

20 Treescape: Trees, animals, landscape, people and 'treetime' 282
LUKE STEER

21 Creation of open woodlands through pasture: Genesis, relevance
as biotopes, value in the landscape and in nature conservation
in south-west Germany 301
MATTIAS RUPP

22 Woodland grazing with cattle: Results from 25 years of grazing
in acidophilus pedunculate oak (*Quercus robur*) woodland 317
RITA M. BUTTENSCHØN AND JON BUTTENSCHØN

23 Ancient trees, grazing landscapes and the conservation of
deadwood and wood decay invertebrates 330
KEITH ALEXANDER

24	The future potential of wood-pastures	339
	IRIS GLIMMERVEEN	
25	A strategic view of the issues for wood-pasture and parkland conservation in England	356
	SUZANNE PERRY	

PART VI
Summary and conclusions — **377**

26	Re-wilding trees for ancients of the future	378
	JILL BUTLER AND KEITH ALEXANDER	
27	Summary and conclusions	389
	IAN D. ROTHERHAM	
	Index	*407*

Notes on Contributors

Keith Alexander: Ancient Tree Forum, Heavitree, Exeter, UK
Rosario Alves: Associação Florestal de Portugal – FORESTIS, Porto, Portugal
Dina Appiah: Bournemouth University, School of Applied Sciences, Poole, UK
Ambroise Baker: Long-term Ecology Laboratory, Biodiversity Institute, Oxford Martin School, Department of Zoology, University of Oxford, Oxford, UK
Caroline Boström: Confédération Européenne des Propriétaires Forestiers (CEPF), Bruxelles, Belgium
Gareth Browning: Forestry Commission, Cockermouth, Cumbria, UK
Armando Buffoni: AMBIENTEITALIA S.R.L., Milan, Italy
Jill Butler: Woodland Trust, Autumn Park, Dysart Road, Grantham, Lincs., NG31 6LL
Jon Buttenschøn: Danish Veterinary and Food Administration, Copenhagen, Denmark
Rita Buttenschøn: Copenhagen University, Forest & Landscape, Frederiksberg, Denmark
Elena Cantarello: Bournemouth University, School of Applied Sciences, Poole, UK
Iris Glimmerveen: Woodland Inspirations Ltd, Nunwick Old Hall, Great Salkeld, Penrith, Cumbria
John Gorst: United Utilities/Wild Ennerdale, Keswick, Cumbria, UK
Martin Goulding: British Wild Boar Association, UK
Ted Green MBE: Ancient Tree Forum, Maidenhead, UK
Richard Gulliver: School of Geographical and Earth Sciences, University of Glasgow, Port Ellen, Isle of Islay, UK
Stephen J.G. Hall: University of Lincolnshire, School of Life Sciences, Faculty of Science, Lincoln, UK
Carmen Hernando Lara: Instituto Nacional de Investigación y Tecnología Agraria y Alimentaria, Grupo de Incendios Forestales (INIA-CIFOR), Madrid, Spain
Della Hooke: University of Birmingham, Birmingham, UK

Marielle Jappiot: Centre National du Machinisme Agricole du Génie Rural des Eaux et des Forêts (CEMAGREF), Le Tholonet, Aix en Provence, France

Melvyn Jones: Sheffield Hallam University, Sheffield, UK

Keith Kirby: Department of Plant Sciences, University of Oxford, UK

Thanasis Kizos Department of Geography, University of the Aegean, Mytilini, Lesvos, Greece

Dries Kuijper: Mammal Research Institute, Polish Academy of Sciences, Białowieża, Poland

Alexander Lovegrove: Bournemouth University, School of Applied Sciences, Poole, UK

Adrian C. Newton: Bournemouth University, School of Applied Sciences, Poole, UK

Lorretta Perrella: Bournemouth University, School of Applied Sciences, Poole, UK

Suzanne Perry: Natural England, Peterborough, UK

Rosa Planelles: Entrenamiento e Información Forestal (EIMFOR), Pozuelo de Alarcón, Madrid, Spain

Tobias Plieninger: Berlin-Brandenburg Academy of Sciences & Humanities, Berlin, Germany

Oliver Rackham: Corpus Christi College, University of Cambridge, Cambridge, UK

Ian D. Rotherham: Sheffield Hallam University, Sheffield, UK

Mattias Rupp: Albert-Ludwigs-University Freiburg, Institute for Landscape Management, Germany

Tomasz Samojlik: Mammal Research Institute, Polish Academy of Sciences, Poland

Jesús San Miguel Ayanz: EC-DG, Joint Research Centre, Institute for Environment and Sustainability, Land Management & Natural Hazards Unit – FOREST (TP 261), Italy

Harald Schaich: Albert-Ludwigs-University Freiburg, Institute for Landscape Management, Germany

Ana Sebastián: GMV Aerospace and Defence S.A.U., Madrid, Spain

Helen Shaw: Lancaster University, Lancaster, UK

Luke Steer: Treescapes Consultancy, Ambleside, Cumbria, UK

Peter Szabo: Institute of Botany, Academy of Sciences Czech Republic, Department of Vegetation Ecology, Brno, Hungary

Frans Vera: Ministry of Agriculture, Nature Management and Fisheries, The Hague, Netherlands

Ian Whyte: Lancaster University, Department of Geography, Lancaster, UK

Derek Yalden: The University of Manchester, Faculty of Life Sciences, Manchester, UK

Illustrations

Figures

1.1	Cattle grazing and browsing parkland	4
1.2	Konik ponies grazing in Poland	6
2.1	A coppice-wood in the second year after felling. The underwood stools are mainly ash. The timber trees are oaks. This dense scatter of small oaks would have been characteristic of a medieval wood. Medieval timber-framed buildings are typically made of hundreds of oak trunks small enough for two men to lift. Such a wood could not operate in the presence of grazing animals, especially since ash is one of the most palatable of trees. Bradfield Woods, Suffolk, August 1986	11
2.2	Uncompartmented wood-pasture. A savanna-like landscape with scattered oaks in grassland (formerly heath, now invaded by bracken). The trees are ancient pollards, at least 450 years old, last cut c.1800. Staverton Park, Suffolk, March 2008	12
2.3	Wood-spurge (*Euphorbia amygdaloides*), one of many plants of ancient woodland that appear from buried seed after each time the wood is felled. Bradfield Woods, Suffolk, June 1987	13
2.4	Boundary of the wood of the great Abbey of Bec, Normandy. On top of the boundary bank is built a massive flint wall. The effort put into defending wood boundaries is a measure of the importance attached to woodland conservation in the middle ages. Le Bec-Hellouin, July 1976	14
2.5	Compartmented wood-pasture. Each coppice was supposed to be cut once in 18 years, then fenced for the first 9 years of regrowth. The plains (blank areas) were always open to livestock and contain pollard trees. Hatfield Forest, Essex	15
2.6	A Vera-escue landscape? The grassy plain is being encroached on by thorn thickets, which in course of time will develop into established woodland like that in the background. Central plain, Hatfield Forest, February 1981	17

2.7	Savanna in a Mediterranean-type climate, analogous to the *montado* of Portugal and the *dehesa* of Spain. Near Ukiah, California, April 2005	18
2.8	Drought-determined savanna. The roots of the live-oak (*Quercus grisea*) extend out at least as far as where the man is standing. Davis Mountains, west Texas, March 1999	18
2.9	Ashwood, severely browsed by three species of deer. Note absence of low cover and absence of herbaceous plants, except the distasteful dog's-mercury. Hempstead Wood, Essex, April 2002	21
3.1	Woodland in South Yorkshire at Domesday	25
3.2	Royal licence to create a deer park at Sandbeck Park, 1637	28
3.3	Tankersley Park c.1730	29
3.4	Sheffield Park, 1637, based on the description in Harrison's survey of the manor of Sheffield	30
4.1	Conifer plantations in Cwmsylfaen near Bontddu, Gwynedd	41
4.2	A wood-pasture scene in Sutton Park near Birmingham	42
4.3	Old oak pollards in Moccas Park, Herefordshire	45
4.4	Pigs being pastured in Wimperhill Wood, Forest of Wyre	47
5.1	The relative importance of pannage in the Carpathian Basin in the fifteenth century expressed as the percentage of settlements with acorn-bearing woods among all settlements in each county	53
5.2	The relative importance of pannage in the Carpathian Basin at the beginning of the eighteenth century expressed as the percentage of settlements with documented pannage among all settlements in each county	54
5.3	Average mast frequencies in years in the villages Co. Nógrád (Hungary) in 1715 superimposed on an altitudinal map	56
5.4	The usage rights of the town of Hodonín in Dúbrava wood at the end of the seventeenth century as described in the 1691 urbarium	58
7.1	Upland coppice alder shadow wood, Peak District, England	75
7.2	Fallow deer	77
7.3	Veteran hawthorn in grazed upland shadow wood, Peak District, England	79
7.4	Cattle browsing parkland	81
9.1	A pre-industrial agricultural (so-called semi- or half-natural) landscape that is commonly used as a baseline for nature conservation and nature management	101
9.2	National Park Białowieża in eastern Poland, considered as the last remnant primeval lowland forest that would once have covered the lowland of Europe before mankind brought it under cultivation by agriculture	102

9.3	The wood pasture the Borkener Paradise in Germany; in the foreground an oak tree grows in the midst of some hawthorns; in the background is a grove surrounded by mantle and fringe vegetation of blackthorn	104
9.4	Dyrhave, adjacent to Copenhagen in Denmark – nineteenth-century image of how the grazing of livestock such as cattle destroyed the primeval forest, but no regeneration took place in the forest and just some senile old oaks are the last witnesses of the once present forest	105
9.5	An oak that grew in openness in the former wood-pasture Sababurg in Germany – after the grazing of livestock was terminated, the park-like wood-pasture changed into a closed-canopy forest dominated by shade-tolerant species beech; light-demanding oak died because of the shade from the trees. The current forest is indicated as an 'Urwald', i.e. it is considered as a modern analogue of the primeval vegetation	106
9.6	The percentage distribution per species of tree in diameter categories per species of pedunculate oak, beech, ash and wych elm in the National Park Dalby Söderkog, a former wood-pasture in southern Sweden	107
9.7	A young oak nursed by blackthorn in the mantle and fringe vegetation of blackthorn that surrounds a grove in the wood-pasture the Borkener Paradise in Germany	109
9.8	The interior of a grove in the wood-pasture Junner Koeland, the Netherlands – the interior lacks a shrub layer as well as regeneration of trees	114
9.9	A grove that changes from the centre onwards into grassland, because of the grazing of cattle, horses and deer in Denny Wood in the New Forest, England.	115
9.10	An oak with its crown still intact next to an oak where the crown died off in Calk Abby Park in England and which formed a second crown low at the trunk	117
10.1	Deadwood beetles	130
10.2	Parkland oaks	132
10.3	Re-grown pollard	133
10.4	Fungi	137
10.5	Ancient park oak	138
11.1	Tree regeneration in grazed woodlands in Borkener Paradies, Germany, according to the grazed woodland hypothesis: thorny bushes establish in grazed vegetation (1) and offer protection for palatable species (e.g. oak) to become established (2); grazing and the lack of light lead to solitary oaks in the landscape (3); after collapse of old oaks, the process starts again with short grazed vegetation (4)	146

11.2	Natural and anthropogenic hedges: ungulate browsing inside forest gaps in the Białowieża Primeval Forest leads to the development of hedge-like structures mainly of hornbeam as in the Strict Reserve (1); this process resembles the constant cutting of hornbeam into a hedge as in this 25 year-old cut hedge at a farm in Varsen, Netherlands (2), demonstrating hornbeam's high tolerance to browsing	149
11.3	Tree regeneration in browsed high forest of the Strict Reserve of BPF: collapse of old trees creates gaps in the canopy (1) and offers regeneration gaps for trees; intensive visitation by ungulates to forest gaps creates shortly browsed trees (2); only browsing-tolerant tree species (hornbeam) can sustain chronic browsing and will eventually escape out of reach of the ungulates (3); gaps are closed within several decades and turned into closed forest (4), which only open when a tree falls down	151
11.4	Watercolour by Jan Henryk Müntz Białowieża Forest – the Bear Hunt (1784); note the dense forest just outside the border of the anthropogenic glade here	156
11.5	Oaks in grazed woodland and browsed high forest: (1) example of a woodland grazed by livestock (horses and cattle), the Borkener Paradies in Germany, with typical large oak trees with low branches and wide crowns; these landscapes often have high aesthetic value as well as harbouring high biodiversity; (2) example of a browsed high forest with wild ungulates only (red deer, roe deer, wild boar, moose and European bison) in BPF in Poland; typical are the tall oak trees with long stem, high branches and narrow crowns – these landscapes have their own beauty and offer hotspots of biodiversity	157
14.1	Four small aspen trees (*Populus tremula*) on a linear crag near the east coast of Colonsay	195
15.1	Silvopastoral oak woodland on formerly arable terraces in Filia municipality, Greece	209
15.2	Map of Greece and location of the plots in the study area	211
15.3	Number of sheep and number of sheep farms on the Aegean Islands and Lesvos, 1950–2001	214
15.4	Size structure of mature a) *Q. macrolepis* (n = 61), b) *Q. pubescens* (n = 64) and c) *Q. cerris* (n = 25) stands (mean ± SD)	216
16.1	Currently open area with scattered woodland to the south side of Loch Affric in Glen Affric	228
16.2	Map of Glen Affric with selected summary pollen diagrams from three selected sites in their locations along the glen, illustrating the differences along the east to west transect of sites	230

16.3	Open grazed landscape of upper Ribblesdale, North Yorkshire	234
16.4	Summary pollen diagram from Wife Park, Ribblesdale	235
17.1	Members of the herd on pasture near Chillingham Burn, extreme east of the park	243
17.2	Two bulls on the Sandy Banks at the north-west of the park	246
17.3	Winter at Chillingham	247
17.4	The herd during winter at Chillingham	248
18.1	Wild boar sow in woodland	257
18.2	Wild boar rooting at a roadside	263
19.1	The Ennerdale Middle Valley	272
19.2	Second Galloway herd grazing the valley bottom	273
19.3	Exclosure showing impact of grazing	278
19.4	Cattle crossing the river Liza	280
20.1	Brothers Water in the Lake District – a mosaic of cropped fertile land; steep, bouldery and infertile valley sides; and exposed leached hilltops	283
20.2	Trees and hawthorn growing on a steep free draining area with thin mineral soil adjacent to a fertile area with deeper moist soil – the vegetation and the differential grazing patterns indicate the fertility and moisture status of the soil	285
20.3	Glen Brittle in Skye – trees are growing on the steep sides of the gorge where sheep graze less intensively than on the more easily traversed land adjacent to it; the soil on the moor is peaty mor whereas the soil in the gorge contains less organic matter and is more free draining	286
20.4	Typical Lake District woodland with rocky, bouldery, thin and infertile soil prone to drought	287
20.5	Alpine meadow in Triglav National Park – here are no fences between the grazed areas and the trees; the trees are growing on steeper ground with less productive soils, often thin and prone to drought	287
20.6	Sherwood Forest – the soil is very sandy, free draining and infertile	288
20.7	A dynamic Lakeland treescape – scrub and woodland 'invading' pasture between Embleton and Cockermouth in the Lake District; reduced stocking numbers are allowing gorse (*Ulex* spp.) and hawthorn (*Crataegus monogyna* Jacq.) to regenerate and these are protecting tree seedlings from grazing animals	289
20.8	A dynamic Slovenian treescape in Triglav National Park – cattle numbers have reduced during the recent past and *Berberis vulgaris* L. and trees are invading the slightly steeper upper area, whereas the remaining cattle are concentrating their feeding in the lower areas	290

20.9	A woodland boundary wall with some isolated trees beyond that may indicate the former extent of the woodland/wood-pasture prior to its enclosure as it naturally expanded and contacted – this is the boundary of the woodland shown on Figure 20.4	295
20.10	Oak tree, Ecclerigg in the Lake District – the tree is incorporated into the field boundary wall that was probably constructed as the result of a parliamentary Enclosure Act; is the tree older than the wall?	297
21.1	Historical wood-pasture in Germany with different influences on vegetation and landscape: Deceleration of natural regeneration, leaf and litter harvest, active herding	303
21.2	Frequency method with 6 x 25 species lists compiled per pasture woodland and per forest (left); and structural method with 3 x 4 vegetation layers and 3 x 100 steps compiled per pasture woodland and per forest (right)	305
21.3	Topographical map of Baden-Wuerttemberg with WPs found in the project and schematic map of Germany, with Baden-Wuerttemberg highlighted	307
21.4	Cow hiding from biting insects; continuous use of shrubs leads to their thinning	309
22.1	Forest cattle grazing in grassland with fragments of scrubland and solitary oak	318
22.2	Density of animal spread woody species (Malus sylvestris, Prunus avium, P.cerasifera, P. serotina, P. spinosa, Rosa spp. and Sorbus aucuparia) in grazed and ungrazed oak woodland	322
22.3	Average annual recruitment (-r) and death (-d) rate in grazed (G) and ungrazed (U) oak woodland	323
22.4	Crested cow-wheat in grazed vegetation at Skovbjerg	326
22.5	Development in species density (average species number per m^2) in grazed and ungrazed oak woodland 11 to 25 years after grazing was initiated	326
24.1	Sloe scrub protecting tree saplings and young trees in Geltsdale's wood-pasture	340
24.2	Hollow ancient rowan, bent over touching lying dead birch, Geltsdale	340
24.3	Swaledale nibbling ash bark in Langdale, Cumbria	342
24.4	Owen Jones creating an oak swill basket at Talkin Treemendous, Cumbria	346
24.5	Burr elm bowl turned and finished by Danny Frost	346
24.6	Hawthorn berries, Geltsdale	348
24.7	Pollarded ash in Langdale's medieval landscape setting	349
24.8	Highland cattle herd in Anloo, Netherlands	351
24.9	Wood-pasture training event at Elan Valley, Wales	353

24.10	Borkener Paradies, north-west Germany	354
25.1	Diagram developed by Neil Sanderson to guide decisions about whether a site is wood-pasture – wood-pastures in good condition are likely to have most factors scoring towards the centre of the circle	357
25.2	Number of trees per hectare in different age classes needed to sustain one veteran tree per hectare	369
26.1	Ancient parkland tree at Windsor Great Park	379
26.2	Trees should be grown from seed in organic soils to benefit their essential mycorrhizal fungi	381
26.3	Browsing animals in wood-pasture	383
26.4	Ash growing in the protection of bramble	384
27.1	Conservation grazing in Devon with Exmoor ponies	390
27.2	Goats grazing and browsing – habitat management, UK	391
27.3	Sheep grazing on heath	393
27.4	Grazed upland shadow wood, Peak District, England	394
27.5	Ancient oak	402

Tables

12.1	Sapling densities (n ha^{-1}) of each tree species, within both inclosed and non-inclosed stands, assessed in the woodlands of the New Forest (Survey 1)	168
12.2	Density of regeneration of woody species associated with thorny shrubs, in areas of woodland expansion	170
12.3	Characteristics of patches of thorny shrubs in relation to the regeneration of tree species, assessed in Survey 2 (data pooled across the three survey plots)	171
12.4	Projected extent of occurrence of principal tree species under different disturbance regimes	174
14.1	Overview of broad trends of susceptibility of Grazing Refuge Habitats to grazing	192
14.2	Overview of broad trends of susceptibility of Grazing Refuge Habitats to environmental factors plus nature of soil	194
14.3	Study species	196
15.1	Farm and land-use statistics for Filia village	213
15.2	Overall occurrence of woody plants in the 70 plots, cover of woody species (where present) and density and basal area of the tree layer (>10cm DBH, where present) (mean values ± SD)	215
15.3	Results of bivariate logistic regression analysis testing for effects of site parameters, vegetation structure and livestock grazing on oak seedling and sapling occurrence	218
21.1	Structure, usage and effects of wood-pastures	308

xviii *Illustrations*

21.2	Conflicts, stakeholders, constructive criticism and proposals for solutions	311
21.3	Floristic comparison between wood-pastures and adjacent forests	313
22.1	Average browsing pressure on important woody species in Skovbjerg	319
22.2	Light requirements for germination of woody species at Skovbjerg assembled from Ellenberg *et al.* (1991) – many of the species germinate some weeks before the sprouting oak-aspen-canopy reaches full cover	321
22.3	Numerical presence of species in grazed and ungrazed oak woodland	324
22.4	Significant increase and decrease (a <0.001, b <0.01 and c <0.5, based on linear regression over time) of the species in young oak woodland based on correlation between time since 1987 and average cover index	325
22.5	Identity tests (t-test and F-test) comparing arrays of grazed and ungrazed field-layer vegetation in 1997 and 2011	327
24.1	Threats to wood-pastures with their associated effects	341
24.2	Effects of shelter on livestock	343
24.3	Grazing harvest from livestock	344
24.4	Management cost comparison of conservation grazing versus strimming for an 80ha site	345
24.5	Estimated timber value of an open-grown tree	345
24.6	Value of some processed tree products	346
24.7	Foodstuffs from trees and bushes for wildlife	349
25.1	List of sites put forward from the UK considered to contain broadleaved wood-pasture and parkland (a case can be made that many of the native pinewoods are also wood-pastures)	360
25.2	Agreed selection criteria for evaluation methodologies – recommended veteran tree site assessment protocol	361
25.3	Listing of protection mechanisms in England	363
25.4	Higher Level Stewardship Scheme agreements supporting wood-pasture and parkland options between 2005 and 2012	370
26.1	Comparison of light requirements of common tree species and the height of the tallest specimen in UK	382

Boxes

20.1	Factors, either singly or in combination, that can allow trees to regenerate in upland grazed landscapes	284
20.2	Factors that can affect the population size of large herbivores for a given site, either singly or in combination	290
25.1	Main threats identified in the UK Habitat Action Plan	366

Part I
Grazed treed landscapes

1 Grazed treed landscapes
Overview and introduction

Ian D. Rotherham

Introduction

The ongoing debates and discussions around the 'Vera hypothesis' of the origins of north-western European landscapes (Vera, 2000) have stimulated academics and others to view the evidence afresh (e.g. Hodder *et al.*, 2005; Hodder and Bullock, 2009; Mitchell, 2005; Whitehouse and Smith, 2010). The issues and research outputs have been considered at a series of major international conferences held in Sheffield, UK; the most recent of these being 'Animals, Man and Treescapes' held in September 2011. The outputs from the meetings have been published (see Rotherham and Green, this volume) and the debates have contributed to raising awareness of the subjects and to developing new research paradigms (Green, this volume). In particular, it seems worthwhile to attempt to place the ideas of Vera's hypothesis (Vera, 2000) into a framework of historic timelines of wooded environments and to consider its relationship to the cultural landscapes from medieval times onwards. A key issue to emerge is the existence historically of populations of 'working trees' mostly in open-grown and in open 'wooded' or 'treed' landscapes.

It is argued here, that the origination of these trees was driven and powered by the ecology of large grazing herbivores (prehistorically and historically) and I would add by the economy of large grazing herbivores too (historically).

Developing the research paradigms

Academics and practitioners alike have been considering the roles of grazing and browsing by large mammalian herbivores in European wooded landscapes of various sorts. One issue to arise from discussions has been what do we recognise as 'woodland' and following from this, how can we evaluate the roles of grazing and browsing animals in these environments – now and in the past. Scholars such as Peterken (1981 and 1996) have addressed the development and dynamics of wooded landscapes from perspectives of both site history and ecology. In terms of the impacts of large herbivores, a natural starting point is to consider known and obviously grazed landscapes with trees and then work back in time to assess their histories and their relationships to other ecological systems and communities. Another approach, which in large part stimulated Vera's own original observations, is to examine presently grazed sites and to interrogate the ecological and landscape implications of grazing impacts. Researchers have noted how ancient parks, originating in medieval or earlier times, provide unique insights into the once great primeval savannahs across much of north-western Europe. Certainly, their remarkable biodiversities provide evidence of such potential lineage. These landscapes present palimpsests of ecology and archaeology that reflect their economically driven origins over 800–1,200 years. Indeed, Since Oliver Rackham's seminal works *Ancient Woodland* (1980) and *The History of the Countryside* (1986), it has been accepted that wood-pasture was once the most abundant type of wooded landscape in north-western Europe. In essence, wood-pasture is a system of land management where trees are grown and grazing by large herbivores (domesticated, semi-domesticated, wild or a combination of stock) is permitted. Wood-pasture in England is well documented for over a thousand years, and the *Domesday Book* (1086) probably records a landscape dominated by the practice. It has been suggested that wood-pasture was an ancient system of management that developed in a multi-functional landscape where woodland was plentiful and where there was little need for formal coppice. The latter is a more intensive and rigorously managed system, intended to ensure vital supplies of wood and timber in a resource-limited landscape (Fowler, 2002; Hayman, 2003; Muir, 2005; Perlin, 1989). The conservation value and importance of pasture woods and of veteran trees was first raised for many people by the work of Harding and Rose in the 1980s (Harding and Rose, 1986). Site-specific studies (e.g. Harding and Wall, 2000) have further confirmed the significance. Pasture-woodland is the older system and in many ways more 'natural'. Significantly, most livestock, wild or domesticated, take leaf fodder or browse if offered (Figure 1.1). This is in preference to grazing (Vera, 2000). These systems of land management, their ecologies and economics, were described by Rotherham (2007a and b). This volume presents recent research that seeks to embed the Vera ideas within a framework of historic cultural ecology (Rotherham, 2011a), and the broader topic of European grazed, wooded landscapes is considered through key contributions by scholars and practitioners from a diversity of academic and professional backgrounds and experiences.

There is considerable interest in the idea of either 'ancient' woods or 'old growth' forest (e.g. Peterken, 1981 and 1996; Rackham, 1976, 1978, 1980, 1986 and 2004) and in relating this to contemporary management. However, many such approaches to wooded landscapes are fundamentally flawed by the absence of an understanding of the historic context of the sites and of their ecologies, or by hands-on practitioner observations of contemporary site management. This volume brings together key works to

Figure 1.1 Cattle grazing and browsing parkland
Source: Ian D. Rotherham

help close the knowledge gap. Recent research in the UK and across Europe seeks to address these issues and to provide a robust interrogation of forest and woodland dynamics that can better inform contemporary management and conservation. This book develops ideas as the evidence base for these assertions. Rotherham (2011a) and Rotherham and Wright (2011) provide models into which historical and ecological information can be placed to assess critically the issues of woodland antiquity and ecological continuity. These help form a useful framework for analysis. These studies also seek to integrate problems of cultural severance (Rotherham, 2008, 2011b) in wooded landscapes. They address the potential for both dynamic change in ecology and the contradictions of long-term spatial stability too. *The Woodland Heritage Manual* (Rotherham *et al.*, 2008) was a major step towards a more integrated approach to the study of wooded landscapes.

Emerging ideas and concepts

The progress reported so far represents an attempt to reconcile to ideas of scholars such as Frans Vera (2000), of an open savannah-like primeval landscape in Europe, with mediaeval and contemporary woodland in countries such as England. Vera's work suggests that large herbivores, browsing and grazing, were drivers in a landscape that mixed successional woodland, big open-grown trees and expansive grasslands or heaths. In attempting to weld the ideas of primeval grazed landscapes to pre-enclosure medieval Europe, the contribution of historians is important. A key idea to emerge from these studies in England, for example, is that of the Act of Commons or Statute of Merton representing watershed in the spatial fixing of 'woods' in their landscape context (Rotherham, 2011a). These named woods are today marked out and identifiable by so-called botanical indicators. Examples to test the ideas were taken from wooded landscapes in England in order to relate known historic timelines to demonstrable changes in ecology, in woodland structure and in pedology (Rotherham, 2011a). The methodology described (Rotherham, 2011a) helps to facilitate an improved understanding of woodland landscape history but it also informs future management and conservation strategies for forest and woodland areas. It also attempts to place wooded landscapes in a wider context of historically grazed ecosystems.

There is still a debate about the nature of this primeval landscape across north-western Europe and this has stimulated cross-disciplinary collaborations to help focus on core evidence-based assessments. To some extent too, these studies have helped to at least provide an idea of what is agreed by most experts and also where there a still areas of dispute and disagreement (e.g. Hodder *et al.*, 2005; Hodder and Bullock, 2009; Mitchell, 2005; Whitehouse and Smith, 2010). The discussions centred on the seminal work of Frans Vera (2000) have helped to clarify how the early landscape might have looked and perhaps aid the understanding of how their biodiversity relates to that which we see today (Hodder *et al.*, 2005; Hodder and Bullock, 2009; Rackham, this volume). Our vision now is not one of wall-to-wall forest, but of plains or savannah that are more open with a rich diversity of other landscape and ecological components too and patches (some very extensive) of more dense closed forest. Much of this variation would be related to basic environmental factors such as climate and waterlogging, but the key factor identified by Vera is the importance alongside these of large grazing herbivores (Figure 1.2). Not everyone agrees with Vera's hypothesis, but the argument is compelling and a refined version of his original vision has a considerable body of supporting evidence (Vera, this volume). Issues of upland/lowland variation and interactions also need to be built into the Vera ideas and these were discussed at the meetings held in Sheffield in 2005 and 2007 (Rotherham, 2005 and 2007c).

A wetter landscape

I suggest that the European landscape had vast expanses of wetland, marsh, fen and peat bog with extensive coastal wetlands too. Over large areas of shallow, wet landscape it is likely that grazing herbivores were ecological

Figure 1.2 Konik ponies grazing in Poland
Source: Ian D. Rotherham

keystone species. Unfortunately, most of the studies so far, in addressing the broader issues of interpretation of palaeoecological or palynological evidence, have rarely considered landscape history specifically. This omission means that some of the interpretations from the rigorous scientific analysis are lacking in their wider applicability. There are attempts now to redress this shortcoming (e.g. Samojlik *et al.*, in prep.). The impacts of widespread and catastrophic drainage across much of the European landscape, and the consequences of post-mediaeval enclosure, for example, are rarely appreciated in full (see for example Rotherham, 2010b and 2013). Many of these redrainage landscapes were well wooded and would have been exploited by grazing herbivores. Pre-enclosure landscapes were connected seamlessly across wide areas and allowed movements of large mammals with seasons and with weather or climate change on a large scale that is unimaginable today.

On drier ground, dense woodland and thickets would grow up in the protection of rings of prickly blackthorn and bramble and here would be the plants and animals that characterise our so-called 'ancient woodlands' today. These plants would include bluebell (*Hyacinthoides non-scripta*), dog's mercury (*Mercurialis perennis*), wood anemone (*Anemone nemorosa*) and yellow archangel (*Galeobdolon luteum*) for example. Outside the prickly halos of the thorns was a wide-open savannah with heath, grassland and giant old trees such as pedunculate oak (*Quercus robur*), which grew to ages of a thousand years or more before collapsing into oblivion. Inspection of giant old trees such as the Major Oak in Sherwood Forest confirm that these trees are not pollards but massive open-grown specimens; relicts from a landscape long-since obliterated by human impacts. Interestingly, photographs of aged oaks in California for example, show exactly the same form and characteristics, again deriving from an era pre-European cultural impacts.

A further powerful driver of succession and change in this landscape would be pre-human fire caused by lightning strikes, taking out the great trees in the open plain. Further north and into the upland zones there would be a change to landscape dominated by great and similarly ancient Scots pines (*Pinus sylvestris*), and again a strong influence of natural, lightning-related fires. Mountain zones would have large areas of disturbance through natural landslip and erosion areas, and all the great rivers would include sometimes-vast floodlands and meandering patterns of erosion and deposition in an ever-changing yet stable landscape. Much of this environment would have limited amounts of available nutrients, especially nitrogen and phosphorous, and this too had a great influence on associated biodiversity. Localised areas such as mountain downwash zones and alluvial fans had higher nutrient levels, and the whole landscape experienced abundant micro-disturbance through natural process, but only limited macro-disturbance. Erosion and deposition areas, animal-related disturbance, fire, successional and lifecycle related changes, such as the collapse of ancient trees, were the key to the ecological dynamics. This landscape provided a template for a richly diverse fauna and flora with regionally district ecologies related to broad climatic influences and localised geological and topographic factors. This was the landscape, the ecology and the biodiversity upon which the footprint of human activity was stamped with increasing effect over the following 5,000–6,000 years.

Interest in ancient 'worked' trees has continued to grow in Britain and across Europe (e.g. Read, 1999). Furthermore, the relationships between parks, forests, woods, commons, heaths, bogs and fens have been explored (e.g. Rotherham 2009a, b and c; Rotherham and Bradley, 2009). This present volume makes a major contribution to these ongoing debates in the six sections of the book:

Part 1   Grazed treed landscapes
Part 2   The lessons of history
Part 3   Landscape dynamics
Part 4   Case studies
Part 5   Conservation, management and wildscapes
Part 6   Summary and conclusions

Each section offers a carefully selected group of essays from key researchers and practitioners across Britain and around Europe on the current issues and debates of this especially interesting and significant subject. The contributions seek to mix ecology, site management, history, archaeology and policy.

Conclusions

The rich academic and practitioner discussions are in part retrospective attempts to understand landscape origins. However, they also look forwards to how we can conserve or even recreate aspects of a functional and sustainable ecology in the future. Chapters presented here address these issues and give thorough accounts of work in progress and work proposed; from research to implementation and monitoring. Examples are provided from across Europe. This volume makes a major contribution to the debates and discussions, and for the first time makes available a thoroughly up-to-date account of the roles of large grazing and browsing herbivores in European wooded landscapes.

References

Fowler, J. (2002) *Landscapes and Lives. The Scottish Forest through the Ages*, Canongate Books, Edinburgh.

Harding, P.T. and Rose, F. (1986) *Pasture-Woodlands in Lowland Britain – A Review of their Importance for Wildlife Conservation*, Institute of Terrestrial Ecology, Monks Wood Experimental Station, Huntingdon.

Harding, P.T. and Wall, T. (eds) (2000) *Moccas: An English Deer Park*, English Nature, Peterborough.

Hayman, R. (2003) *Trees. Woodlands and Western Civilization*, Hambledon and London, London.

Hodder, K.H. and Bullock, J.M. (2009) 'Really wild? Naturalistic grazing in modern landscapes', *British Wildlife*, 20 (5): 37–43.

Hodder, K.H., Bullock, J.M., Buckland, P.C. and Kirby, K.J. (2005) 'Large herbivores in the wildwood and in modern naturalistic grazing systems', *English Nature Research Reports*, Report Number 648, English Nature, Northminster House, Peterborough.

Mitchell, F.J.G. (2005) 'How open were European primeval forests? Hypothesis testing using palaeoecological data', *Journal of Ecology*, 93: 168–77.

Muir, R. (2005) *Ancient Trees Living Landscapes*, Tempus, Stroud.

Perlin, J. (1989) *A Forest Journey*, Harvard University Press, Massachusetts.

Peterken, G.F. (1981) *Woodland Conservation and Management*, Chapman & Hall, London.

Peterken, G.F. (1996) *Natural Woodland: Ecology and Conservation in Northern Temperate Regions*, Cambridge University Press, Cambridge.

Rackham, O. (1976) *Trees and Woodland in the British Landscape*, J.M. Dent & Sons Ltd, London.

Rackham, O. (1978) 'Archaeology and land-use history' in Corke, D. (ed.) *Epping Forest – the Natural Aspect?*, Essex Nat., N.S. 2, pp16–57.

Rackham, O. (1980) *Ancient Woodland; its History, Vegetation and Uses in England*, Arnold, London.
Rackham, O. (1986) *The History of the Countryside*, J.M. Dent & Sons Ltd, London.
Rackham, O. (2004) 'Pre-existing trees and woods in country-house parks', *Landscapes*, 5(2):1–16.
Read, H. (1999) *Veteran Trees: A Guide to Good Management*, English Nature, Peterborough.
Rotherham, I.D. (ed.) (2005) 'Crisis and continuum in the shaping of landscapes', *Landscape Archaeology and Ecology*, 5: 117pp.
Rotherham, I.D. (2007a) 'The ecology and economics of medieval deer parks', *Landscape Archaeology and Ecology*, 6: 86–102.
Rotherham, I.D. (ed.) (2007b) *The History, Ecology and Archaeology of Medieval Parks and Parklands*, Wildtrack Publishing, Sheffield.
Rotherham, I.D. (2007c) 'The implications of perceptions and cultural knowledge loss for the management of wooded landscapes: A UK case-study', *Forest Ecology and Management*, 249: 100–15.
Rotherham, I.D. (2008) *'The importance of cultural severance in landscape ecology research'*, in Dupont, A. and Jacobs, H. (eds) (2008) *Landscape Ecology Research Trends*, Nova Science Publishers Inc., New York, pp71–87.
Rotherham, I.D. (2009a) 'Hanging by a thread – a brief overview of the heaths and commons of the north-east midlands of England', in Rotherham, I.D. and Bradley, J. (eds) (2009) *Lowland Heaths: Ecology, History, Restoration and Management*, Wildtrack Publishing, Sheffield, pp30–47.
Rotherham, I.D. (2009b) 'Habitat fragmentation and isolation in relict urban heaths – the ecological consequences and future potential', in Rotherham, I.D. and Bradley, J. (eds) (2009) *Lowland Heaths: Ecology, History, Restoration and Management*, Wildtrack Publishing, Sheffield, pp106–15.
Rotherham, I.D. (2009c) 'Cultural severance in landscapes and the causes and consequences for lowland heaths', in Rotherham, I.D. and Bradley, J. (eds) (2009) *Lowland Heaths: Ecology, History, Restoration and Management*, Wildtrack Publishing, Sheffield, pp130–43.
Rotherham, I.D. (2010a) 'Cultural severance and the end of tradition', *Landscape Archaeology and Ecology*, 8: 178–99.
Rotherham, I.D. (2010b) *Yorkshire's Forgotten Fenlands*, Pen and Sword Books Limited, Barnsley.
Rotherham, I.D. (2011a) 'A landscape history approach to the assessment of ancient woodlands', in Wallace, E.B. (ed.) *Woodlands: Ecology, Management and Conservation*. Nova Science Publishers Inc., USA, pp161–84.
Rotherham, I.D. (2011b) 'The implications of cultural severance in managing vegetation for conservation', *Aspects of Applied Biology*, 108: 95–104.
Rotherham, I.D. (2013) *The Lost Fens: England's Greatest Ecological Disaster*, The History Press Ltd, Stroud.
Rotherham, I.D. and Bradley, J. (eds) (2009) *Lowland Heaths: Ecology, History, Restoration and Management*, Wildtrack Publishing, Sheffield.
Rotherham, I.D. and Wight, B. (2011) 'Assessing woodland history and management using vascular plant indicators', *Aspects of Applied Biology*, 108: 105–12.
Rotherham, I.D., Jones, M., Smith, L. and Handley, C. (eds) (2008) *The Woodland Heritage Manual: A Guide to Investigating Wooded Landscapes*, Wildtrack Publishing, Sheffield.

Samojlik, T., Jędrzejewska, B. and Rotherham, I.D. (in prep.) 'What shaped the forest? A conceptual model to estimate and quantify historic human impacts on forest environments – a case study in Europe's last primeval lowland forest', (paper in preparation).

Vera, F.W.M. (2000) *Grazing Ecology and Forest History*, CABI Publishing, Oxon, UK.

Whitehouse, N.J. and Smith, D. (2010) 'How fragmented was the British Holocene wildwood? Perspectives on the "Vera" grazing debate from the fossil beetle record', *Quaternary Science Reviews*, 29(3–4): 539–53.

2 Woodland and wood-pasture

Oliver Rackham

Landscapes with trees are of many kinds. Omitting gardens, orchards, hedgerow trees and trees in fields, the remaining treed landscapes are of two main types. In one type grazing and browsing large mammals are part of the normal dynamics of landscape; in the other they are not.

In England there is a distinction, which goes back over a thousand years, between *woodland*, in which there is a tradition of not having significant grazing, and *wood-pasture*, which combines trees and grazing animals (Figure 2.1). These are distinct land uses: the shade of trees spoils the

Figure 2.1 A coppice-wood in the second year after felling. The underwood stools are mainly ash. The timber trees are oaks. This dense scatter of small oaks would have been characteristic of a medieval wood. Medieval timber-framed buildings are typically made of hundreds of oak trunks small enough for two men to lift. Such a wood could not operate in the presence of grazing animals, especially since ash is one of the most palatable of trees. Bradfield Woods, Suffolk, August 1986

pasture, and browsing animals eat tree seedlings and regrowth sprouts (Rackham 1976). Wood-pasture, which now survives less often than woodland, is associated with pollarding – cutting trees high enough for the livestock not to reach the sprouts – and thus with ancient trees (Figure 2.2).

This is part of the worldwide distinction between *forest* (in the forester's sense) and *savanna*. Savannas, corresponding to one form of English wood-pasture, are grassland (or heath) with scattered trees (Grove & Rackham (2001) ch. 12). Savannas have had a poor press: foresters regard them as not proper forest, agronomists and grassland ecologists as not proper grassland. There is a vague general consensus (not based on historical evidence) that savannas outside the tropics are no more than artefacts of land management and thus – in some schools of ecological thought – not worth consideration; it is incorrect to take an interest in the ecology or conservation of savanna.

For the purpose of this article, savanna is where the trees are widely enough spaced to have non-shade-bearing vegetation between them.

Figure 2.2 Uncompartmented wood-pasture. A savanna-like landscape with scattered oaks in grassland (formerly heath, now invaded by bracken). The trees are ancient pollards, at least 450 years old, last cut *c.*1800. Staverton Park, Suffolk, March 2008

Woodland

> In a region absolutely covered with trees, human life could not long be sustained... The depths of the forest seldom furnish either bulb or fruit suited to the nourishment of man; and the fowls and beasts on which he feeds are scarcely seen except upon the margin of the wood, for here only grow the shrubs and grasses, and here only are found the seeds and insects, which form the sustenance of the non-carnivorous birds and quadrupeds.
>
> George Perkins Marsh (1864)

In woodland, the shade of the trees allows little pasture to grow. Livestock – here I deal with cows and deer, not monkeys – like leaves and tree-bark, but these once eaten do not grow again at a height for the animals to reach them. Shade-bearing plants are attenuated, do not produce tons per hectare of biomass and many are distasteful or poisonous. To sustain large numbers of animals, domestic or wild, the trees must not form a continuous canopy.

Most woodland is cut down from time to time. Its ecology depends on periodic cycles of felling and regrowth; much of the flora of English woodland, including plants indicative of ancient woods, is visible only in the years after felling and passes the years between fellings as buried seed

Figure 2.3 Wood-spurge (*Euphorbia amygdaloides*), one of many plants of ancient woodland that appear from buried seed after each time the wood is felled. Bradfield Woods, Suffolk, June 1987

(Figure 2.3). Grazing animals eat the regrowth shoots of the trees and eat many of the woodland herbs that come up after each felling. Woods that are periodically felled are destroyed after only two or three fellings if exposed to even moderate densities of herbivores. Historically people have gone to great trouble to keep livestock out of felled woods (Figure 2.4).

There is a partial exception in woods of alder (*Alnus*) in Wales, which sometimes are not fenced off from fields but nevertheless survive sheep-grazing, because shoots of alder are rather unpalatable to sheep.

From prehistoric times onwards woods have fluctuated: they have been grubbed out to make fields and pastures, but have returned in times of agricultural recession. By the Middle Ages coppice-woods, mostly private land, were permanent features of the landscape. A good proportion of the medieval woodland of England still survives, if often neglected or degraded by replanting (for example about 38 of the 150 woods that belonged to Bury St Edmunds Abbey, and 17 of the 50 woods of the Bishops of Ely (Rackham 1998a, 2000).

Wood-pasture

Wood-pasture was the main component of the wooded commons, parks and wooded Forests of medieval and later England. There were two kinds,

Figure 2.4 Boundary of the wood of the great Abbey of Bec, Normandy. On top of the boundary bank is built a massive flint wall. The effort put into defending wood boundaries is a measure of the importance attached to woodland conservation in the middle ages. Le Bec-Hellouin, July 1976

one being *compartmentalised*. A compartmentalised wood-pasture was divided into *coppices* and *plains*. Each coppice was periodically felled and then fenced, typically for six to nine years, until the regrowth had become tall enough to escape serious browsing; domestic livestock or deer were then admitted for the rest of the cycle (Figure 2.5). Examples are Hatfield Forest, Monks' Park (Bradfield Woods, Suffolk), and the complex of 23 woods now called Wetmoor Woods, Gloucestershire. This practice may be peculiarly English.

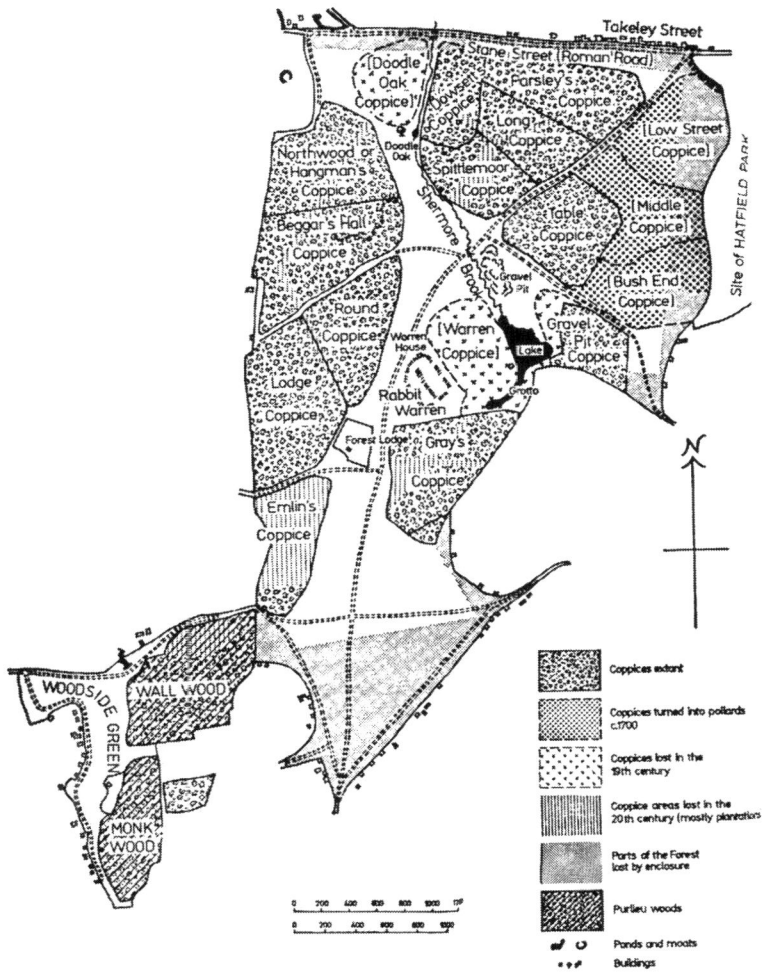

Figure 2.5 Compartmented wood-pasture. Each coppice was supposed to be cut once in 18 years, then fenced for the first 9 years of regrowth. The plains (blank areas) were always open to livestock and contain pollard trees. Hatfield Forest, Essex

Uncompartmental wood-pastures tend to become trees plus grass or heath, for example, Epping Forest, Bradgate Park (Leicestershire) and Ashtead Common (Surrey). The trees are either pollards or are allowed to stand longer than in woodland because of the difficulty of replacing them in the face of grazing animals. (Historically, especially large timber trees were valued for purposes such as mill posts, cranes, ships' keels and cathedral roofs. If left to get too big they could become of little use because of the difficulty and expense of felling, moving and subdividing them in an age without power tools.) Pollards and other wood-pasture trees thus tend to become veteran and then ancient trees, the home of hole-nesting birds, bat roosts, ancient-tree insects and 'woodland continuity' lichens.

Survival of ancient wood-pasture is rarer than of woodland. In the England of *Domesday Book* (1086), wood-pastures probably formed the majority of the 15 per cent of the country recorded as 'woodland'. Savanna formed much of the 'woodland', even of historic Scotland, where Glen Finglas, with its hundreds of ancient pollard alders and hazels, appears to be the only extensive survival.

How can wood-pastures last for more than one generation of trees?

Ecological conservationists like to provide for a stable model in which, out of a thousand trees, every year (let us say) five trees die and a hundred seedlings arise of which five will grow into trees. Wood-pasture seldom works like this, because in every normal year the livestock find and eat every one of those hundred seedlings. One cannot expect a new generation of trees to arise regularly every year, or even every 50 years. (Moreover, in Britain many trees arise not from seed but from the stumps or root-sprouts of their predecessors.)

A wood-pasture needs to be somehow discontinuous in space or episodic in time. There may be thorny thickets of hawthorn or blackthorn, in whose protection a new generation of trees can arise (Figure 2.6). Or the grazing may fluctuate: every century or so the cattle die from disease or get eaten by invaders, and there is a phase of regeneration of trees (as in north-west Greece following the disruption of World War II). That is what evidently happened in the past, but it is difficult for conservationists to legislate for. Can rinderpest or Viking raiders be written into a management plan? Yet without something like them – with constant grazing – the trees will in the long term die out.

Savanna in other countries

Savanna, uncompartmental wood-pasture, happens where some factor allows trees to grow but not forests. This may be grazing, drought (as in inland Australia), cold (as in the larch and pine meadows of the Alps) or fire (as in the interior of North America). These factors may interact: in

Figure 2.6 A Vera-esque landscape? The grassy plain is being encroached on by thorn thickets, which in course of time will develop into established woodland like that in the background. Central plain, Hatfield Forest, February 1981

drier climates, where vegetation grows more slowly, it takes less intensive grazing to make savanna.

In semi-arid regions trees, and indeed shrubs or grass tussocks, may grow widely spaced above ground, expanding their roots to fill all the space below ground and thus to catch rain that falls between the trees as well as on them. Drought-determined savannas are widespread in North America, Africa, Australia and in southern Europe (Figures 2.7, 2.8). In Britain all savannahs appear to be cultural, with the possible exception of Staverton Park, Suffolk, which is in the driest part of the country and where the soil, blown sand, retains little moisture.

Human activity interacts with other factors: if tree growth is already restricted by drought it takes less intensive grazing to make savannah. In mid-Texas savanna takes the form of *motts*, clonal patches of trees (*Quercus fusiformis, Ulmus crassifolia* etc.) rather than single stems. The motts appear to be centuries old and have passed through four human cultures, two Native American and two European; yet if grazing and burning are abandoned the gaps between the oak and elm motts infill with the fire-sensitive *Juniperus ashei*.

In the world as a whole, it is often difficult to tell where natural savannahs end and cultural savannas begin. In Australia, much of the eucalyptus

18 *Oliver Rackham*

Figure 2.7 Savanna in a Mediterranean-type climate, analogous to the *montado* of Portugal and the *dehesa* of Spain. Near Ukiah, California, April 2005

Figure 2.8 Drought-determined savanna. The roots of the live-oak (*Quercus grisea*) extend out at least as far as where the man is standing. Davis Mountains, west Texas, March 1999

savanna was maintained by Aboriginal burning, which favoured kangaroos and other edible beasts. When Europeans came, who distrusted fire, fire suppression allowed rainforest trees to spread and form an understorey to the scattered big eucalypts with their charred bases. Much of the supposedly 'primeval' forest of West Africa is secondary to cultural savanna maintained by a formerly dense human population; after slavers murdered or carried off the people, savanna turned into forest (Fairhead & Leach 1998).

Most of the world's ancient trees are in savanna rather than forest; they can pass through prolonged or intermittent periods of decline without incurring the competition of neighbouring trees.

Forest and savanna in prehistory

The theory that Britain and Europe in pre-Neolithic times was covered from coast to coast in wildwood of continuous, unbroken trees, unaffected by the activities of earlier peoples, is harder to sustain than it used to be. I criticised it in an article in 1998 called 'Savanna in Europe' (Rackham, 1998). Francis Vera proposes a rival theory of wildwood, that it was a savanna-like mosaic of areas of grassland and areas of trees, changing on a timescale of centuries as woods expanded at their edges into grassland, giving rise to new grassland as the woods declined and died out in the middle.

This is not the place to discuss in detail which is right. At present the weight of pollen-analytical opinion is against Vera, but his theory is unlikely to disappear. It allows for the pre-Neolithic existence of grassland and heath with their characteristic plants. Wildwood contained long-lived plants, like *Succisa pratensis*, that do not flower in shade, so it must have had at least modest permanent open areas. Continuous forest could hardly have sustained large numbers of ground-living herbivores. Nor could it have attracted people to take up agriculture; why should they abandon hunting and gathering and put in years of hard labour digging up trees before they could grow anything? What could they have lived on while doing it? In a 'Vera world' all they needed to do was to dig up grassland, leaving the trees till later when they had more manpower.

One of the biggest effects of *Homo sapiens* on natural ecosystems was apparently the extermination of the super-elephants and other giant herbivores that lived in previous interglacials. The loss of these living bulldozers could not fail to have had profound consequences. Not surprisingly, the evidence for savanna is stronger in past interglacials.

I envisage wildwood as containing at least modest open areas, providing a habitat for deer and wild oxen and a site for veteran trees and the special creatures that live on them. In southern Europe, where tree growth is further restricted by drought, savanna and veteran trees would have been abundant, as locally they still are.

An unsolved problem, in this reconstruction, is the fate of what are now the grazing-sensitive plants of ancient woodland. What could have prevented herbivores from entering long-established wooded areas and eating up *Primula elatior* and its like? Was there something in wildwood that had a similar effect to the compartmentation of medieval wooded Forests?

Significance for conservation and the future

Ian Rotherham speaking at the 2011 Sheffield conference remarked on the 'nature-culture complexity of woodland origins'. This complexity is fundamental to understanding the difference between woodland and wood-pasture and how each functions.

In Britain, the most widespread conservation problem, apart from globalisation of tree diseases, is the loss of the distinction between woodland and wood-pasture. In the lowlands the problem is large numbers of deer (many more than were present historically), especially now that woods are not fenced as they used to be. The deer eat everything edible in the wood, and when that is exhausted they go out and eat the farmers' fields. Far more deer thus live in the wood than it can support; the wood loses its specifically woodland plants and turns into trees plus grass. In the north and west, woods suffer from sheep getting in as the historic boundary walls of woods decay.

Wood-pastures, conversely, often suffer from too little grazing, which leads to the savanna infilling and turning into undistinguished recent woodland. Or alternatively the ancient trees may survive, but the grassland or heath is replaced either by agricultural-type grassland or by bracken. Wood-pastures very rarely retain both the ancient trees and the non-tree component.

There is (or was) a school of conservationists who believe it their duty to drop everything else they are doing and devote all their energies to restoring wildwood as it was before it was corrupted by civilized humanity (Hamblert Speight 1995). This is unrealistic, not least because not enough is known about the structure of wildwood to form a basis for restoration; also because many changes, like the loss of lime, the introduction of grey squirrel and the globalisation of plant diseases, are beyond human ingenuity to reverse.

Woodland and wood-pasture both result in part from human intervention, but they have been in operation for so long – for so many generations of plants and animals – that those activities now form part of the normal dynamics and ecosystems have become adapted to them. Woodland has distinctive herbaceous plants, which are not necessarily shade-bearing: many of them require the wood to be cut down periodically. Wood-pasture has the distinctive plants and animals that live on veteran trees, and those that require both trees and open ground. Each has its own distinctive set of birds.

Conservationists should normally stick to conservation: trying to preserve what is there, rather than recover what might have been. It is all too easy to set up a simplified and idealised model of what an ecosystem ought to be, and to try to make real ecosystems conform to it. There is an unfortunate tendency for conservationists, like foresters before them, to devise a 'sound' system of management and apply it to all land that they control – rather than understanding and perpetuating the features that make each site unique and different from other sites. A site should not be forced to turn into something that it is not.

The distinction between woodland and wood-pasture needs to be recognised and maintained. Treating woodland as if it were wood-pasture, through allowing deer or sheep to invade it, loses the low cover and most of the ground vegetation, and will ultimately lose the wood. Treating wood-pasture – cultural savanna – as if it were woodland will cause it to infill into forest, which will lose the veteran trees (through the competition of younger trees) and all that goes with them, and will also lose the grassland or heath component.

Figure 2.9 Ashwood, severely browsed by three species of deer. Note absence of low cover and absence of herbaceous plants, except the distasteful dog's-mercury. Hempstead Wood, Essex, April 2002

In wood-pasture it is important to try to find out how the trees replaced themselves, and to replicate the instability that seems to be an essential mechanism. Every year must not be the same as every other year. Compromises between woodland and wood-pasture, with a constant reduced level of grazing, should be avoided. It is sometimes claimed that there is an 'appropriate grazing regime' that conservationists ought to establish and stick to. This is unrealistic. If there are large herbivores at all, they will have definite likes and dislikes; it is they that will decide what and how much they eat. It does not take many deer or cattle to eat the bottom out of a wood (Figure 2.9).

References

Fairhead, J. & Leach, M. (1998) *Reframing Deforestation* Routledge, London
Grove, A.T. & Rackham, O. (2001) *The Nature of Southern Europe: an ecological history,* Yale University Press
Hambler, C. & Speight, M.R. (1995) 'Biodiversity conservation in Britain: science replacing tradition' *British Wildlife* 6 137–47
Perkins Marsh, G. (1964) *Man and Nature,* Charles Scribner, Yew York.
Rackham, O. (1976) *Trees and Woodland in the British Landscape* Dent
Rackham, O. (1980) *Ancient woodland: its history, vegetation and uses in England* Edward Arnold, London [2nd ed. 2003, Castlepoint Press, Dalbeattie]
Rackham, O. (1998a) 'The Abbey woods' in *Bury St Edmunds: medieval art, architecture, archaeology and economy* ed. A. Gransden, British Archaeological Association, London. 139–60, Plates XXXIII–XXXIV.
Rackham, O. (1998b) 'Savanna in Europe', in Kirby, K.J. and Watkins, C. (eds) *The Ecological History of European Forests,* CAB International, Wallingford, Oxon, 1–24.
Rackham, O. (2000) 'Woodland in the Ely Coucher Book' *Nature in Cambridgeshire* **42** 37–44, 61–7
Rackham, O. (2006) *Woodlands* (New Naturalist vol. 100), Harper Collins, London

Part II
The lessons of history

3 Woods, trees and animals

A perspective from South Yorkshire, England

Melvyn Jones

Introduction

Stock rearing and woodland management went hand in hand in the past in South Yorkshire in deer parks, on wooded commons and in coppice-woods. And this strong relationship is well documented over the last thousand years. This chapter uses surviving documentary sources to survey the relationship between woods, trees and animals in the region over the last millennium.

Silva pastilis in South Yorkshire at Domesday

If we take the results of William the Conqueror's great national survey of 1086 at face value then woodland cover had been drastically reduced by the late eleventh century and the countryside was not covered by the boundless wildwood of people's imagination. Rackham has calculated that the Domesday survey of 1086 covered 27 million acres of land of which 4.1 million were wooded, that is 15 per cent of the surveyed area. His figure for the West Riding of Yorkshire is 16 per cent (Rackham, 1980). My own calculation for South Yorkshire is just under 13 per cent (Jones, 2009). By way of comparison, woods today, including plantations, cover just over 6 per cent of the region. What this means is that in the eleventh century, South Yorkshire was relatively sparsely wooded, even by today's standards.

Domesday woodland in South Yorkshire was described in four main ways: as *silva*, *silva modica*, *silva minuta* and *silva pastilis*. *Silva* is simply woodland; the meaning of *silva modica* is not clear; *silva minuta* is coppice; and *silva pastilis* is wood-pasture. Of the 111 manors in which woodland was recorded, 102 had wood-pastures and 7 had coppices (Figure 3.1). All seven occurrences of coppice woods were in the eastern half of the region, two in the eastern part of the Coal Measures (at Little Houghton and Barnburgh) and five on the magnesian limestone (at Adwick-le-Street, Hampole, Marr, High Melton and a manor incorporating land at Sprotbrough, Cusworth and Balby). However, although wood-pastures were found throughout South Yorkshire they were very extensive and the only

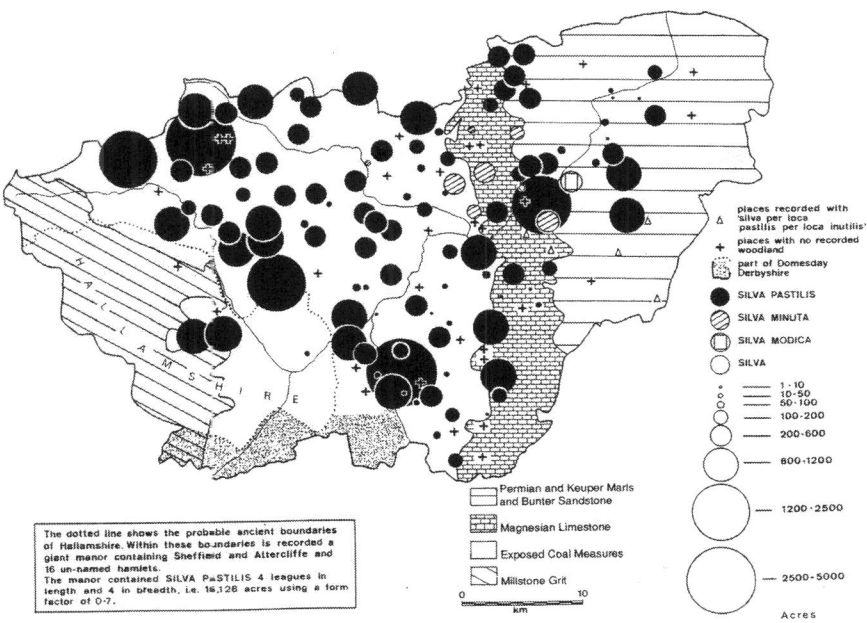

Figure 3.1 Woodland in South Yorkshire at Domesday
Source: Jones (2009)

type of woodland found in the millstone grit country and throughout most of the Coal Measures in the central and western parts of the region.

Wooded commons

Silva pastilis in the form of wooded commons continued to be a widely distributed type of land use until at least the mid-eighteenth century in the central third of South Yorkshire. The wooded commons were in great contrast to the moorland commons of the Dark Peak zone in the extreme west and the marshland commons of the Humberhead Levels in the east. Wooded commons were unfenced woods in which underwood and timber were harvested but in which the animals of commoners, i.e. those who had certain rights on the common land of a manor, were allowed to graze freely. Commoners usually also had the right of the underwood and dead wood but the timber trees usually belonged to the lord of the manor. In the manor of Sheffield as late as 1637 there were 21,000 acres (8,499ha) of common land, much of it wooded (Ronksley, 1908).

The first record of a wooded common in South Yorkshire after the Domesday survey was in 1161 when the monks of the abbey of St Wandrille (who built Ecclesfield Priory) were given permission by the lord of Hallamshire, Richard de Louvetot, to pasture their flocks from January to Easter, their swine in the autumn, and to take dead wood in a large wood stretching from Birley Edge down to the Don, all the way from Wardsend to Oughtibridge. In an undated charter of about 1297, Thomas de Furnival, the lord of the manor of Hallamshire, made a number of grants of common in the largely uninhabited uplands and river valleys in the western part of the manor. He gave to the men of Stannington, Morewood, Hallam and Fulwood the right of herbage and foliage (to graze their stock and to gather green and dry wood) throughout his forest of Riveling (i.e. Rivelin Chase, his private hunting forest) all the way from Malin Bridge at the confluence of the rivers Rivelin and Loxley to Stanage Edge at a height of more than 1,421 feet (435m) more than seven miles (11km) to the south-west. In an inquisition *post mortem* of 1332 on the death of Thomas de Furnival it was recorded that his properties in the manor of Sheffield included pastures in, among other places, 'Greno, Billy Wood, Ryvelyngden and Baldwinhousteads' (Curtis, 1918). Ryvelyngden was the valley of the Rivelin occupied by his private hunting forest, which continued in part to be wood pasture until the parliamentary Enclosures of the eighteenth and early nineteenth centuries but by 1600 the other three, Greno Wood, Beeley Wood and Bowden Housteads Wood, were all enclosed coppice-woods. But some wood-pastures continued in existence for more than two centuries longer.

There are graphic descriptions of a number of wooded commons in the late seventeenth and early eighteenth centuries. In 1650, Loxley Chase was referred as 'one Great wood called Loxley the herbage common and consisteth of great Oake timber'. About ten years earlier, another wooded common, Walkley Bank, was said to have 'a great store of rough Oake trees & some Bircke (birch) woods'. In the same year Stannington Wood, formerly part of Rivelin Chase and that covered 217 acres (88ha) in 1637, was said to consist of 'pt of rough Timber and part of Springe wood'.

Perhaps the best documented and interesting of the few surviving former South Yorkshire wooded commons is Loxley Common and the coterminous Wadsley Common in north-western Sheffield. The area in question covers the valley side to the north of the River Loxley with Loxley Edge, a sandstone escarpment outcropping in the eastern half of the site and with a boulder-strewn slope below it. Loxley Common is variously referred to in historical documents as Loxley Common, Loxley Chase and Loxley Firth. Loxley Common was mentioned in a late thirteenth-century document in which Thomas de Furnival, the lord of Hallamshire, granted common rights to the inhabitants of the area. In the inquisition *post mortem* in 1332 on the death of Thomas de Furnival it was recorded that the pannage of the woods (*pannagio boscorum*), i.e. the pasturing of pigs on acorns, at 'Rivelyngden and Lokesley' was worth forty shillings (Curtis, 1918). Just over 300 years later in 1637 it was recorded in John Harrison's survey of the manor of Sheffield as 'A Common Called Loxley wood & ffirth' and covered 1,517 acres (614ha) (Ronksley, 1908). As already noted, in 1650 the common was referred to as 'one Great wood called Loxley the herbage common and consisteth of great Oake timber'. Two old oak pollards still survive on the site.

Deer parks

Medieval deer parks were symbols of status and wealth. In South Yorkshire they were created by the nobility and they were also attached to monasteries. There were also two royal deer parks: Conisbrough Park, formerly the property of the de Warenne family that reverted to the Crown in the fourteenth century, and Kimberworth Park that became Crown property for a period in the late fifteenth century. As all deer were deemed to belong to the Crown, from the beginning of the thirteenth century landowners were supposed to obtain a licence from the king to create a park, although this appears not to have been necessary if the proposed park was not near a royal forest. The medieval parks at Conisbrough and Sheffield predated the issuing of royal licences and so must have been of twelfth century or even earlier, possibly Saxon, origin.

The Crown issued rights of free warren that gave a landowner the right to hunt certain animals – such as game birds, hare, fox, badger, wild cat, polecat and pine marten – within a prescribed area. This was often the forerunner to the fencing of demesne land to create a deer park. Searches of parish histories, principally Hunter's two-volume *South Yorkshire* (Hunter, 1828–31) reveal that more than 80 grants of free warren were given in the medieval period in South Yorkshire and that in nearly a third of the cases, a deer park is known to have been subsequently created.

Nationally the great age of park creation was the century and a half between 1200 and 1350, a period of growing population and agricultural prosperity. Landowners had surplus wealth and there were still sufficient areas of waste on which to create parks. In South Yorkshire, the majority of grants of free warren that, as already noted, were often the forerunners of the creation of deer parks, were given in the period from 1250 to 1325 when 44 grants were made. Significantly, no grants of free warren were given for 30 years following the Black Death (1349), but then there were 21 grants between 1379 and 1400. The last-known medieval royal licence to create a deer park was given in 1491–2 when Brian Sandford was granted permission to create a park at Thorpe Salvin. This grant is also notable for the fact that it was accompanied by a gift of 12 does from the king's park at Conisbrough 'towards the storing of his parc at Thorp' (Jones, 1996). The last-known local licence was granted to the 2nd Viscount Castleton in 1637 by King Charles I to create a deer park at Sandbeck (Figure 3.2). The licence states that Viscount Castleton was given permission to make separate with pales, walls or hedges 500 acres or thereabouts of land, meadow,

28 *Melvyn Jones*

Figure 3.2 Royal licence to create a deer park at Sandbeck Park, 1637
Source: from the original in Sandbeck Park estate office with permission

pasture, gorse, heath, wood, underwood, woodland tenements and hereditaments to make a park where deer and other wild animals might be grazed and kept (Rodgers, 1998).

The deer in medieval parks were carefully farmed (Birrell, 1992). Besides their status-symbol role, the main functions of parks were to provide for their owners a reliable source of food for the table, supplies of wood and timber, and in some cases quarried stone, coal and ironstone. They were, therefore, an integral part of the local economy. Besides deer, hares, rabbits (also introduced by the Normans and kept in burrows in artificially made mounds) and game birds were kept in the medieval parks of South Yorkshire. Herds of cattle, flocks of sheep and pigs were also grazed there. Fish ponds were also important features to provide an alternative to meat in Lent and on fast days.

To use the modern term a deer park was also usually part of the manorial 'forestry' operation. Although there are records of parks without trees, deer parks usually consisted of woodland and areas largely cleared of trees. The park livestock could graze in the open areas and find cover in the wooded areas. The cleared areas, called launds or plains, consisted of grassland or heath with scattered trees. The king's park keeper at Conisbrough Park in the second half of the fifteenth century was referred to in a document written in French as *Laundier et Palisser de n're park de Connesburgh*, i.e. keeper of the unwooded areas and the park boundaries at our park at

Conisbrough (Hunter, 1828). Many of the trees in the launds would have been pollarded, i.e. trees cut at least 6ft from the ground leaving a massive lower trunk called a bolling above which a continuous crop of new growth sprouted out of reach of the grazing deer, sheep and cattle. In the launds, regeneration of trees was restricted because of continual grazing and new trees were only able to grow in the protection of thickets of hawthorn and holly. Some of the unpollarded trees might reach a great age and size and were much sought after for major building projects.

The woods within deer parks were managed in different ways. Some woods were 'holted', i.e. they consisted of single-stemmed trees grown for their timber, like in a modern plantation. Most park woods were coppiced and were surrounded by a bank or wall to keep out the grazing animals during the early years of re-growth. Later in the coppice cycle the deer would have been allowed into the coppice-woods. There were also in South Yorkshire's deer parks, separate woods or special compartments within coppice-woods in which the dominant tree was holly and that were called holly hags. In Tankersley Park, an engraving of c.1730 shows two holly hags. In the engraving (Figure 3.3) one, the Far Hollings, is on the extreme left surrounded by a stone wall and the other is the small square-shaped wood near the park wall on the extreme right. The holly was cut in winter for the deer and other park livestock. I return to the subject of holly hags below.

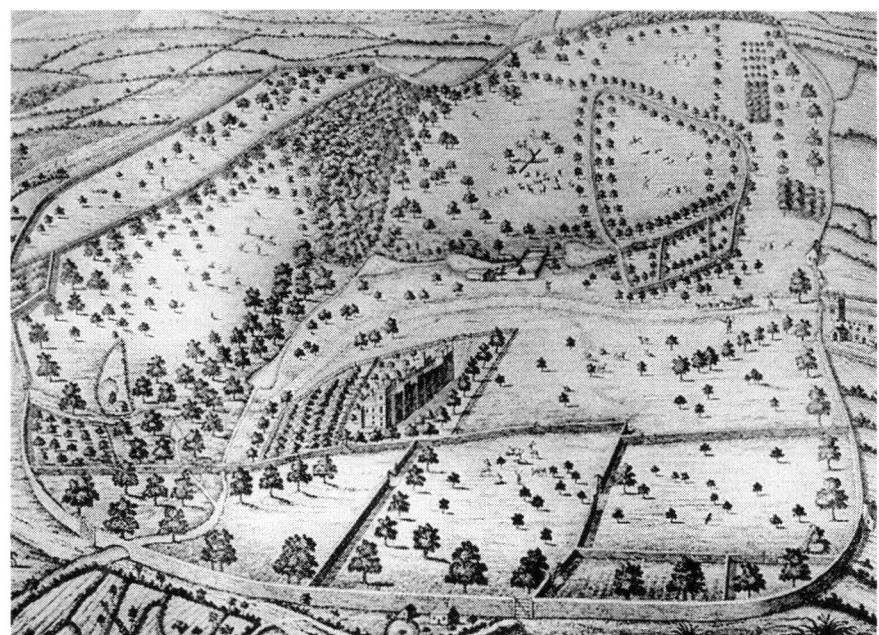

Figure 3.3 Tankersley Park c.1730
Source: from the original in St Peter's church, Tankersley

Sheffield Park is particularly well documented and a clear picture emerges of the woods and trees within it and the way they were used in the medieval period. This deer park, which at its greatest extent covered 2,462 acres (nearly 1,000ha) and was 8 miles (13km) in circumference, came right up to the eastern edge of the town of Sheffield (Figure 3.4). Harrison in his survey of the manor of Sheffield in 1637 named the various parts of the park including some with woodland names including Arbor Thorn Hirst and Stone Hirst (hyrst is a wooded hill) but they would only have been covered with scrub woods of hawthorn and holly. The cleared areas within the park are also precisely named in Harrison's survey: 'ye Lands', 'Cundit Plaine', 'Blacko Plaine' and 'Bellhouse Plaine'. Ye Lands is probably a corruption of laund. These launds or plains contained large, aged oak trees in the seventeenth century, that would have already been very large trees two or three hundred years earlier in the late medieval period. They were described in great detail by John Evelyn in his book, *Silva*, first published in 1670. Evelyn said that in 1646 there were a hundred trees in the park whose combined value was £1,000. He described one oak tree in the park whose trunk was 13ft in diameter and another which was 10 yards in circumference. He also described another massive oak that when cut down yielded 1,400 'wairs', which were planks 2 yards long and 1 yard wide, and

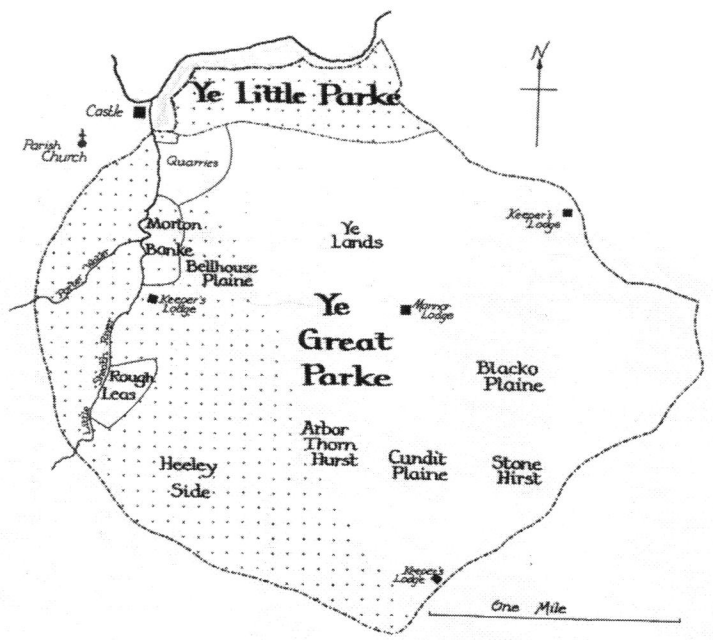

Figure 3.4 Sheffield Park, 1637, based on the description in Harrison's survey of the manor of Sheffield

20 cords from its branches. A cord was a pile of wood 4ft high, 4ft wide and 8ft long. He described another oak, that when felled and lying on its side was so massive that two men on horseback on either side of it could not see each other's hat crowns. On Conduit Plain (the Cundit Plaine of Harrison's 1637 survey), Evelyn reported that there was one oak tree whose boughs were so far spreading that he estimated (giving all his calculations) that 251 horses could stand in its shade (Evelyn, 1706).

These mighty veteran and ancient oaks had had a multiplicity of uses for centuries. When felled they provided not only timber for building projects, but also charcoal (from their branches) and wood for a multiplicity of crafts and industries. Standing, live, veteran oak trees also had an important function. These open-grown enormously branched trees (the one on Conduit Plain described by Evelyn had branches extending for 15½ yards (14m) in every direction) produced, unlike most woodland grown oaks, burgeoning crops of acorns. These not only provided the food for fattening pigs during the pannage season but also for keeping the deer population in good heart in preparation for the long winter.

A 'Feet of Fines' document of 1268 and the mid-fifteenth-century manorial rolls give glimpses of the uses to which wood and timber in Sheffield Park were put in the late thirteenth and mid-fifteenth centuries respectively. A Feet of Fines document was one relating to a judgment on a title to land. This one relates to the dowry of Berta, the widowed wife of Thomas de Furnival, lord of Hallamshire (Thomas, 1924). The judgment stated that Berta de Furnival had rights to a third of the income from certain activities in the park. This income included that derived from pannage and herbage. Herbage (also known as agistment) was the renting of grazing space for cattle, sheep and horses and pannage (*pannagio porcorum*) was the autumnal grazing of pigs on fallen acorns.

The manorial roll of the 1440s (YAS, 1921) shows that besides its role as a food larder for the lord, the park in the mid-fifteenth century supplied firewood for the castle, timber for building repairs at the castle stables and the chapel in the castle. Pollarded oaks were also felled to make scaffolding and hurdles for repairs at the castle. Oaks in the park were felled to make posts, rails and palings for fencing the castle garden. Income was also derived from allowing holly trees to be cropped (for fodder), from the pannage of pigs, the sale of felled trees and a parcel of underwood, and from charcoal made from the branches of trees where they were being cleared to make a new pasture.

Between the late fifteenth and eighteenth centuries many medieval deer parks either changed their function and hence their appearance, or, more commonly, disappeared altogether. Well-wooded parks often simply became large coppice-woods. Examples of the reversion to managed woodland in South Yorkshire are Cowley Park, Hesley Park, Shirecliffe Park, Tinsley Park and what is now Ecclesall Woods.

Holly hags

Holly hags, special woods of holly, have already been mentioned in connection with feeding deer in deer parks in winter but they were once also features of the wider farming countryside where they were used as fodder for sheep until at least the eighteenth century. The use of holly for winter fodder was recorded in the Conisbrough manorial rolls in 1319 when two people were fined for cutting holly for feeding to their animals (Conisbrough Court Rolls in Doncaster Archives: c/1/8-15 and c/1/8-31). Another medieval record was in the Sheffield area in 1442 when the lord of Hallamshire's forester at Bradfield noted in his accounts payment for holly sold for animal fodder in winter (Thomas, 1924). John Harrison in his survey of the Manor of Sheffield in 1637 recorded 27 separate 'Hollin Hagges' that were rented by farm tenants from the earl of Arundel (Ronksley, 1908). The use of holly as fodder in the Sheffield area was also graphically described by two early diarists. In 1696 Abraham de la Pryme wrote that:

> In south-west Yorkshire at and about Bradfield and in Derbyshire they feed all their sheep in winter with holly leaves and bark, which they eat more greedily than any grass. To every farm there is so many holly trees{...}care is taken to plant great numbers of them in all farms hereabouts.
>
> (Surtees Society, 1870)

Abraham de la Pryme went on to say that when the sheep see the shepherd come with his axe they all follow him to the first tree and stand and wait for the first bough to fall to the ground. He said that when they have eaten all the leaves they begin to devour the bark.

In 1725, 29 years after de la Pryme had reported on the custom around Bradfield,, a party headed by the Earl of Oxford travelled through Sheffield in a south-easterly direction across an area of common land still called Birley Moor and Hollinsend. It was noted that they travelled:

> through the greatest number of wild stunted holly trees that I ever saw together. They extend themselves on the common for a considerable way. This tract of ground that they grow upon is called the Burley Hollins{...}[*They have*] their branches lopped off every winter for the support of the sheep which browse upon them, and at the same time are sheltered by the stunted part that is left standing.
>
> (Historical Manuscripts Commission, 1901)

That holly was considered a valuable crop and had to be protected is illustrated by an entry in the accounts of the Duke of Norfolk's woods in the early eighteenth century. In the winter of 1710, the duke of Norfolk's woodward, noted in his accounts that he had paid Henry Bromhead 'for him and horse going 2 days in ye great snow to see if anyone croped holling' (Arundel Castle Manuscripts in Sheffield Archives: S283). The impression given is that Bromhead would have had a blunderbuss over his saddle!

Bull Wood, the deer park holly hag at Tankersley, still exists. Writing in 1977, two local researchers, Spray and Smith, believed that there were at least the remains of five holly hags in the country to the west of Sheffield. They mention Fox Hagg and Coppice Wood in the Rivelin Valley, the holly bushes still surviving at Holly Edge and at Holly Busk near Bolsterstone and a location near the edge of Loxley Common (Spray and Smith, 1977).

Agistment

Tenants' animals were allowed access to coppice woods for grazing on the woodland grasses and herbs once the coppice was well grown and beyond possible damage from browsing animals. There are records of horses and cattle being grazed in local coppice woods. The practice was known as agistment or herbage. For example, in the surviving agistment records for Ecclesall Woods in Sheffield between 1709 and 1714, horses, mares, foals, cows, calves, heifers and stirks (bullocks or heifers between one and two years of age) were agisted. They were there for varying periods of time, from a full agistment (a 'whole gist') of four months, for a 'halfe gist', for a month or for just a few weeks (Bright Papers in sheffield Archives: Br 100). Two records demonstrate the importance of only allowing grazing animals into coppice-woods that were well grown. In 1710, for example, Joseph Ashmore, the duke of Norfolk's woodward, charged himself two shillings for 'My Mare & fole in Woolley Woods this Spring a month' adding, just to make it absolutely clear that he was not contravening the normal custom, 'its old Cutt' (Arundel Castle Manuscripts: S283). Eight years later in 1718 the vicar of Ecclesfield, just after coppicing had taken place in Greno Wood, was paid twopence for giving notice to tenants at a Sunday service that they should 'take care that their cattle do no longer Continue to Graise in Greno Wood for Spoyling ye young sprouts' (Arundel Castle Manuscripts: S283).

Animals were likely to stray from pastures and commons into neighbouring coppice-woods. When detected they were impounded and the owner fined. In 1718, Enoch Moor was fined one shilling when nine of his sheep were pounded out of Greno Wood and in 1720 two men were paid three shillings and sixpence for their trouble in 'pounding 5 sheep belonging to Mr Watts that was trespassing in Little Hall Wood' (Arundel Castle Manuscripts: S283). The village pinfold in which these animals would have been impounded still survives in Grenoside village.

Postscript: Snigging

There is another type of record of the relationship between animals and woodland management in South Yorkshire. Records of coppice management in the period from 1600 to 1850 regularly mention the use of horses in the management of the woods. The term used for the work that horses did was *snigging*, sometimes also known as *tushing*, the pulling of felled trees out of a wood. And in recent years the practice has been reinstated to protect the ground flora and archaeology when group felling and thinning are taking place in Sheffield's and Rotherham's ancient woods.

References

Birrell, J. (1992) 'Deer and deer farming in medieval England', *The Agricultural History Review*, 40: 112–26.

Curtis, E. (1918) 'Sheffield in the fourteenth century: Two Furnival inquisitions', *Transactions of the Hunter Archaeological Society*, 1: 41.

Evelyn, J. (1706 edition) *Silva or a Discourse of Forest Trees*, publisher unknown, 229–30.

Historical Manuscripts Commission (1901) Lord Edward Harley's (later earl of Oxford) *Journies and Tours in the Eastern counties 1723–1738*, Portland Mss, Vol. 6.

Hunter, J. (1828–31) *South Yorkshire: The History and Topography of the Deanery of Doncaster*, 2 volumes, J.B. Nicols and Son, London, Vol. 1.

Jones, M. (1996) 'Deer in South Yorkshire: An historical perspective' in Jones, M., Rotherham, I.D. and McCarthy, A.J. (eds) *Deer or the New Woodlands?*, *The Journal of Practical Ecology and Conservation*, Special Publication, No 1: 11–26.

Jones, M. (2009) *Sheffield's Woodland Heritage*, 4th edition, Wildtrack Publishing, Sheffield.

Rackham, O. (1980) *Ancient Woodland; its History, Vegetation and Uses in England*, Arnold, London.

Rodgers, A. (1998) 'Deer parks in the Maltby area' in Jones, M. (ed.) *Aspects of Rotherham: Discovering Local History*, Vol. 3, Wharncliffe Publishing, Barnsley.

Ronksley, J.G. (ed.) (1908) *An Exact and Perfect Survey of the Manor of Sheffield and other Lands by John Harrison, 1637*, Robert White & Co, Worksop.

Spray, M. and Smith, D.J. (1977) 'The rise and fall of holly in the Sheffield region', *Transactions of the Hunter Archaeological Society*, 10: 239–51.

Surtees Society (1870) *The Diary of Abraham de la Pryme, the Yorkshire Antiquary*, Surtees Society, Durham, Vol. 54.

Thomas, A.H. (1924) 'Some Hallamshire rolls of the fifteenth century', *Transactions of the Hunter Archaeological Society*, 2: 74.

YAS (Yorkshire Archaeological Society) (1921) *Feet of Fines for the County of York 1218–1231*, Record Series, LXII.

4 Re-wilding the landscape

Some observations on landscape history

Della Hooke

Introduction

It is argued here that knowledge of landscape history must underpin any future conservation plans. Landscapes have been subject to change over time and the recreation of wood-pasture, not to mention areas of 're-wilding', can now only be applied to fairly limited areas of countryside. The use of seasonal pasture played an important economic and social role throughout history, and was maintained in Norman forests – although hunting was an elitist activity it paradoxically sustained those of lower status in the community. Today, grazing remains essential if this type of traditional woodland landscape is to be conserved.

The concept of re-wilding

A good many years ago (probably in the 1960s or 1970s), the concept of turning parts of Snowdonia into a 'wilderness' by permitting relatively unrestrained tree growth was broached. But there was a very real danger of thus masking or even wiping out the evidence of many centuries of settlement and land-use change. Almost every upland valley in Snowdonia still has such remains, even if prehistoric remains are visually few. The footprints of the medieval *hafodydd* can still be detected on the *ffridd* pastures that were often enclosed around the sixteenth century, the encroachments around the lowland and upland commons, upland farm units that were abandoned later with the break-up of some large estates, short-lived mining communities exploring the mineral potential, and all their associated field systems and boundaries – such examples are manifold (Hooke, 1997a, 2003). It was even possible to suggest the different periods of stone walling that accompanied this ebb and flow of settlement (Hooke, 1997b). Blanket uncontrolled tree growth would inevitably destroy this kind of evidence and the field patterns themselves. Of course, there were historically more woods and trees, as I was able to show in a study of the parishes of Caerhun and Llanbedrycennin in the Conwy valley in the late 1990s – with settlements bearing such names such as *coed* 'wood', *llwyn* and *gelli* 'grove', or *goitre* 'homestead in a wood', especially in the tributary valleys of the Conwy where woodland had diminished but not disappeared by the nineteenth century. Virtually every upland region of England and Wales is rich in similar kinds of archaeological remains, testament to a region's changing land-use history and, as such, worthy of preservation.

The concept of 're-wilding' came again to the fore in the 1990s, but this time as an approach that was rather better thought out. Essentially this involves the reintroduction of native animal species that are no longer present, plus the restoration of the habitats necessary for their survival. It is clear that any such moves would require careful management and since then various projects and experiments have tested the water so that further steps can be controlled. The idea that minimal human intervention should leave nature in control might just be a possibility in other parts of the world but not in Britain, and cultural implications always deserve consideration. Indeed, it is widely recognised that the ubiquity of human disturbance forces us to 'confront the fact that we cannot have wilderness that is truly wild or natural' (Cole, 2001, cited by Hodder and Bullock, 2009: 41). However, large-scale core wilderness areas with connecting corridors might still be achievable in the long run if local support could be gained.

Plans for the reintroduction of such species as moose, brown bear, elk, wild boar, lynx and wolves on an estate in Scotland have so far proved highly contentious (e.g. MCofS, n.d.). Indeed, for most of Britain it would be a dramatic reversal of our country's history. The white-tailed eagle has been successfully reintroduced on certain isles off the west coast of Scotland but rejected in Suffolk and currently beavers are being reintroduced – but carefully monitored – on the Allandale estate, also in Scotland. Arguments have been presented for the reintroduction of the lynx, one of the most endangered European mammals, and it has been estimated that this country could support about 450 lynx (Hetherington, 2008). Lynx mostly take deer but sheep would, for safety, have to be pastured some distance away from woodlands. Some of our native animals (i.e. those present after Britain had been severed by rising sea levels from the Continent) were lost at an early date. The aurochs and elk, for instance, seem to have virtually disappeared during the Bronze Age and the brown bear much later, probably in the early medieval period. But the beaver, lynx, wild boar and wolf were still present in medieval times, the wolf surviving in Scotland until 1743 and slightly longer in Ireland (Yalden, 2003). The beaver and wild boar were hunted out but the latter has returned from escaped stock, as in the Gloucestershire Forest of Dean. Some other mammals have experienced diminishing populations but the polecat is recovering in numbers, especially in Wales.

From a cultural point of view the concept of 'wilderness' is attractive to some who relish the 'appearance' of wildness and a sense of remoteness

– but there are always others who oppose landscape intervention of this kind for a variety of reasons. Woods are seen as 'savage' and 'threatening' and they give rise to the fears expressed in fairy tales and mythology. Throughout the early medieval and medieval period wilderness in literature and legend is regarded as a place of demons, unknown perils and uncertainty (Hooke, 2010a: 69–75; Neville, 1999: 127).

Landscapes managed for ecology are not necessarily the (to some) tortured landscapes of Renaissance gardens or even the manicured lawns of later garden designers – 'wild' landscapes were not to be appreciated until the idea of 'romantic' landscapes raised its head with a new body of artists later in the eighteenth century. Wilson's *Cader Idris*, for instance, painted in the 1760s depicts a rough landscape of primitive simplicity, a retreat from an ever-more complicated world. The attraction of wild landscapes was further expressed the writings of such people as John Muir in the later nineteenth century and early twentieth century (e.g. Muir, 1911). But still, for some, 'wild' landscapes are not necessarily producing the 'managed' landscapes that imply order and civilisation, the 'tidy' landscape of suburbia. And some would argue that they represent 'abandoned wildernesses' as opposed to the preferred landscapes of agricultural productivity – as Arthur Gibbs, in his *A Cotswold Village*, expressed his idea of the rural idyll. His favoured view was of :

> A wide of undulating downland, divided into fifty-acre fields by means of loose, uncemented walls of grey stone. The grass is green for the time of the year, and scattered about are horses, cattle, and sheep, contentedly nibbling the short fine turf.
> (Gibbs, 1868–1899/1988: 98–9).

It has been said that the English prefer a countryside that is '*tamed and inhabited, warm, comfortable, humanized*' rather than wild (Lowenthal and Prince, 1965: 190).

Whatever changes are made to restore former habitats would obviously have considerable impact on the landscape. Re-wilding may represent a possibility where large tracts of open countryside survive, as in parts of Scotland, but elsewhere will probably have to be restricted to rather smaller areas. One has to approach the subject from a number of viewpoints and one would hope that these might include the historical as well as the ecological. Historical landscapes are more than mere ecosystems: they illustrate ways in which man has adapted wild nature for his own use in ways that have been sustainable for generations. While future changes in land management and farming are inevitable it will be a challenge to accommodate these while preserving – and actually improving – the habitats available for our native fauna and flora.

Traditional landscapes

If re-wilding is accepted as an aim for some places, no single era can assume any immediate right to supremacy as the preferred objective. Change has contributed to the character of the landscape we see today. The landscape has never been static and one might question just what kind of landscape we wish to conserve, and for whose benefit, but the historical development of the landscape needs to be fully understood. Our flora and fauna have indeed survived and thrived in the *traditional* landscapes of Britain, produced by man's land-use practices over thousands of years. Such landscapes contribute vastly towards the maintenance of 'countryside character' as well. It is not just the animals of the 'wilderness' we need to consider but our general fauna, especially as we appear to be producing increasingly damaged habitats. Insect life is jeopardised by over-reliance upon the widespread use of insecticides and many bird species are diminishing in number as food supplies are reduced; some small mammals are equally endangered. Thus problems arise not only from competing introduced species such as mink and for a short time, coypu, but from present-day land-use practices. Traditional landscapes are potentially just as, if not more, important than 'wilderness' and commendable moves towards the conservation maintenance of such landscapes have been taken under the Countryside Stewardship and later Farm Environment schemes, largely in response to the vast changes that took place in the second half of the twentieth century. Within them, trees and woods often play significant roles but the impacts of people have also left evidence of historical significance:

- Lowland heath was one such 'traditional landscape', much of which was destroyed in Dorset, for instance, by programmes of conifer planting after the two World Wars. Ironically, military usage helped to preserve this valuable wildlife habitat in some areas, as on the tank ranges to the north-east of East Lulworth. Yet, there have sometimes been problems with heathland restoration – the scraping away of topsoil can destroy the archaeological evidence of flint scatters or other prehistoric remains; the total removal of trees ignores the historic nature of many such landscapes. Elsewhere too, other open habitats and 'unproductive scrub' have been converted to productive plantations that generally support much more limited ecosystems. Plantations are not necessarily 'bad', and populations of hen harriers, owls, sparrow hawks and crossbill for example, may increase. However, they do create a very different kind of countryside, only occasionally or temporarily provided with the open areas that encourage many other species, and cannot match deciduous woodland in this respect.
- Open moorlands also are not secure from change. On the flanks of upland hills and on infertile granite or similar outcrops in the lowlands, moorlands were greatly extended in prehistoric times. This was especially

the case as grazing was extended from the Bronze Age onwards, with cattle being a major sign of wealth. Tree regeneration was hampered and deteriorating climatic conditions at the end of the Bronze Age further damaged fragile soils. In many areas, there was a change from a tree level as high as 700–800 metres to the open barren views of today; leaching or peat formation prevented further tree growth. However, the moorlands became a habitat for such birds as the merlin and hen-harrier, which could more easily catch their prey over open ground. With the predatory birds targeted by nineteenth- and twentieth-century gamekeepers, anxious to protect their black grouse for their shoots, many have become threatened species and the dangers are not yet over. It has also been shown how raptors can help to keep game bird populations healthy by removing ill and infected specimens (Davies, 2005).

There are other conflicts: today conservationists welcome the presence of 'wild' ponies such as the Exmoor and Dartmoor breeds. Yet these animals are not really 'wild' and farmers, who now derive little economic benefit from their presence, are advocating culling young foals to avoid the cost of winter provisions. The animals help to keep the gorse – no longer needed for roofing etc. – from invading open habitats.

- The extension of ploughland, again stimulated by wartime experiences and made possible by the addition of vast amounts of chemical fertilisers, irrevocably changed much of our chalk downland. Again, Celtic field systems that had survived for over a thousand years were destroyed by a few ploughings.

Other traditional farmed landscapes have suffered considerable decline over the last 50 years or so. Indeed, open-field landscapes are now mere fragments following eighteenth- and nineteenth-century enclosure. However, the grass baulks of some of these fragments, as at Forrabury in Cornwall, still carry wild flowers, as do ancient grassland pastures, many of which arose from Tudor enclosure; many of these species-rich old grasslands have now been ploughed up. Over 63 per cent of our old orchards, too, have been grubbed up since the 1950s, leading to pollen loss. These orchards historically helped to maintain insect and bee populations.

- Our water meadows were an essential way of providing hay for winter feed throughout history. Although advanced methods of irrigating them were a relatively late development, these were aimed at stimulating the early growth of grass for grazing sheep. The meadows, which would originally have been subjected periodically to natural flooding, also conserved a rich flora. However, they have been much encroached upon by building and roads, the latter frequently choosing to follow relatively flat valleys. Brian (1993) shows how our common meadows were decimated by such changes. The loss of our wild flowers can only have had a detrimental effect upon bee populations, essential for pollination. Some also still show evidence of the channels that represent

attempts to irrigate or 'flood' them, mainly in western and southern England constructed in the seventeenth and eighteenth centuries (Williamson, 2002: 59–62).
- Many people are aware of hedgerow loss – hundreds of miles of hedgerows, some ancient, have been removed to make way for extended arable farming with large machines. The patterns of old hedgerows were a significant feature of the historical character of many regions, not only fossilising field systems that in some areas can be traced back to Roman times, but also acting as valuable corridors for wildlife and providing homes for nesting birds. But even here there are caveats – while few would question their benefit as wildlife corridors, deliberate hedgerow creation has its drawbacks. If one knows the historical character of the countryside, there are some regions that have been open arable for well over a thousand years – such as the arable lands of parts of Cambridgeshire – and here the introduction of hedgerows would be historically quite wrong. Equally beneficial results might be achieved with unsown field margins, beetle banks or patches left for wild plants and grasses.
- Wetlands, once important for fishing and fowling and the gathering of rushes and reeds for, among other uses, roofing thatch, have often been destroyed by being drained for agriculture. This has been particularly detrimental to our wild bird populations and at an early stage led to the loss of species such as the stork, crane and spoonbill. However, the spoonbill, as a bird migrating to the UK in summer, successfully nested at Martin Mere in Lancashire in 1999 and is an occasional visitor (WWT Martin Mere, 2010) while the crane is being reintroduced on the Somerset Levels and Moors (RSPB, 2011), and has re-established in East Anglia and in South Yorkshire (Ian Rotherham, pers. comm.). In the Fenlands, extensive reclamation followed the adoption of wide-scale schemes often first carried out by Dutch engineers in the seventeenth century. Today an attempt is being made to conserve or recreate area of fenland, often associated with areas of willow and alder carr, largely for the benefit of bird populations, as at Wicken Fen in Cambridgeshire.
- Upland landscapes are of many kinds but indiscriminate grazing, as encouraged in recent decades by subsidies based upon numbers of stock kept, not only destroys trees but damages variegated grassland. This has the effect of suppressing the ground flora with the number of sheep increasing nearly fourfold in the twentieth century. With some 25 per cent of the Welsh upland affected by the spread of conifer plantations since 1945 this has meant increased pressure on the remaining open land (Green, 2003: 408; Wildlife Trusts, 1996) (Figure 4.1). Further, 'improvement' by the drainage and reseeding of upland pastures further undermines the historical character of the countryside as well as being destructive to the richness of ground flora.

Figure 4.1 Conifer plantations in Cwmsylfaen near Bontddu, Gwynedd
Source: D. Hooke

Upland landscapes are indeed 'on the edge' of change. With upland farming providing such poor returns it has been suggested that eco-tourism might be a better source of finance. Certainly a decline in sheep numbers would be ecologically beneficial and there is undoubtedly scope for the extension of woodland in upland valleys. But there is also a need for eco-tourism to be carefully managed if unsuitable activities are not to take place.

Wood-pasture landscapes

The author's specific interest lies with wood-pasture landscapes. Many upland pastures still show traces of an earlier wood-pasture usage and are generally species rich but many wood-pasture regions were also characteristic of lowland Britain (Figure 4.2).

This is a type of landscape that does not fit readily into modern farming but is one of our most treasured traditional landscapes. From at least the late Iron Age into the early medieval period, wooded regions played an important economic and social role. The pasturing of domestic stock, often initially driven for considerable distances to their summer pastures, was a basic feature of the rural economy. The woodland was thus maintained as open in character but well scattered with large native trees, especially those that survive well in such conditions – like the oak, the young shoots of

Figure 4.2 A wood-pasture scene in Sutton Park near Birmingham
Source: D. Hooke

which are unpalatable to stock and which can grow again from a deep root once the top growth has been bitten-off (Rackham, 2003: 293). The distribution of oak – and ash, which readily regrows if stocking numbers are diminished – can be mapped from early medieval place names and charters and clearly shows marked concentrations in the regions known to have been well-wooded areas of seasonal pasture in pre-Conquest times (Hooke, 2010a: 192 Figure 14, 193–200).

The pig was perhaps the most important animal pastured in this way, although cattle, horses, sheep and even goats are mentioned in contemporary documents. The fact that the pigs rooted around the trees helped to bring in light and ensure good crops of acorns (Green, 2010: 57). In the south of England, beech-mast was an equally important source of seasonal food. Pre-Conquest charters offer the earliest detailed documentary accounts of this practice, especially for the Weald of south-eastern England (Hooke, 2011). Here domestic animals – pigs, cattle/plough-beasts, sheep, horses or goats – were driven annually from estates in the surrounding lowlands to seasonal pastures called dens. Sometimes information is also provided about the numbers of pigs involved – such as the 120 pigs and 50 cattle pastured in two dens belonging to estates held by Christ Church, Canterbury (Sawyer, 1968, S 323). Domesday Book records that the Warwickshire estate of Stoneleigh in its amassed woods had pasture for 2,000 pigs (Plaister, 1976: 1, 4).

Forests after the Norman Conquest helped to preserve deciduous woodland across much of England; these maintained populations of deer, wild boar and other woodland animals, the fallow deer being reintroduced in large numbers. The use of seasonal pasture was kept alive in the Norman forests, even if sometimes it led to conflict – the open nature actually helping huntsmen to follow their quarry. Neither should the social role be forgotten – although hunting was an elitist activity it paradoxically sustained those of lower status in the community. Real conflict occurred when woods were increasingly taken into private ownership, as on the Paget estates on Cannock Chase in Staffordshire in the eighteenth century. Here, attempts to enclose coppices to provide wood for the family's ironworks reduced the commoners' rights to take loppings from trees and their rights of pasture for their stock, leading to riots in the sixteenth century (Harrison, 1999: 103). Riots also occurred when enclosure affected the remaining open commons in the eighteenth and nineteenth centuries, as at Ogley Hay in south Staffordshire. This was one of the small administrative divisions known as hays that had survived the breakup of the former Cannock Forest, but following enclosure in 1838/39 the access roads used by surrounding communities to take their animals into the woods were deliberately stopped up, leading to unrest. Patches of former forests and chases that escaped desecration at the hands of eighteenth- and nineteenth-century agricultural improvers may on occasions still be blessed with wood-pasture landscapes.

Woodlands are rich in archaeology. On occasions they have apparently regrown over earlier landscapes – linear dykes are found within Wychwood in Oxfordshire and at Blunt's Green in the Warwickshire Arden that appear to represent Iron Age enclosures constructed close to territorial frontiers. Trees have colonised abandoned Iron Age hill forts in the Welsh Borderland, and at Welshbury hill fort, in the Forest of Dean, the limes are probably growing from a rootstock as ancient as the earthworks themselves. In northern Hampshire, in the woods of Faccombe and Crux Easton, it is still possible to trace the so-called 'Celtic' field systems that were in use in late Iron Age and Roman times; these were abandoned in post-Roman times and the woods became important areas of royal game reserve (Hooke, 1988). Trees soon move in if fields or buildings are neglected, as wild nature regains control; thorns rapidly colonise abandoned fields and pastures and trees sprout on crumbling walls, like the gnarled yews that grow around and through the walls of Craswell Priory in Herefordshire.

Other archaeological relics reveal former woodland usage: woodbanks can be traced that are testament to changing management, as in Sutton Park near Birmingham. Here some of them bounded the hollins that were a valuable source of leaf fodder in the Middle Ages. Since coal measure rocks are usually infertile, these have often remained wooded in the midland region. In the Forest of Dean, in Gloucestershire, the coal measures overlie iron ore beds that supported ironworking by late Iron Age times and continued use of these iron resources has often left a landscape of heavily disturbed rough ground within woodland, partly arising from the mining of iron from shallow outcrops and eroded caves systems (for the origin of such scowles, see Gloucestershire City Council, 2004, 2008), plus the associated features of charcoal-burning platforms. The iron ore was generally roasted prior to smelting to remove moisture and impurities and was then refined by further heating in a bloomery, which required about 2 acres of wood annually for each ton of iron produced (Cleere and Crossley, 1985: 100). The charcoal-fired blast furnace replaced the bloomery as continental technology was introduced into England in the mid-fifteenth century and by the mid-seventeenth century Dean had the greatest concentration of ironworks in the country with 20 blast furnaces in the region (Riden, 1993); it was not until 1795 that coke-fired furnaces were introduced at the Cinderford Ironworks, so plentiful were the supplies of charcoal. In Dean, the Weald of south-east England and other early ironworking regions, the charcoal-burning areas may be revealed by earthen platforms, still often floored with charcoal, that are often connected by systems of trackways. The industrial features that were located close to these supplies of charcoal – the bloomeries, blast furnaces and forges – have often left significant archaeological remains, added to which are the hammer-ponds or their dams and leats, wheel-pits, masonry and quantities of slag and cinder. In some woods, coal mining has given rise to both early bell-pits and later features associated with shaft mining. Built by people who frequently combined mining with small-scale farming, cottages were often built around the edges of woodland commons in the post-medieval period, as in Dean, giving rise to straggling hamlets or isolated settlements, some now long abandoned but still traceable on the ground (Hooke, 2010b). These and other indicators of ancient crafts, such as sawpits, make woodlands vulnerable areas for any extension of the woods.

Wood-pasture landscapes have often been conserved in historic parklands, as at Staverton in Suffolk, and here trees were often pollarded to encourage the growth of timber out of the reach of grazing animals, giving rise to some of our most valuable veteran trees. Today, the importance of managing these pollard trees, so much a feature of ancient wood-pasture, is fully recognised. Parks are actually some of our most precarious landscapes and have undergone change throughout history. Medieval parks were initially game reserves, with deer kept for hunting and a source of venison, but as the fashion began to have them as an ornamental feature

Re-wilding the landscape 45

Figure 4.3 Old oak pollards in Moccas Park, Herefordshire
Source: D. Hooke

around a large Tudor mansion so they became expressions of good taste, beauty and wealth. Parkland design changed with fashion taste but the 'natural' landscapes favoured by Brown and Repton preserved something of their earlier appearance – at Moccas in Herefordshire the ancient pollards were retained by Brown, possibly inherited from an earlier existence within the Forest or park of Dorstone (Figure 4.3). Many parks were made, however, over former farms and fields; in an otherwise praiseworthy wish to replace trees as ornamental features in their own right, I have seen the remains of deserted medieval villages, with their house platforms and tofts, almost totally destroyed by injudicious planting.

Woodland management

Both upland and lowland woods were heavily planted with conifers after the two World Wars in an attempt to make Britain more self-reliant and less dependent upon imported materials. This led again to the loss of deciduous woodland and wood-pasture. While conifer plantations may support their own kind of wildlife, including the goshawk and some species of owl, and may eventually revive mycorrhizas in the soil, the extension of native deciduous woodland seems to be the preferred objective. Not only does this support a rich wildlife but people like deciduous trees:

> Deciduous trees please the English because they are delicately patterned, softly outlined, varied in form and color, scrambled in texture, seasonal in foliage, tolerant of undergrowth, and generally older than conifers. [Whereas] Conifers are considered gloomy, harsh, and oppressive, partly because many of them are strictly commercial.
>
> (Lowenthal and Prince, 1965: 197)

In Ennerdale Valley, Cumbria, native woodlands are being allowed to spread to promote the objective of a wilder landscape there and Galloway (highland) cattle have replaced sheep. The Brecon Beacons Management Plan draft (2009) also noted how :

> The point is fast approaching whereby what biodiversity remains will no longer be sufficient to provide a sustainable future for both habitats and species. What remains today is only a fraction of what was here in the past.
>
> (BBNP, 2009: 59)

As result it was recommended that, while commercial forestry must continue, and could complement the open landscapes of unimproved upland grazing at higher altitudes, the lower valley native trees should be allowed to expand further onto the higher slopes where they are already an existing feature, and also to migrate into the commercial plantations with additional broadleaf planting. In the Mynydd Du Forest, broadleaf planting should increase from 7 per cent of the woodland cover to some 33 per cent by a combination of felling and replanting and natural regeneration management (BBNP, 2011: 102).

If woods are not pastured – or if the effects of pasture are not deliberately copied – then there is a danger that they become a tangle of trees, bracken and bramble. One area of woodland, Lady Park Wood in the Wye Valley, has been kept with the minimum of intervention and has remained relatively open through natural tree fall and browsing by deer, but the latter, given that no higher carnivores survive today, are now having to be culled (Peterken, 2005) so that no real 'natural' habitat seems possible today. By removing predators, deer may now increase in number so as to cause considerable tree damage.

In general, historical management methods of managing woods have also often faded away – the creation of varying habitats by coppicing, for instance, has been neglected, to the detriment to many species. Coppices were of enormous value to iron manufacturers in the seventeenth century, as in Dean or around Coalbrookdale. They were especially important for producing wood for the charcoal furnaces, and a wide range of usable wood for hurdles, building etc. and other commodities such as hazel nuts. The regular cycle of light and shade, together with the grazing being restricted when coppices were first cut, favoured such woodland plants as wood anemones, bluebells and foxgloves; the glades and scrub were good for nesting birds and insects together with many species of butterfly. In the Forest of Wyre on the Worcestershire/Shropshire border, pigs are again being introduced in Wimperhill Wood to clear scrub from part of a (possible) prehistoric enclosure (ex inf. Adam Mindykowski, Worcestershire Archaeology Unit) (Figure 4.4).

Figure 4.4 Pigs being pastured in Wimperhill Wood, Forest of Wyre
Source: Adam Mindykowski

There are many present-day schemes for managing woodlands and heathlands by pasturing stock – whether the animals are Longhorn cattle, as in Windsor Forest or Epping Forest, or, on the Malvern Hills on the Herefordshire/Worcestershire border, where hardier breeds are required, Highland or Belted Galloways (sheep, eating shorter grass, are also put onto the hills to complement the cattle grazing). Elsewhere, as in Wyre, pigs are used for selected purposes.

It is widely recognised today that woodland has an added recreational value, a concept that has been taken on board by such projects as the National Forest and various community forests and may be a strong feature in most upland reafforestation schemes. In all of these, it is recognised that woodland management should ideally incorporate pasturing by a range of animals. Even if wood-pasture can only be sustained in limited regions, such controlled grazing remains essential if this type of traditional woodland landscape is to be conserved.

Conclusions

Re-wilding on a large scale for the introduction of now extinct species may have only limited potential over much of Britain. However, the reinstatement or conservation of traditional landscapes has a major role to play in both preserving the habitats of our native fauna and flora and also in preserving the regional character of our countryside. All attempts to manage the landscape should pay close attention to man's contribution towards creating these landscapes in the first place and ensure that any surviving historical evidence is not destroyed. In a situation in which many habitats are becoming degraded and where the quality of landscape is also being diminished, the recreation of treescapes, especially the kind of open woodland that once characterised wood-pasture regions, has much to offer from almost every angle: timber resources, beneficial ecosystems and wildlife habitats, and places for recreation or personal wellbeing. In them animals, man and nature combine to produce a rich milieu of both biodiversity and beauty.

Today there are worrying pressures: not least the new emphasis upon more intensive food production. There is a danger that environmental concerns may be given lower priority in the political agenda; this is something we all need to be aware of.

References

BBNP (Brecon Beacons National Park) (2009) 'Brecon Beacons National Park Management Plan 2010–2011, Draft Version 2', www.breconbeacons.org/the-authority/planning/strategy

BBNP (2011) 'Brecon Beacons Management Plan 2010–2011', www.breconbeacons.org/the-authority/planning/strategy

Brian, A. (1993) 'Lammas meadows', *Landscape History*, 15: 57–69.

Cleere, H.F. and Crossley, D. (1985) *The Iron Industry of the Weald*, Leicester University Press, Leicester.

Cole, D.N. (2001) 'Management dilemmas that will shape wilderness in the 21st century', *Journal of Forestry*, 99: 4–8.

Davies, R. (2005) *British Wildlife*, 16(5): 339–47.

Gibbs, J.A. (1868–99) *A Cotswold Village*, Reprinted 1988, Allan Sutton Publishing, Stroud.

Gloucestershire City Council (2008) 'Scowles', www.gloucestershire.gov.uk/index.
Gloucestershire City Council (2004) 'The scowles of the Forest of Dean', Archaeology Service, www.gloucestershire.gov.uk/utilities
Green, M. (2003) 'The Welsh uplands – past, present and future', *British Wildlife*, 14(6): 403–12.
Green, T. (2010) 'The importance of open-grown trees – from acorn to ancient', *British Wildlife*, 21(5): 334–8.
Harrison, C. (1999) 'Fire on the Chase: Rural riots in sixteenth-century Staffordshire', in P. Morgan and A.D.M. Phillips (eds) *Staffordshire Histories. Essays in Honour of Michael Greenslade*, Staffordshire Record Society & Centre for Local History, University of Keele, Keele, pp97–126.
Hetherington, D. (2008) 'The history of the Eurasian lynx in Britain and the potential for its reintroducton', *British Wildlife*, 20(2): 77–86.
Hodder, K.H. and Bullock, J.M. (2009) 'Really wild? Naturalistic grazing in modern landscapes', *British Wildlife*, 20(5): 37–43.
Hooke, D. (1988) 'Regional variation in southern and central England in the Anglo-Saxon period and its relationship to land units and settlement', in D. Hooke (ed.) *Anglo-Saxon Settlements*, Basil Blackwell, Oxford, pp123–52.
Hooke, D. (1997a) 'Place-names and vegetation history as a key to understanding settlement in the Conwy valley', in N. Edwards (ed.) *Landscape and Settlement in Medieval Wales*, Oxbow Mongr, 81, Oxford, pp79–95.
Hooke, D. (1997b) 'The effect of English settlement in medieval north Wales', in G. de Boe and F. Verhaeghe (eds) *Rural Settlements in Medieval Europe*, I.A.P. Rapporten 6, Zellik, pp331–44.
Hooke, D. (2003) 'Place-names and land use in coastal Ardudwy, with comparisons with the Conwy valley in north Wales', in T. Unwin and T. Spek (eds) *European Landscapes: From Mountain to Sea*, Huma Publishers, Tallinn, Estonia, pp139–45.
Hooke, D. (2010a) *Trees in Anglo-Saxon England: Literature, Lore and Landscape*, Boydell Press, Woodbridge.
Hooke, D. (2010b) 'Early wood commons and beyond', *Landscape Archaeology and Ecology*, 8(1): 107–20.
Hooke, D. (2011) 'The woodland landscape of early medieval England', in N.J. Higham and M.J. Ryan (eds) *Place-Names, Language and the Anglo-Saxon Landscape*, Boydell Press, Woodbridge, pp143–74.
Lowenthal, D. and Prince, H.C. (1965) 'English landscape tastes', *Geographical Review*, 55(2): 186–222.
MCofS (Mountaineering Council of Scotland) (n.d.) 'Position statement on Alladale Project', www.mcofs.org.uk/assets/access
Muir, J. (1911) *My First Summer in the Sierra* (London), reprinted in *The Eight Wilderness-Discovery Books* (1992), Diadem Books, London.
Neville, J. (1999) *Representations of the Natural World in Old English Poetry*, Cambridge University Press, Cambridge.
Peterken, G.F. (2005) 'Natural woodland reserves – 60 years of trying at Lady Park Wood', *British Wildlife*, 17(1): 7–16.
Plaister, J. (1976) *Domesday Book, 23, Warwickshire*, Phillimore, Chichester.
Rackham, O. (2003) *Ancient Woodland, its History, Vegetation and Uses in England*, new edition, Castlepoint Press, Dalbeattie, Kirkcudbrightshire.
Riden, P. (1993) *A Gazetteer of Charcoal-Fired Blast Furnaces in Great Britain in Use since 1660*, Merton Priory Press, Cardiff.

RSPB (Royal Society for the Protection of Birds) (2011) 'The great crane project', www.rspb.org.uk/supporting/campaigns/greatcraneproject/project.aspx

Sawyer, P.H . (1968) *Anglo-Saxon Charters, an Annotated List and Bibliography*, Royal Historical Society, London.

Wildlife Trusts (1996) *Crisis in the Hills – Overgrazing in the Uplands*, The Wildlife Trusts, Lincoln.

Williamson, T. (2002) *The Transformation of Rural England. Farming and the landscape 1700–1870*, University of Exeter Press, Exeter.

WWT Martin Mere (Wildlife & Wetlands Trust) (2010) 'Saving wetlands for wildlife and people', www.org.uk/visit-us/martin-mere/news

Yalden, D.W. (2003) 'Mammals in Britain – a historical perspective', *British Wildlife*, 14(4): 243–51.

5 Rethinking pannage

Historical interactions between oak and swine

Péter Szabó

Introduction

Pannage is the 'fattening of domestic pigs under [oak and sometimes beech] trees when there is a crop of mast' (Rackham, 2003). In Europe, pigs have been driven into woods since prehistory (Grigson, 1982; Hamilton *et al.*, 2009). Along with pigs, pannage was also exported to the US (Shaw, 1940). The period of pannage (with some local variation) was from September to December. Although an important resource, it is necessary to stress that mast was not enough to rear pigs. Good mast years do not happen every year, and because of the restricted time available for the average domestic pig, 'not all pigs lived to experience a mast year' (Bruun and Fritzbøger, 2002). Even when there is good mast, acorns do not provide a full diet for pigs. In bad mast years, as well as off season (in other words most of the time), pigs had to be fed some other food. The point in pannage was the opportunity to fatten up pigs quickly before slaughtering – it was an occasional but welcome bonus in a pig husbandry that was not dependent on it. Pannage declined in importance in the modern period, but it persisted in, for example, southern Europe or the US until recently (Shaw, 1940).

Pannage is profusely discussed in historical written sources starting from the Roman period and the early Middle Ages (Wickham, 1994). In medieval and early modern Europe, pigs were a taxable asset and generally belonged to the sphere of interest of the landowner. Even when firewood and timber were freely accessible to peasants, acorns and beech-mast remained the exclusive property of landowners. As a result, pannage left many legislative records (records of customary law and of disputes over pannage rights) (Birrell, 1987), as well as financial documentation (records of money received for allowing pigs into woodlands). The fact that masting does not occur every year was well-known, and medieval documents usually referred to this by including a formula, such as '*cum acciderunt*' ('when [acorns] happen'). The number of pigs a certain wood could support was sometimes used as a rough-and-ready unit of measurement to describe the size of the wood. The best known example of this is some parts (but not all) of the *Domesday Book* (1086) in England (Rackham,

2003), but the practice was known elsewhere as well. In addition, pannage features in many high and late medieval books of hours, the widespread and popular calendars of the period, which have been often used in research about medieval land management (Wieck, 1988).

Because of its copious documentation, historians tend to view pannage as a highly important part of medieval and early modern woodland economy (Darby, 1950; Wilson, 2003). However, pannage is usually discussed in general terms and we know relatively little about its everyday details, geographical distribution pattern and ecological consequences. Based mostly on Central European medieval and early modern material, this chapter sets out to examine some of the issues connected to the history of pannage that have so far received less attention in scholarly literature. In particular, I focus on 1) possible changes in the geographical distribution of pannage through time, 2) sources on the frequency of mast years in the past, and 3) possible connections between the current distribution of oak woods and historical pannage.

Changes in the geographical distribution of pannage

In many regions of Europe, pannage was practiced continuously for centuries. The Carpathian Basin is one of these regions. Several oak species are native here (*Quercus robur, Q. petraea, Q. pubescens, Q. cerris*) and oaks grow almost everywhere except in the highest mountains. To establish in which areas of the Carpathian Basin pannage was the most important and whether there was any change in the pattern observed, I processed two kinds of sources from the fifteenth and the eighteenth centuries.

In the fifteenth century in the Kingdom of Hungary (which roughly coincided with the Carpathian Basin) in certain legal issues (e.g. estate divisions, dowry and arrest upon high treason) authorities were obliged to find out how much money someone's possessions were worth (Szabó, 2003). Possessions included plots of peasants, land, buildings, livestock, mills, etc. In such cases, an *estimation* was commissioned. Among other things, estimations recorded different kinds of woodland, one of which was *silva glandinosa*, or acorn-bearing wood. In practice this term referred to a woodland with pannage (Szabó, 2005). With the exception of a few larger sources that cover dozens of settlements, most estimations describe a single village. Estimations are elusive sources and have to be searched for individually. I managed to find data for 360 settlements or c.300,000ha. Although this represents only a small portion of the Carpathian Basin, the geographical coverage is relatively satisfactory: 33 of the 63 counties in the Kingdom of Hungary (as they stood at the end of the fifteenth century, excluding counties in Transylvania) are illuminated by the estimation of at least one settlement. In Figure 5.1, the importance of pannage in each country is expressed as the percentage of settlements with acorn-bearing woods among all settlements for which data are available.

The other source I used was the *Regnicolaris conscriptio* of Hungary from 1715. This is the first general taxation survey of the kingdom, commissioned by the Habsburgs shortly after the expulsion of the Ottomans from Hungary (it is therefore in certain aspects similar to the *Domesday Book*) (Dávid, 1957; Magyar Országos Levéltár (Hungarian National Archives) (MOL) N 78). In each settlement the kinds of woodland were recorded, including information on whether pannage was available or not. Pigs were pannaged in 1,653 settlements out of the total 6,918, i.e. approximately every fourth settlement had pannage. We should also note that 1,462 settlements had no woodland at all, which means that c.30 per cent of those settlements that had woodland practised pannage. Figure 5.2 shows the percentage of settlements with pannage among all settlements in each county.

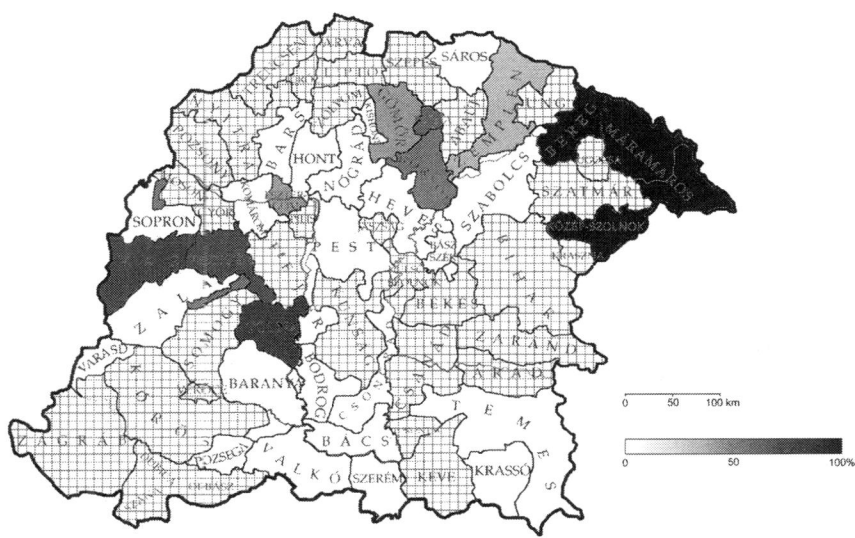

Figure 5.1 The relative importance of pannage in the Carpathian Basin in the fifteenth century expressed as the percentage of settlements with acorn-bearing woods among all settlements in each county
Note: Hatched areas have no data.

Figures 5.1 and 5.2 are separated by around two and a half centuries. Although the medieval dataset is rather fragmentary, the comparison offers relevant conclusions. Two permanent strongholds of pannage in the Carpathian Basin are apparent: one in the north-east (around today's border area between Romania, Ukraine, Slovakia and Hungary), and the other in central western Hungary around Lake Balaton. In one area there

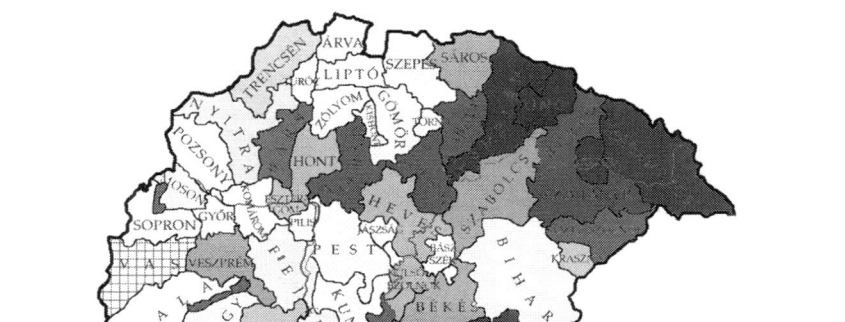

Figure 5.2 The relative importance of pannage in the Carpathian Basin at the beginning of the eighteenth century expressed as the percentage of settlements with documented pannage among all settlements in each county
Note: Hatched areas have no data.

was change: counties Bars, Hont and Nógrád in central northern Hungary had little pannage in the Middle Ages but significant amounts in the eighteenth century. The limitations of the medieval data are demonstrated by the region of counties Csanád, Zaránd and Arad in the south, which appears to have been a significant pannage area in the eighteenth century but has no data for the fifteenth century; therefore I cannot assess whether continuity or change was characteristic here.

Data show that pannage has its own spatial dynamics. In some regions it could be practised with the same intensity for centuries, in others its significance may have changed. The reasons behind this can only be guessed at. The occurrence of mast years is driven to a certain extent by climate, and climate change can have different consequences in regions with differing natural conditions. Oak woods themselves may fluctuate with time, causing changes in pannage patterns. Socioeconomic changes cannot be left of consideration either (Szabó and Hédl, 2012). Pannage may disappear if people choose to rear cattle instead of pigs. In other cases, pannage will disappear only from the sources, for example from account books, if it becomes free and stops generating income. Separating actual changes in pannage from the illusion of change created by historical sources can be a considerable challenge.

Sources on the frequency of mast years in the past

There are two ways to study the frequency of mast years in periods not covered by modern observational studies: dendroecology and archival research. Dendroecology can identify the reduction in tree-ring width caused by masting once other factors have been filtered out (Drobyshev et al., 2010; Eis et al., 1965; Holmsgaard, 1958; Speer, 2001). For archival research, many types of written sources can be used from forestry statistics to yearly records of pannage incomes. Such research usually focuses on collecting information on masting events, from which frequency statistics are created (Hilton and Packham, 2003; Lindquist, 1931; Májer, 1982; Matthews, 1955; Maurer, 1964; Övergaard et al., 2007; Paar et al., 2011; Schenk, 1994; Wachter, 1964).

However, there is an important source type that has so far not been used at all. It is different from the sources most often used in that it records average masting patterns. In places where pigs were taxable, authorities sometimes wanted to find out how often mast years happened on average in order to establish yearly land taxes. In 1876, the Hungarian forestry journal *Erdészeti Lapok* published guidelines for foresters as to how to establish masting patterns: the past 25 years were to be taken into account either with the help of data from account books or by interviewing the locals (Anonymous, 1876). The above-mentioned *Regnicolaris conscriptio* of Hungary from 1715 contains information on average masting patterns as well. For example, the entry on the village of Garáb reads '[the villagers] have acorn-bearing woods, but these produce acorns hardly once in ten years, nonetheless in time of acorns the villagers hope to pasture ca. 50 pigs, which makes an income of 12.50 florins' (translation from Latin original) (MOL N 78, téka 18). Records of average intervals between mast years are highly significant, because taxation documents are the only sources that record past observations on the frequency of mast years. Furthermore, because taxation documents typically cover larger geographical areas, they provide data on regional differences as well. It has been noted in literature that masting patterns are different in different regions (Hilton and Packham, 2003; Schenk, 1994), however, few studies have examined the possible driving factors for such differences (Drobyshev et al., 2010; Packham et al., 2008), and to my knowledge no research exists on regional masting patterns in the past. Figure 5.3 illustrates how geographical patterns in masting can be studied with the help of taxation documents. It shows average mast frequencies as reported by the villagers in Co. Nógrád in Hungary in 1715 (MOL N 78, téka 18) superimposed on a basic altitudinal map. High heterogeneity (in values as well as in distribution) in this map makes it clear that mast frequency was a complex phenomenon at this time and place; it was certainly not driven by altitude. Because very different values occur next to each other, one can assume that climate played little role. Factors to explain heterogeneity would have to be local, such as soil conditions, microclimate or management history.

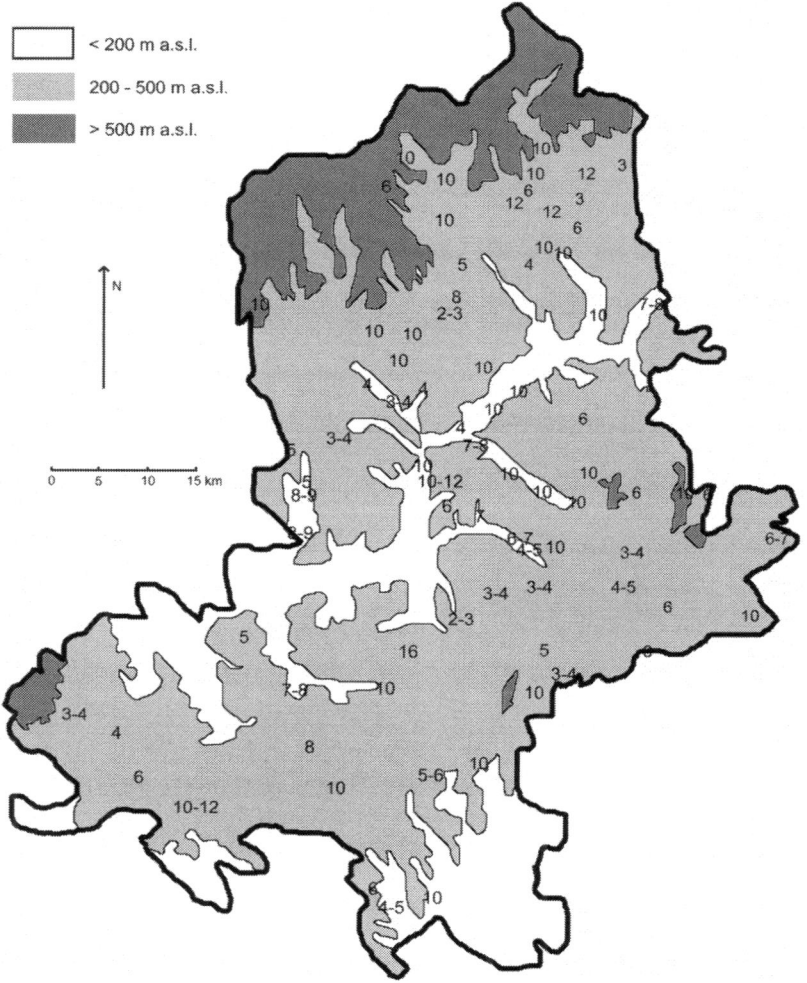

Figure 5.3 Average mast frequencies in years in the villages Co. Nógrád (Hungary) in 1715 superimposed on an altitudinal map

Pannage and oakwoods

Oakwoods form an important part of many European landscapes. The distribution of oakwoods is traditionally explained with the help of natural conditions, such as climate, elevation and soil. While this is certainly a useful approach, other factors could also be considered. Traditional management, in contrast to modern forestry, is not normally seen as a driving factor of the distribution of various forest types. Oakwoods might prove to be a counterexample. Because of the economic potential of acorns, woodland owners may have decided to conserve oakwoods or even to transform other types of forests into oakwoods.

I will use the example of Hodonínská Dúbrava, a large ancient oakwood in the south-eastern Czech Republic to illustrate my point. Dúbrava is a subcontinental oakwood on blown sand with underlying basic rock. It is situated on the right bank of the river Morava at the Czech-Austrian-Slovak border. The site is gently sloping towards the south-west, with the lowest and highest points at 164m and 242m. The basic rock and the acidic soils, the varied micro-topography characteristic of blown sand, as well as the closeness of the river Morava create a fine vegetation mosaic with many wet hollows. Dúbrava is the core site of a specific type of sub-continental oakwood, *Carici fritschii-Quercetum roboris* (Chytrý, 1997; Roleček, 2007) – a community endemic to this and a few adjacent sites. Other vegetation types, such as alluvial forests (*Alnion incanae*) and alder carrs (*Alnion glutinosae*) also occur.

In 2008–9, two small forest hollows inside the wood were sampled for pollen. The analysis, among other things, showed that oak pollen was present in small numbers until the mid-fourteenth century, when it suddenly increased to dominance at the expense of alder and hazel (for more on this issue, see Jamrichová *et al.*, 2012). A comparison with other pollen cores in the region showed no similar developments, therefore climate change or other non-local driving forces could be excluded. At the same time, archival research brought to light several documents that could be connected to this change in vegetation. The first one is an interpolated charter from 1350, which gave the citizens of Hodonín the right to take dry wood and grass in the wood but forbade them to fell living oaks (Boček, 1839: 204–5; for a discussion on the dating of the charter, see M ínský and Šmerda, 2008). They also had the right to pasture their animals in Dúbrava. The other document is the foundation charter of the Augustinian monastery in Brno from AD1370. It mentions that the tenants of the monastery had the right to cut timber and firewood in Dúbrava 'with the exception of oak trees, which they must not cut down at all' (translation from Latin original Moravský zemský archiv (Moravian Archives) (MZA) F5 karton 11 inv. č. 744, fol. 25–32). The ban on cutting oaks was included in a number of privileges in later periods as well (e.g. 1531 – MZA F5 karton 3 inv. č. 29; 1600 – MZA F5 kniha 1a). That the vegetation changed in the fourteenth century is also shown by a change in the name of the wood. Originally it was called Klečka, meaning 'a place with shrubs'. This name was gradually replaced by Dúbrava ['oakwood ']. 'Dúbrava' was first used as a common noun specifying tree composition in the wood (e.g. 1370: 'the oakwood [dúbrava] that is called Klečka' – MZA F5 karton 11 inv. č. 744, fol. 25–32), and only later became a geographical name. The last occurrence of the name Klečka is from 1531 (MZA F5 karton 3 inv. č. 29); after that only Dúbrava was used.

The reason for protecting the oaks is apparent from later documents, mostly from the *urbaria* of AD1600 and 1691 (MZA F5 kniha 1a, 3), which provide detailed information on the management of Dúbrava. The most characteristic feature was multiple use, which included wood-pasture, pannage, hay cutting, firewood cutting and the collecting of strawberries and oak galls. Financially, pannage was second in importance to the selling of firewood, creating c.20 per cent of all woodland incomes. The usage system was complex but not random: every use, including pannage, was carefully regulated. Figure 5.4 illustrates a sample of this complex system: the usage rights of the town of Hodonín in various parts of Dúbrava at the end of the seventeenth century (MZA F5 kniha 3). As can be seen, Hodonín had pannage rights free of charge in approximately one fourth of the wood. Other nearby settlements could also drive their pigs into parts of Dúbrava, creating considerable pressure on acorn crops. According to the pollen results, oak remained dominant in the wood after the fourteenth century. Pannage itself was practised until the beginning of the nineteenth century, when Dúbrava was divided into woodland-only and pasture-only parts.

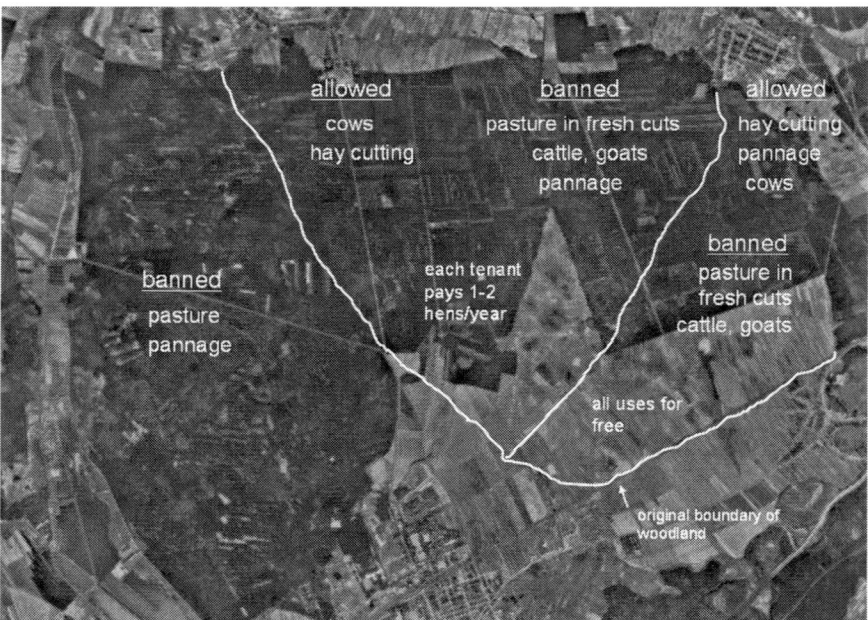

Figure 5.4 The usage rights of the town of Hodonín in Dúbrava wood at the end of the seventeenth century as described in the 1691 urbarium
Note: The aerial photograph is from 1953. The white lines indicate boundaries of various usage rights and the original woodland boundary. Note that in the south-east the wood used to be slightly larger.

Three types of evidence (pollen, linguistic and archival) concur that deliberate management measures from the fourteenth century onwards changed the vegetation of Dúbrava wood – it is assumed that the strongest motivation behind this was the opportunity for pannage. In the modern period, the oakwood vegetation survived the abandonment of pannage. In this particular case, pannage appears to have been a major driving force of vegetation development and directly modified the distribution of oakwood s in the area.

Conclusions

Pannage has been discussed in historical ecology for more than a century. In this short chapter I have tried to demonstrate that in spite of the relatively rich literature on the subject, several issues deserve a fresh look. Even the most obvious subject, the frequency of mast years in the past, would benefit from looking for and utilising new types of sources. Other areas, such as changes in the geographical distribution of pannage or the possible connections between pannage and current oak-wood distribution, have so far not received any attention at all. New research will help us understand the ecology of oakwoods, a highly sensitive issue because in many regions oaks no longer regenerate (e.g. Doležal *et al.*, 2010; Helama *et al.*, 2009; Ragazzi *et al.*, 1989). Historical ecological knowledge also has the potential to contribute to nature conservation efforts.

References

Anonymous (1876) 'Az államerd kben alkalmazandó erd rendezési eljárás alapelveir l', *Erdészeti Lapok*, 15: 368–84.
Birrel, J. (1987) 'Common rights in the medieval forest: Disputes and conflicts in the thirteenth century', *Past and Present*, 117: 22–49.
Boček, A. (ed.) (1839) *Codex Diplomaticus et Epistolaris Moraviae*, tom. 2, Olomouc, Czech Republic.
Bruun, H.H. and Fritzbøger, B. (2002) 'The past impact of livestock husbandry on dispersal of plant seeds in the landscape of Denmark', *Ambio*, 31(5): 425–31.
Chytrý, M. (1997) 'Thermophilous oak forests in the Czech Republic: Syntaxonomical revision of the *Quercetalia pubescenti-petraeae*', *Folia Geobotanica*, 32: 221–58.
Darby, H.C. (1950) 'Domesday woodland', *Economic History Review*, 3: 21–43.
Dávid, Z. (1957) 'Az 1715-20. évi összeírás', in J. Kovacsics (ed.) *A történeti statisztika forrásai*, Közgazdasági és jogi könyvkiadó, Budapest, Hungary.
Doležal, J., Maz rek, P. and Klimešová, J. (2010) 'Oak decline in southern Moravia: The association between climate change and early and late wood formation in oaks', *Preslia*, 82: 289–306.
Drobyshev, I., Övergaard, R., Saygin, I., Niklasson, M., Hickler, T., Karlsson, M. and Sykes, M.T. (2010) 'Masting behaviour and dendrochronology of European beech (*Fagus sylvatica* L.) in southern Sweden', *Forest Ecology and Management*, 259: 2160–71.

Eis, S., Garman, E.H. and Ebell, L.F. (1965) 'Relation between cone production and diameter increment of Douglas fir (*Pseudotsuga menziesii* (Mirb.) Franco), grand fir (*Abies grandis* (Dougl.) Lindl.), and western white pine (*Pinus monticola* Dougl.)', *Canadian Journal of Botany*, 43: 1553–9.

Grigson, C. (1982) 'Porridge and pannage: Pig husbandry in Neolithic England', in S. Limbrey and M. Bell (eds) *Archaeological aspects of woodland ecology* (British Archaeological Reports International Series 146), Archaeopress, Oxford, UK.

Hamilton, J., Hedges, R.E.M. and Robinson, M. (2009) 'Rooting for pigfruit: Pig feeding in Neolithic and Iron Age Britain compared', *Antiquity*, 83: 998–1011.

Helama, S., Laanelaid, A., Raisio, J. and Tuomenvirta, H. (2009) 'Oak decline in Helsinki portrayed by tree-rings, climate and soil data', *Plant and Soil*, 319: 163–74.

Hilton, G.M. and Packham, J.R. (2003) 'Variation in the masting of common beech (*Fagus sylvatica* L.) in northern Europe over two centuries (1800–2001)', *Forestry*, 76: 319–28.

Holmsgaard, E. (1958) 'Effect of seed-bearing on the increment of European beech (*Fagus sylvatica* L.) and Norway spruce (*Picea abies* (L.) Karst)', in *Papers of the Twelfth Congress of the International Union of Forestry Research Organizations*, 3, International Union Forestry Research Organizations, Oxford.

Jamrichová, E., Szabó, P., Hédl, R., Kuneš, P., Bobek, P. and Pelánková, B. (2012) 'Continuity and change in the vegetation of a Central European oakwood', *Holocene*, in press.

Lindquist, B. (1931) 'Den Skandinaviska bokskogens biologi', *Svenska skogsvårdsföreningens tidskrift*, 29: 175–532.

Májer, A. (1982) 'A bükkösök makktermésének id szakossága', *Erd*, 31: 388–92.

Matthews, J.D. (1955) 'The influence of weather on the frequency of beech mast years in England', *Forestry*, 28: 107–16.

Maurer, E. (1964) 'Buchen- und Eichensamenjahre in Unterfranken während der letzten 100 Jahre', *Allgemeine Forstzeitschrift*, 31: 469–70.

M ínský, Z. and Šmerda, V. (2008) 'Svítání st edov ku (Doba slovanská a p emyslovských knížat)', in M. Plaček (ed.) *Hodonín: d jiny m sta do roku 1948*, Hodonín, Czech Republic.

Övergaard, R., Gemmel, P. and Karlsson, M. (2007) 'Effects of weather conditions on mast year frequency in beech (Fagus sylvatica L.) in Sweden', *Forestry*, 80: 555–65.

Paar, U., Guckland, A., Dammann, I., Albrecht, M. and Eichhorn, J. (2011) 'Häufigkeit und Intensität der Fruktifikation der Buche', *AFL-Der Wald*, 2011/6: 26–9.

Packham J.R., Thomas P.A., Lageard J.G.A and Hilton G.M. (2008) 'The English beech masting survey 1980–2007: Variation in the fruiting of the common beech (*Fagus sylvatica* L.) and its effects on woodland ecosystems', *Arboricultural Journal*, 31: 189–214.

Rackham, O. (2003) *Ancient Woodland: Its History, Vegetation and Uses in England*, 2nd edition, Castlepoint Press, Colvend, UK.

Ragazzi, A., Dellavalle Fedi, I. and Mesturino, L. (1989) 'The oak decline: A new problem in Italy', *European Journal of Forest Pathology*, 19(2): 105–10.

Roleček, J. (2007) 'Vegetace subkontinentálních doubrav ve st ední a východní Evrop ', PhD dissertation, Masaryk University, Brno, Czech Republic.

Schenk, W. (1994) 'Eichelmastdaten aus 350 Jahren für Mainfranken – Probleme der Erfassung und Ansätze für umweltgeschichtliche Interpretation', *Allgemeine Forst und Jagdzeitung*, 165: 122–31.

Shaw, E.B. (1940) 'Geography of mast feeding', *Economic Geography*, 16: 233–49.

Speer, J.H. (2001) 'Oak mast history from Dendrochronology: A new technique demonstrated in the Southern Appalachian region', unpublished PhD dissertation, University of Tennessee, Knoxville, USA.

Szabó, P. (2003) 'Sources for the historian of medieval woodland', in J. Laszlovszky and P. Szabó (eds) *People and Nature in Historical Perspective*, CEU Medieval Studies and Archaeolingua, Budapest, Hungary.

Szabó, P. (2005) *Woodland and Forests in Medieval Hungary*, Archaeopress, Oxford.

Szabó, P. and Hédl, R. (2012) 'Socio-economic demands, ecological conditions and the power of tradition: Past woodland management decisions in a Central European landscape', *Landscape Research*, in press.

Wachter, H. (1964) 'Über die Beziehungen zwischen Witterung und Buchenmastjahren', *Forstarchiv*, 35(4): 69–78.

Wickham, C. (1994) 'European forests in the early Middle Ages: Landscape and land clearance', in *Land and Power: Studies in Italian and European Social History, 400–1200*, British School at Rome, London.

Wieck, R.S. (1988) *The Book of Hours in Medieval Art and Life*, Sotheby's Publications, London.

Wilson, D. (2003) 'Implications of feeding pigs in the Anglo-Norman forest', in L. Jeleček, P. Chromý, H. Jan , J. Miškovský and L. Uhlíová (eds) *Dealing with Diversity: Proceedings of the 2nd International Conference of the ESEH*, Charles University, Prague, Czech Republic.

6 The post-glacial history of grazing animals in Europe

Derek Yalden

Introduction

Europe, especially Western Europe, has a limited range of large mammals that might be termed grazers. This is in part a consequence of its isolation to the south by the Mediterranean, restricting possible influxes of any part of the large African mammal diversity. It also reflects the extent to which the last glaciation swept most of the fauna out of northern Europe, restricting it either to southern refuges (notably Iberia, Italy and the Balkans) or pushing it well to the east. It also reflects, in part, the evident replacement of open habitats, extensive steppe and tundra, by equally extensive woodland in the post-glacial period, say 11,000 to 5,500 years ago. Only in the farmed landscapes created progressively since 5.5ka (kiloannum or thousand years ago) have grazing mammals recovered their dominance in the landscape, but those grazers are mostly introduced Middle Eastern species.

'Grazing mammals' implies those that eat grass, as opposed to browsers, which eat shrubs, twigs and leaves. In practice, the dichotomy is unrealistic: most grazers eat some browse, as well as herbs mixed with the grass, and most browsers eat some grass. The mountain hare, *Lepus timidus*, whose diet is up to 70 per cent grass in summer, switches to heather, *Calluna vulgaris*, a dwarf shrub, in winter when most grass is dead or eaten. The horse, *Equus caballus*, perhaps the epitome of a grazer, switches from its 80 per cent grass diet to browse when drought in summer or snow cover make grass unavailable. Roe deer, *Capreolus capreolus*, the most specialised browser among the native ungulates, nevertheless eats up to 46 per cent herbs and 7 per cent grass in summer (Harris and Yalden, 2008), though bramble is its favoured food. Smaller mammals such as *Microtus* voles might be less adaptable than larger species, but their history is less well documented by archaeological data, and this account is limited to larger species.

Knowledge of the history of the large mammal fauna of the British Isles was reviewed by Yalden (1999), and has been updated and expanded for several species by contributors to O'Connor and Sykes (2010). A concise, comprehensive account of the Danish fauna has been provided by Aaris-Sorensøn (2009) and Poortvleit (1993) offers an imaginative, illustrated popular account from a Dutch perspective. Summaries of the available information on some of the important species, continent-wide, have also been presented (e.g. *Cervus elaphus*, Sommer and Benecke (2006); *Sus scrofa*, Scandura *et al.* (2011); *Capreolus capreolus*, Sommer *et al.* (2009); *Bos primigenius*, van Vuure (2005)). Importantly, Sommer and Benecke (2006) summarised the occurrences of large mammals in Late Glacial refugia in southern Europe. These accounts rely on both direct evidence of archaeological remains and on phylogeographical studies. In combination, these two sets of information are contributing an increasingly coherent story of the recent history of the European fauna.

Time and faunas

There are five obvious time periods with distinctive mammal faunas during the last 100,000 years. These are: 1) before the Last Glacial Maximum (LGM), (around 115–23ka; 2) at the Last Glacial Maximum, (23–16ka); 3) the Late Glacial (16–11.5ka); 4) the Post-glacial warm period, (from 11.5ka to about 5.5ka; and 5) the farming period (from 5.5ka to now).

Before the Last Glacial Maximum

The first is a somewhat shadowy period of mostly cold, dry conditions interspersed with short, slightly warmer intervals. Most of Western Europe had an open landscape, very suitable for grazing mammals. Large herds of bison, *Bison priscus*, horse, *Equus ferus*, and reindeer, *Rangifer tarandus*, with woolly mammoth, *Mammuthus primigenius*, woolly rhinoceros, *Coelodonta antiquitatis* and Irish elk, *Megaloceros giganteus*, roamed over the continent from at least the Pyrenees to southern Scandinavia. There are a few records of muskox, *Ovibos moschatus*, now confined to North America, from this period.

Last Glacial Maximum

At the LGM, ice cover expanded from Scandinavia and Scotland, creating northern ice-sheets reaching as far south as the mouth of the River Shannon, the Gower coast and the Wash, in the west, and to northern Germany and Poland in the east; these ice-sheets probably fused across what is now the North Sea. A smaller ice sheet covered the Alps, while the Pyrenees were at least capped by ice. Between the northern and southern ice-sheets, Western Europe was a cold open tundra-like landscape, too cold for most mammals. Even woolly mammoth seem to have been absent from this area for about 2,000 years at the maximum of the cold (Stuart *et al.*, 2004). Species that needed warmer conditions, or forested habitats, moved south or east. Identification of these refugia has been a major research topic in recent years. There were red deer populations in Iberia and south-west France, in Italy, in the Balkans and within the arc of the Carpathian Mountains. Roe deer and aurochs were present in the same refuges (Sommer and Nadachowski, 2006; Sommer et al., 2008, 2009). The rabbit, *Oryctolagus cuniculus*, persisted only in the Iberian/south-west France refugium (Hardy et al., 1995), while the brown hare, *Lepus europaeus*, survived only in the Balkans and around the Black Sea (Kasapidis et al., 2005). By contrast, the cold-adapted mountain hare, *Lepus timidus*, occurred widely across the steppe-tundra from southern Ireland and southern England, south as far as the Pyrenees and Catalonia, and eastwards across the north European plain, north of the Alps and Carpathians (Melo-Ferreira et al., 2007).

Late Glacial

The Late Glacial period saw some amelioration in climate, and birch scrub re-colonising southern Britain. Woolly mammoth, reindeer, Irish elk and horse also reinvaded as far north as Denmark and England. This is the period when Irish elk became numerous in Ireland and the Isle of Man; elk *Alces alces* also appeared in Britain and Denmark, though it apparently never reached Ireland. Mammoth was scarce by this stage, but Campbell (1977) points out that horse and reindeer were the common prey of late palaeolithic human hunters in most British sites. At Robin Hood Cave, though, they were primarily hunting mountain hare (Charles and Jacobi, 1994). At the mildest, most wooded, part of this interval, red deer returned briefly to southern Britain and Ireland, and there is now one record of roe deer in North Wales from this interval. However, right at the end of the Late Glacial, the climate reverted to glacial conditions for a relatively short (perhaps 600-year) period (the Younger Dryas), and this was apparently sufficiently severe to kill off the Irish elk, not only in Ireland but across the whole of Western Europe. The woolly mammoth also disappeared from Western Europe at this time, though it survived in Siberia for a few thousand years longer. Other temperate species that had spread northwards retreated south again, including humans, elk and red deer, and a fauna characteristic of cold dry conditions spread west as far as England, where saiga, *Saiga tartarica*, appeared briefly along with steppe pika, *Ochotona pusilla*. Horse and reindeer were the common large grazers in this period, but there were apparently no humans to hunt them and mammal remains from this interval are fewer than before or since.

Further south in Europe, to which the temperate fauna retreated, humans in Cantabria were still able to hunt red deer as their main prey, along with locally available alpine ibex, *Capra ibex* (Straus, 1999). A few roe deer, wild boar, horse and chamois, *Rupicapra rupicapra*, were also taken, with rarely reindeer. In Aquitaine, to the north of the Pyrenees, reindeer were the main prey, along with horse, but some red deer were available and scarce roe deer and wild boar hint at some woodland (Straus, 1999). In the uplands of southern Belgium, it seems to have been cold enough for musk ox, as well as reindeer, horse, bison, chamois and ibex (Straus, 1999).

Post-glacial

The end of the Younger Dryas, marking the end of the last glaciation, was abrupt. It has been estimated that summer temperatures rose about 8°C in around 50 years. The tundra vegetation was over-run quite quickly by scrub of juniper, *Juniperus communis*, and then birch, *Betula* spp. Within 500 years, birch woodland dominated much of southern Britain. Moreover, the mammals responded equally quickly to this change. Horse and reindeer, characteristic of open conditions, seem to have disappeared from England within 300 and 500 years, respectively, though the reindeer survived longer in the north, until about 1,100 years into the post-glacial in Sutherland. Conversely, the mammals typical of boreal and temperate conditions returned northwards very quickly. In Aquitaine, Straus (1999) documents the replacement of reindeer by red deer as the main human prey, and in Belgium, equally, the cold-adapted fauna was replaced by roe and red deer, wild boar and aurochs. In England, at Starr Carr and Thatcham, dated to around 11.0ka and within 500 years of the end of the Late Glacial, humans were likewise hunting elk, red and roe deer, aurochs and wild boar. In Denmark, the wild boar arrived by about 11.4ka. Although regarded as rooters and harvesters of autumn fruits and nuts, wild boar are largely grazers through the summer; however, they cannot cope in winter with snow deeper than around 20cm, making them a very typical member of the temperate woodland fauna and limiting the northern edge of their range (Yalden, 2001). While the three deer are largely browsers when possible, the aurochs was certainly a grazer, needing the bulk of grass provided by extensive open areas. Like wild boar it arrived in Denmark about 11.4ka. Within about 2,000 years of the end of the Late Glacial, pollen records suggest that southern Britain was covered by deciduous woodland of oak, *Quercus*, elm, *Ulmus*, and alder, *Alnus*, with a hazel, *Corylus*, understorey. The persistence, indeed (to judge from the frequency of its remains, abundance) of aurochs through to about 4.0ka in England, and to about 3.0ka in Denmark, poses a problem to those who have envisaged the Mesolithic landscape of Western Europe as continuous tree cover from coast to coast across England. It seems that it lived particularly on lower ground, in river valleys and coastal grasslands (Hall, 2008), but it is reported to be a woodland animal by historians. It seems likely that the woodland was rather more open than the bald pollen rain figures suggest, and allowed the aurochs access to sufficient grazing, but red deer and wild boar would also have needed some grassland to make up their full diets.

Most surprising is the absence of bison in the post-glacial fauna in the western-most parts of Europe. While numerous in both Britain and Denmark during the early period, pre-LGM, and present in Eastern

Europe into recent times, it appears not to have returned to the lowlands of the north-west. Perhaps the woodland was too dense to sustain them, and they survived only on higher ground, for instance in the Massif Central and the Alpine foreland. There is some evidence that tarpan (wild horse) may have survived in such areas through to Neolithic times (Boyle, 2006), as it did in Denmark, yet it was no longer present in England in this period.

Farming

The arrival of Neolithic farming culture in north-west Europe around 5.5ka is marked by a decline in elm pollen, perhaps due to the felling of trees to create small fields, and by the arrival of domestic sheep, *Ovis aries*, goats, *Capra hircus*, cattle, *Bos taurus*, and pigs, *Sus domesticus*. Cattle and pigs could conceivably have been domesticated from the native aurochs and wild boar, but genetic evidence says they were not; like the sheep and goats, which were never native to Europe, they were brought, already domesticated, and along with the cereals wheat and barley, from the Middle East. Hunting was quickly and almost completely replaced by harvesting of these domestic mammals as a source of meat, so they dominate archaeological samples (Yalden, 1999). Only red deer, used as an important source of antlers for tool-making, retains some importance. Horse, domesticated in southern Russia, arrived somewhat later in Western Europe, around the end of the Neolithic. Either deliberate persecution, or the indirect effects of habitat loss as more and more woodland cover was cleared, resulted in the decline and extinction of the larger wild mammals from much of Western Europe, most starkly from the British Isles (Yalden, 1999), but also, for instance, from the Danish islands (Aaris-Sorensøn, 2009). Elk were lost by around 4.0ka and aurochs by 3.0ka, in both Britain and Denmark. Wild boar survived longer, to about AD1400 in Britain and AD1700 in Denmark. Elsewhere they survived longer, aurochs reputedly to AD1687 in Poland, wild boar through to modern times in Germany and further east. In England, even red and roe deer became extinct around AD1700, though they survived in most other countries, including Scotland.

The converse has been the enormous expansion in the numbers of domestic grazers. The current (adult) mammal fauna of Great Britain is believed to include around 21 million sheep and 4 million cattle, compared with perhaps 500,000 roe deer and 360,000 red deer (Harris *et al.*, 1995; Yalden, 2003). There are also around 750,000 horses, 853,000 pigs and a few reintroduced wild boar, maybe numbering 1,000. Thus large domestic grazers vastly outnumber wild ones. Smaller grazers, such as rabbit and brown hare, may make up some missing numbers, but not much biomass.

Discussion

In respect of the impact grazing mammals might have on recent woodland history in Western Europe, and our interpretation of it, this potted history makes three main points.

First, in the Late Glacial, with an open landscape, herds of reindeer and horses might have suppressed tree growth, but probably not so much as the prevailing climate. They certainly do not seem to have slowed the spread of woody vegetation when climatic conditions improved. On the contrary, it seems that climate and consequent vegetation determined the fauna, probably both its diversity and its abundance.

Second, in the Mesolithic wooded period, the most abundant and widespread large mammals, roe deer, red deer and wild boar were largely browsers, though the two larger species certainly graze, especially in summer, and imply an open wooded landscape with a good ground flora. This is most evidently indicated by the extensive range and frequency of aurochs, surely like modern cattle essentially a grazer. Its historical attribution to woodland probably indicates a preference for seclusion from humans, rather than a strict habitat preference, just as the last European bison, *Bison bonasus*, also frequented the Białowieża Forest. At least coastal and river-valley grasslands must have been available, but probably, again, the woodland contained a good ground flora, perhaps grassy clearings as well. In winter, horses in the New Forest, as well as deer, browse evergreen species such as holly, *Ilex*, gorse, *Ulex*, and ivy, *Hedera* (Putman, 1986). It is less certain that these large mammals were sufficiently numerous to drive the woodland–grassland cycle suggested by Vera (2000). Mitchell (2005) argues that the woodland invasion of Ireland, facing grazing only by wild boar and Irish hare, was not apparently faster than in England, with four more large grazers. Rackham (2003) cautions that Mesolithic humans, by selective burning to attract game towards new growth, might have promoted the grassland-woodland mosaic, and the tree-felling activities of beavers would have added meadows and glades near water (Coles, 2006).

Third, since the Neolithic, the increasing level of grazing from domestic ungulates has certainly been accompanied by a decrease in the quantity of woodland, and the level of grazing imposed by the combination of wild and domestic ungulates in the New Forest, for example, is certainly sufficient to prevent woodland regeneration – except when exclosures deliberately permit it (Putman, 1986). Farmers have traditionally used ivy and holly as emergency winter food, and it has also been suggested that cutting branches of elm, an early sprouting tree whose young shoots are rich in phosphorus, contributed to the 'elm-decline' that marked the evident arrival of farmers about 5.5ka (Godwin, 1975; Rackham, 2003). It is difficult to compare the rates of grazing attributable to wild and domestic grazers then and now. If the Mesolithic ungulate fauna of Britain can be plausibly estimated by comparison with the equivalent woodland fauna of modern Białowieża Forest, the

Mesolithic fauna might have been about twice as numerous as the wild mammals are now. However, the domestic mammals are about eight times as numerous, and it is hard to believe that the smaller Mesolithic fauna could have created as much pressure on the vegetation. Moreover, grazing pressure by the Mesolithic ungulates would have been lessened by the intervention of predators. This effect is demonstrated in Yellowstone Park, where aspen, *Populus tremuloides*, groves are regenerating now that they no longer provide an undisturbed sanctuary for large herds of wintering wapiti. Reintroduced wolves ensure that they disperse (Ripple *et al.*, 2001).

References

Aaris-Sorensøn, K. (2009) 'Diversity and dynamics of the mammalian fauna of Denmark throughout the last glacial-interglacial cycle, 115-0 kyr BP', *Fossils and Strata*, 57: 1–59.

Boyle, K.V. (2006) 'Neolithic wild game animals in Western Europe: the question of hunting', in Serjeantson, D. and Field, D. (eds) *Animals in the Neolithic of Britain and Europe*, Oxbow, Oxford, pp10–23.

Campbell, B. (1977) *The Late Palaeolithic of Britain*, Clarendon Press, Oxford.

Charles, R. and Jacobi, R.M. (1994) 'The lateglacial fauna from Robin Hood Cave, Creswell: A re-assessment', *Oxford Journal of Archaeology*, 13: 1–32.

Coles, B. (2006) *Beavers in Britain's Past*, Oxbow Books, Oxford.

Godwin, H. (1975) *The History of the British Flora*, 2nd edition, Cambridge University Press, Cambridge.

Hall, S.J.G. (2008) 'A comparative analysis of the habitat of the extinct aurochs and other prehistoric mammals in Britain', *Ecography*, 31: 187–90.

Hardy, C., Callou, C., Vigne, J.-D., Casane, D., Dennebouy, N., Mounolou, J.-C. and Monneret, M. (1995) 'Rabbit mitochondrial DNA diversity from prehistoric to modern times', *Journal of Molecular Evolution*, 40: 227–37.

Harris, S. and Yalden, D.W (eds) (2008) *Mammals of the British Isles: Handbook*, 4th edition, Mammal Society, Southampton.

Harris, S., Morris, P., Wray, S. and Yalden, D. (1995). *A Review of British Mammals: Population Estimates and Conservation Status of British Mammals Other than Cetaceans*, JNCC, Peterborough.

Kasapidis, P., Suchentruck, F., Magoulas, A. and Kotoulas, G. (2005) 'The shaping of mitochondrial DNA phylogeographic patterns of the brown hare (*Lepus europaeus*) under the combined influence of Late Pleistocene climatic fluctuations and anthropogenic translocations', *Molecular Phylogenetics and Evolution*, 34: 55–66.

Melo-Ferreira, J., Boursot, P., Randi, E., Kryukov, A., Suchentruck, F., Ferrand, N. and Alves, P.C. (2007) 'The rise and fall of the mountain hare (*Lepus timidus*) during Pleistocene glaciations: Expansion and retreat with hybridization in the Iberian Peninsula', *Molecular Ecology*, 16: 605–18.

Mitchell, F.J.G. (2005) 'How open were European primeval forests? Hypothesis testing using palaeoecological data', *Journal of Ecology*, 93: 168–77.

O'Connor, T. and Sykes, N. (2010) *Extinctions and Invasions. A Social History of British Fauna*, Windgather Press, Oxford.

Poortvleit, R. (1993) *Journey to the Ice Age*, Harry N. Abrams, New York.

Putman, R.J. (1986) *Grazing in Temperate Ecosystems. Large Herbivores and the Ecology of the New Forest*, Croom Helm, London.

Rackham, O. (2003) *Ancient Woodland: Its History, Vegetation and Uses in England*, New Edition, Castlepoint Press, Colvend.

Ripple, W.J., Larsen, E.J., Renkin, R.A. and Smith, D.W. (2001) 'Trophic cascades among wolves, elk and aspen on Yellowstone National Park's northern range', *Biological Conservation*, 102: 227–34.

Scandura, M., Iacolina, L. and Apollonio, M. (2011) 'Genetic diversity in the European wild boar *Sus scrofa*: Phylogeography, population structure and wild x domestic hybridization', *Mammal Review*, 41: 125–37.

Sommer, R.S. and Benecke, N. (2006) 'Glacial refugia of mammals in Europe: Evidence from fossil records', *Mammal Review*, 36: 251–66.

Sommer, R.S., Zachos, F.E., Street, M., Jöris, O., Skog, A. and Benecke, N. (2008) 'Late Quaternary distribution dynamics and phylogeography of the red deer (*Cervus elaphus*) in Europe', *Quaternary Science Reviews*, 27: 714–33.

Sommer, R.S., Fahlke, J.M., Schmölke, U., Benecke, N. and Zachos, F.E. (2009) 'Quaternary history of the European roe deer *Capreolus capreolus*', *Mammal Review*, 39: 1–16.

Straus, L.G. (1999) 'High resolution archaeofaunal records across the Pleistocene Holocene transition on a transect between 43 and 51 degrees north latitude in western Europe', in Dxcykx, J.C. (ed.) *Zooarchaeology of the Pleistocene/Holocene Boundary*, BAR International Series 800, Archaeopress, Oxford, pp21–9.

Stuart, A.J., Kosintsov, P.A., Higham, T.F.G. and Lister, A.M. (2004) 'Pleistocene to Holocene extinction dynamics in giant deer and woolly mammoth', *Nature*, 431: 684–9.

van Vuure, C. (2005) *Retracing the Aurochs: History, Morphology and Ecology of an Extinct Wild Ox*, Pensoft, Sofia-Moscow.

Vera, F.W.M. (2000) *Grazing Ecology and Forest History*, CABI, Wallingford, Oxon.

Yalden, D.W. (1999) *The History of British Mammals*, T. and A.D. Poyser, London.

Yalden, D.W. (2001) 'Mammals as climatic indicators', in Brothwell, D.R. and Pollard, A.M. (eds) *Handbook of Archaeological Sciences*, Wiley, Chichester, pp147–54.

Yalden, D.W. (2003) 'Mammals in Britain – a Historical Perspective', *British Wildlife*, 14: 243–51.

Part III
Landscape dynamics

7 Reinterpreting wooded landscapes, shadow woods and the impacts of grazing

Ian D. Rotherham

Introduction

Human colonisation of the north-western European landscape brought gradual, though sometimes abrupt changes. From templates of savannah and forest, human activities created 'cultural landscapes' of wood, forest, meadow, field, pasture, heath, common, fen and bog (Rotherham, 2011a and b). The habitats and species associations of the original landscape found their niches within the emerging cultural ecology of the medieval period. Evidence for such lineage is extensive and complex, but some simple ecological observations point the way. Several key aspects of grassland ecological diversity suggest a longstanding evolution of communities and complexity in ecosystems with a significant presence in the primeval landscapes. First, we have a rich biodiversity of flowering plants and of invertebrates associated with pastures, meadows and heaths. Second, these communities include many species with complex mutualistic relationships between the plants and a range of fungi, but especially the vesicular-arbuscular mycorrhizae. Such interactions include multiples of interconnections between fungi and various plant individuals and species. Finally, there are incredibly complex interrelationships evolved, such as between the large blue butterfly and the ants that play host to the butterfly larvae. It is clear that all these examples suggest long-term stability of open communities in early European landscapes and the evolution of intimate and complex interrelationships. However, it is not suggested that this landscape was static and unchanging. These were extensive mosaics of interconnected ecosystems with dynamic stability yet fluid and changing. The natural dynamics of water and land, of lightning-strike fires, and of Vera's herbivores would ensure that disturbance and successional change were key drivers in this ancient and evolving ecology. These impacts were mimicked and mirrored in the effects of human usage of the cultural landscape.

However, most of the earlier subsistence landscapes were swept away by improvement and parallel revolutions in industry and in agriculture. Rising human populations and urbanisation drove the changes and from the 1700s onwards, European landscapes and ecologies were changed forever. Habitats and communities that had ebbed and flowed in the preindustrial environment struggled to adapt to eutrophication and macro-disturbance. In a little over 200 years, the extensive and seamless savannah landscapes of Frans Vera's vision were reduced to isolated fragments in a sea of inhospitable anthropogenic conditions. In contemporary landscapes, almost all the original template has been erased and eroded, and in environments of urban-dwelling, industry and intensive farming and forestry, it is hard to see the past through the present. Massive landscape transformations of the last 300 years have changed things almost beyond recognition. Yet the ecology is surprisingly persistent; sometimes retaining an almost desperate hold as species persist in tiny relict areas or as opportunities arise, moving to new sites. As humanity changed early European landscapes and ecology, some of the complexity evolved in Vera's wide-open spaces between the dense forests and extensive wet woodlands, survived and adapted. Over countless centuries these animals, plants and fungi, and those of the wooded zones, became the ecologies of the cultural landscape of the early medieval period. In the absence of petrochemically-driven power and inorganic fertilisers, the species and communities dovetailed into niches in habitats similar to those in which they had evolved. Extensive heaths, commons, meadows, sheep walk, fens, bogs, woods and forests were ideal analogues of their origins.

These issues are addressed in detail by Rotherham (2011a and b) and Rotherham and Handley (2011). Today, we see shadows and imprints of this ancient ecology in the modern landscape. Old meadows and pastures, ancient heaths, medieval coppice-woods and similar features bear testimony to this remarkable lineage. Ancient parks are the most visibly obvious remnants of formerly extensive grazed wooded landscapes. However, even where deer parks survive (and this is rare), they do so as unique landscapes separated in time and function from their origins. In many cases, these areas reflect the landscapes of the time and place they were imparked, and changes of economic function and ecology over a long lifespan. When created, carved out of a working medieval or earlier landscape, their ecologies were driven by multifunctional systems of economic utilisation. Over time, as purpose changed so did ecology, each new phase incorporating, preserving or removing, and modifying those that preceded it. It is argued by Frans Vera (2000), Ted Green and others, that some of these great parks are landscapes that originated in medieval or earlier times, and give unique insights into the former extensive primeval savannahs of north-western Europe.

These observations provide a starting point for further investigation. In particular, medieval parks are of great interest since they, like the primeval savannah, were landscapes that mixed trees and grazing or browsing mammals. It is suggested that parks were one facet of a landscape complex that included wood-pasture, wooded commons and forests, perhaps as relicts of what was probably in prehistory a great wooded savannah across much of north-western Europe. In both origins and ecology, parks are essentially a form of 'pasture-woodland', related to forests, heaths, moors and some commons, with grazing animals and variable tree cover (Rotherham, 2007a and b). Aside from the obvious external enclosure, these landscapes were often essentially unenclosed grazing lands. In considering their ecology, it is important to establish origins and relationships to other wildlife habitats, and to recognise too the range from areas of only a few hectares to massive sites extending over many kilometres of unenclosed ground. In the two centuries following the Norman Conquest, numbers of parks in England increased dramatically to perhaps 3,000, with possibly 50 in Wales and 80 in Scotland. From the early thirteenth century, a royal licence was technically necessary to create a park in areas of royal forest; though Cummins (1988) notes that in both England and Scotland baronial parks were sometimes created without licence. Where documents survive, they provide invaluable reference materials for a now vanished age, giving insight into landscape and ecology. The average English medieval park was around 50ha, but size varied considerably. The date of establishment, the area enclosed, the functions of the park and the interplay between enclosed and unenclosed areas all influenced the ecology of these landscapes (Jones, 1996; Jones *et al.*, 1996; Rotherham, 2007a and b). As lands were imparked, at the same time other areas were set aside as 'woods' and for other uses by the lord or by the commoners. The interactions between these various landscape components, and the implications of legal recognition and of enclosure, are central to the theme of this research.

Ancient wooded landscapes

In Britain, it is generally accepted that there are two broad distinctions in 'ancient woodland' landscapes. First, there are coppice-woods, often managed since the medieval period as simple coppice, or more frequently 'coppice-with-standards'. These sites often have relatively few large trees, but possess strikingly rich and sometimes diverse ground floras (Figure 7.1). Second, there are parklands, which may have historic links back to their use as medieval parks. These areas generally have poorer ground floras due to grazing livestock, and are frequently notable for massive and ancient trees, chiefly 'pollards'. Sometimes the park may have no ancient trees, but often this is because they have been removed for financial reasons at some time in the site's long history. Many of these landscapes, both ancient coppice and mediaeval parklands, have significant evidence of human impacts and settlements within them (see Beswick and Rotherham, 1993). I suggest here that reality is more complex and that these are two clearly identifiable types within a broad complex of wooded landscapes.

Research over the last 20 years has demonstrated that many assumptions about these wooded landscapes are incorrect or naïve in their interpretation. Researchers such as Paul Harding stimulated interests in British

Figure 7.1 Upland coppice alder shadow wood, Peak District, England
Source: Ian D. Rotherham

pasture-woodlands, and both Frans Vera and Ted Green have challenged many accepted 'truths' of woodland history. Together, the new approaches help to place park and other grazed landscapes in their wider ecological context. Indeed, parks were juxtaposed with, but different from, medieval coppice-woods, although the bigger parks often included 'woods' within them. The surviving landscapes present unique resources for conservation and provide rich insights into ecological history (Rollins, 2003). As part of this process, research by scholars such as Keith Alexander and Roger Key has transformed the understanding of the importance of parks for invertebrates. Ted Green has awakened interest in ancient tree fungi and with Jill Butler and Helen Read especially, has emphasised the significance of the trees themselves.

To further develop the context of our current studies it is important to note the seminal and influential writings of authorities such as Oliver Rackham (1976 and 1986), George Peterken (1981 and 1996), and Donald Pigott (1993). Together these forge coherent visions of woodland landscape ecology, with parks and other grazed areas representing an important component. However, it is of significance that for many years medieval parks were not considered 'ancient woodland' by conservation agencies.

They were in effect the 'Cinderellas' of nature conservation. Now, and from a broader 'woodland' perspective, it is possible to assess the historical ecology of medieval parks and other grazed wooded areas and place them in their landscape context. Parks have trees (usually but not always), and large (and sometimes smaller) grazing mammals, and to survive, trees need protection. Some parkland trees are ornamental and others are managed '*working*' trees, with fundamental differences in species and structures associated with these different functions. Taigel and Williamson (1993) and Bettey (1993) give useful introductions to the complexities of these landscapes. Such historical contributions are important since ecologists must understand history and historians the ecosystem. The potential of cross-fertilisation is considerable and Rackham (2004) provides an eloquent exposition on the evolution of park landscapes and of their trees in particular. In this chapter I suggest that the interpretation of these landscapes may require revisiting in order to appreciate more fully the complexities of the resource. In particular, I suggest there were extensive, grazed, wooded landscapes remaining at the time of the early medieval period, and these linked backwards in time to Vera's primeval environment. However, and importantly, when manorial coppice 'woods' were enclosed, either triggered by or more likely reflected in, the Statute of Merton (1235), much of this wooded landscape remained unenclosed and importantly for research purposes, unnamed. Not enclosed or named in the manorial records, these wooded landscapes were in the wider commons, including fens, moors, heaths and wastes.

The palimpsests of wooded landscapes

In terms of ecology, we need to unravel these complex relationships, yet it is difficult to separate parks from enclosed areas and associated grazing landscapes (Rotherham, 2007a and b). They share features with other unenclosed grazed landscapes with trees and woods (chases, forests, moors, heaths, commons and some fens). Furthermore, many parks 'took in' significant elements of earlier landscapes when they were enclosed often from 'waste' or 'forest'. Sometimes park management allowed parts of this ancient ecology to survive and parks also include ecology from periods of active management, from subsequent abandonment or changed use. Each phase preserves, modifies or removes earlier ecology as these working landscapes evolved through time. To understand today's ecology requires awareness of changes through pulses and periods of both management and neglect.

Ecological research often fails to differentiate between contrasting parkland origins and histories, and for many ecologists a park is a park. However, the reality is very different and we often lack any reliable historical framework. Furthermore, there is little actual information on their ecology as working parks so assumptions today are made retrospectively from modern observations or information gleaned from household and estate accounts. Cantor and Hatherly (1979) observe that medieval parks were very different from today's parks and were often areas of rough, uncultivated landscape, usually wooded, and frequently on the edge of manors away from cultivation. Owned by the lord of the manor, these lands were managed as hunting parks, stocked with deer and other game (Figure 7.2), and provided food and sport in varying balance. Our vision of a working medieval park should be set in a landscape of open field, waste, common, heath, bog and fen, woodland and royal forest. The ecologies of these different components were inexorably linked and intimately intertwined and form a complex of 'wooded landscapes' or 'treescapes'.

Figure 7.2 Fallow deer
Source: Ian D. Rotherham

The extensive medieval landscapes, including parks, provided hunting, foodstuffs, wood and timber for building, and fuel. Alongside deer, medieval parks contained wild boar, hares, rabbits, game birds, fish in fishponds and grazing for cattle and sheep. For some, such as Bradgate Park, pannage (feeding pigs on acorns) from the oaks provided revenue in rents. Medieval parks generally had large areas of heath or grassland (called launds or plains) dotted with trees, along with woods (called holts or coppices, and if for holly (*Ilex aquifolium*) hollins). The launds provided food for animals in summer, and the hollins provided it through the winter. Parks may have held and maintained deer (fallow (*Dama dama*) and red (*Cervus elaphus*)) for the table and the hunt. In the latter case, this sometimes involved release beyond the park pale and into the chase beyond (Whitehead, 1964 and 1980). Their ecologies varied as larger parks took in and maintained more of the earlier wilderness, waste and associated ecology. Hunting in, and the ecologies of, parks, forest and chase, were closely associated.

With socioeconomic changes, the fashions for parks and the means for their upkeep fluctuated. Most deer parks were created from AD1200 to 1350. They then declined following the impact of the Black Death (Mileson, 2005). Subsequently, boundaries moved. Small parks were enlarged or replaced by new creations, as parks and their relationships to great houses, changed with time and fashion. Originally an enclosed area at a small distance from the main house, perhaps containing hunting lodges, later parks were increasingly the settings for houses and gardens. The house moved to the park, or the park was moved or modified to envelop the house. Expensive and difficult to maintain, many deer parks fell from fashion, abandoned and destroyed. Between the fifteenth and eighteenth centuries, medieval deer parks were deliberately removed (disparkment), to become large, compartmented coppice-woods, or farmland. As the rural economy changed so did the values and costs of a park. Many were abandoned during the English Civil War (1642–9), and few survived intact as the wave of agricultural improvement swept through the landscape from 1600 onwards. The greater medieval parks may have shared common origins from the legacy of Vera's primeval savannah. However, other areas, upland moors and moorland fringe, and lowland heaths, commons and downs, probably reflect this same lineage; even today, many of these lands are grazed, and many have ancient albeit small, trees. These are the lands unenclosed in medieval times and linked, albeit tenuously, to Vera's open, fluid primeval landscape. For example, some of the species-rich grasslands such as the Derbyshire Dales limestone pastures are in effect, the remains of the open areas of Vera's landscape. Here are anciently complex, species-rich grasslands within landscapes of hazel and patches of ancient woodland (identified by Pigott in the 1960s). These and other wooded sites are now being recognised as 'shadow woods' or 'ghosts woods' (Figure 7.3); either relicts from once obviously ancient woodland sites, or perhaps more excitingly, ancient wooded landscapes until now overlooked.

The ecology of parks and other grazed wooded landscapes

The ecologies of working parks and other grazed wooded landscapes reflect the factors described above, and what survives today, mirrors these events and pressures (Rotherham, 2007a, b and 2010a). Park landscapes had unimproved grassland across much of the grazed area; species and communities

Reinterpreting wooded landscapes 79

Figure 7.2 Veteran hawthorn in grazed upland shadow wood, Peak District, England
Source: Ian D. Rotherham

varying with grazing intensity, soil type and wetness. Many grassland plants and associated invertebrates cannot cope with short swards and intensive grazing. However, if grazing levels were low or areas seasonally protected from livestock, the vegetation would grow tall, flower and set seed; similar to modern unimproved pasture and hay meadow. Such areas would be rich in wild flowers and in associated invertebrates such as butterflies, bees and hoverflies; a patchwork of shorter grass, bare ground, and in acidic locations, heath. Wet areas (valley bottoms or land with impeded drainage) had extensive moist grassland, marsh or bog. The typical plants of ancient woodlands (such as dog's mercury (*Mercurialis perennis*), wood anemone (*Anemone nemorosa*), primrose (*Primula vulgaris*) and bluebell (*Hyacinthoides non-scripta*)) would have been restricted and found only in enclosed woods, copses, lane sides, hedgerows or stream sides. They survived perhaps in areas of less intensive grazing, and in the protection of prickly bramble and blackthorn.

Keystone species in the park were deer, with other grazing mammals of varying domestication; these animals being the main drivers in the deer-park ecosystem. Other important ecological components were fungi in the unimproved grasslands, and associated with extensive animal dunging. There would have been a rich fungal flora of both mycorrhizal associates of both trees (ectomycorrhizas) and of grasses and forbes in the sward (vesicular-arbuscular mycorrhizas). These would present as both individual groups of toadstool fruiting bodies, as can be seen today with the dung-associated species such as the shaggy ink caps (*Coprinus* sp.), and as spectacular '*fairy rings*'. Associated with animal dunging would be rich faunas of coprophagous and predatory flies and dung beetles. It can be assumed that high numbers of animals would lead to carcases and faunas of species such as burying beetles. With the high numbers of mammals were rich faunas of parasites such as mites, ticks and biting or egg-laying flies, and further associated food chains of complexity.

Imparking sometimes included deliberate or accidental preservation of domesticated, semi-domesticated or wild grazing mammals within the enclosure. The white park cattle are a case in point, with the Chillingham Park herd in Northumberland perhaps the best example; aside from a small herd established some distance away as a precaution against foot-and-mouth disease, this unique breed of ancient cattle survives at only one location. Whitaker in 1892 described the park as 1,500 acres, well wooded, and with moor and wild grounds (Whitaker, 1892). This ancient and extensive park enclosed and encapsulated an entire ecosystem that has been maintained ever since, though with considerable modifications (Stephen Hall, 2007, and this volume). Outside the park, species including the cattle disappeared long ago. Enclosing extensive semi-natural landscape was not exclusive to deer parks. Seventeenth- and eighteenth-century ornamental parks often involved similar scales of enclosure, sometimes from common fields but often from 'waste'. This may have included marshes, grasslands, heaths and extensive bogs.

Large oaks were grown for timber with trunks and boughs carefully nurtured to form particular shapes and sizes for specific functions. Careful planning and management over many decades are key aspects of park historical ecology. The records of great estates often give precise details of the removal of trees, their price, and destination. Park trees mixed with

Reinterpreting wooded landscapes 81

timber trees, enclosed when the park was formed, and others deliberately planted. Parks such as Chatsworth, Derbyshire, include later additions through conversion of field systems and hedgerow trees, now veterans in the contemporary landscape. Most of the very old trees, often oak (*Quercus robur*), are specimens that have been actively managed for at least several centuries and then abandoned. Large trees performed many functions in working parks, providing shelter in winter and shade in summer for cattle and deer. Importantly, they provided herbage to feed to the livestock; most deer and cattle preferring to browse leaves and shoots, than graze grass (Figure 7.4). To ensure a continuous supply of branches and leaves, the trees were cut high, several metres above ground, keeping re-growth out of the reach of the grazing animals, until the parker cut it for fodder. The

Figure 7.2 Cattle browsing parkland
   Source: Ian D. Rotherham

technique was known as pollarding and is in effect a high coppice. Furthermore, the provision of special hollins and hags ensured that herbage was provided for livestock throughout the winter. For several months of the year, and longer during colder periods, grass does not grow in Britain and livestock depend on stores of hay, a valuable and often-scarce commodity, and cut branches of evergreen holly. Pollarding extended the lifespan of trees beyond that normally achieved and in so doing ensured a major supply and continuity of deadwood, a highly important wildlife habitat. Other anciently grazed wooded landscapes also have old trees and deadwood but these are generally less obvious.

Parks are the most obvious landscapes that mix trees and grazing animals. However, once one starts to examine the landscape more critically, it is apparent that many other systems have a similar approach. The once-vast lowland fens of eastern England (Rotherham, 2010b) had extensive woodland and large numbers of grazing animals. Heaths, commons and unenclosed pastures (such as Longshaw, north Derbyshire), mix ancient trees and open grazing lawns with long-term continuity of management to match that of the nearby Chatsworth Park. A major difference is that ancient trees in these landscapes are generally small and may be species such as hawthorns, which are often overlooked. Examining ecology and pedology in these wider landscapes reveals the imprint or '*shadow*' of former 'woodland' status; they are '*ancient wooded landscapes*'. The origins or at least the recognition of, the components of these ancient landscapes are apparent in medieval legislation. In particular, the Statute of Merton (Act of Commons) (1235) provides a window into a watershed moment for these landscapes.

The Statute of Merton: A Magna Carta of the landscape

1235 Henry III

Whereas in a Statute made at Merton, it was granted that Lords of Wastes, Woods, and Pastures, might approve the said Wastes, Woods, and Pastures, notwithstanding the contradiction of their Tenants, so that the tenants had sufficient Pasture to their Tenements with free Egress and regress to the same; and Forasmuch as no Mention was made between Neighbours and Neighbours, many Lords of Wastes, Woods, and Pastures, have been hindered heretofore by the Contradiction of Neighbours having sufficient Pasture; and because foreign Tenants have no more right to Common in the Wastes, Woods, and Pastures of any Lord than the Lord's own Tenants; It is Ordained, that the Statute of Merton, provided between the Lord and his tenants, from henceforth shall hold place between Lords of Wastes, Woods, and Pastures, may make Approvement of the residue …………….. from henceforth no Man shall be grieved by Assise of Novel Disseisin for Common of Pasture.

The act probably reflected what was happening and provided legal recognition setting down rights of land use and function at manorial level; what was previously very fluid and extensive, became fixed and localised. A named 'wood' was now set in its landscape, bounded by fence, wall, hedge or ditch, and given a name. Similarly, common, heath, fen, field and waste were marked and recorded. This process transformed Vera's landscape to what we see today, including 'wooded' areas left outside the 'woods'. Greater, early medieval deer parks enclosed the landscape, as it existed before Merton, including woods and other wooded features and so these are good places to search for the ecological shadows of once extensive wooded landscapes.

Transformation and fashion

Rackham (1986) notes that parks were troublesome, precarious enterprises with boundaries expensive to maintain, especially for large parks. Owners were often absent for much or all of the year, a situation that could lead to mismanagement and neglect. Even well-run parks faced ongoing problems of maintenance and many smaller parks were short-lived; some were already out of use by the thirteenth century. Sometimes a park was retained but moved within the manor, which clearly affected the ecology.

During the sixteenth century, park function shifted from game preserve and source of wood and timber, to grand country house setting. Disused deer parks might revert to woodland through neglect or deliberate replanting or be converted to farmland. Some like Trelowarren in Cornwall retained the park pale to enclose the newly formed fields. If changes allowed habitat continuity, then some original ecology such as rare deadwood insects might hang on. New ornamental parklands were in demand but were not created from blank canvases but generally adapted and imposed on an earlier landscape that sometimes gives historic connectivity to the older ecology (Rotherham, 2007a and b). The landscape archaeology may include early non-park features and, lacking some ancient deer park indicators, they hold species of medieval woodlands, of hedgerows, and perhaps veteran pollard. Again, this gives '*acquired antiquity*', as elements normally associated with genuinely ancient sites are acquired or 'borrowed' from earlier periods.

Many other woodland shadows or ghosts were lost or destroyed by land 'improvement', in Britain for example, during the 1700s and 1800s, through parliamentary enclosures, and then especially during the post-World War II drive for production. Others remain but are abandoned and severed from customary or traditional uses. In all these landscapes grazing animals were important economic and ecological drivers. However, it is also important for conservation purposes that we recognise the other human-driven cultural impacts too, such as cutting of fuelwood and constructional timber, turf- and peat-getting, furze cutting and other harvesting. These complex processes operating over centuries accidentally created and then sustained much of the ecology that we value today. In landscapes occupied by people, grazing was important but it did not operate in isolation. Without such recognition, there is only limited awareness of the serious challenges for ecological sustainability in the twenty-first century.

Conclusions

Grazed wooded landscapes are more widespread than generally recognised, with many ancient areas overlooked. Some sites may have uniquely long timelines of grazing impacts and influences to complement those of historic parks and forests. It is possible to develop visions of historical ecology joining primeval savannah ideas of Frans Vera to the landscape and countryside history of Oliver Rackham. The evidence is there if we have the eyes to see. How we find, preserve and conserve this complex and often-intangible heritage is a challenge with no single approach or correct answer. Much that remains is threatened by cultural severance and the ending of traditional management (Rotherham, 2008, 2009a, b and c, and 2010a).

To raise awareness, identify sites and to encourage good management, involving local people and engaging with local communities is vital. This approach was used to develop the *Woodland Heritage Manual* (Rotherham *et al.*, 2008). Ongoing studies in the Peak District and surrounding areas are proving helpful in placing ideas and theories into frameworks reflecting specialist disciplines such as palaeoecology, palynolgy and archaeology.

It is suggested that remnants of medieval parks are vestiges of very ancient landscapes; albeit transformed and manipulated by human hand over the centuries. These may precede human domination and agriculture, with Vera's vision of forested savannah indicating a lineage to great primeval origins of the European forest. Harking back evocatively to the past, this view also informs the future. The vision of landscapes is freed from anthropogenic constraints of medieval agricultural and pastoral scenes, setting new challenges for deeply embedded precepts of nature conservation.

Bibliography and references

Beswick, P. and Rotherham, I.D. (eds) (1993) 'Ancient woodlands – their archaeology and ecology – a coincidence of interest', *Landscape Archaeology and Ecology*, 1.
Bettey, J.H. (1993) *Estates and the English Countryside*, Batsford, London.
Cantor, L.M. and Hatherly, J. (1979) 'The medieval parks of England', *Geography*, 64: 71–85.
Cummins, J. (1988) *The Hound and the Hawk*, Weidenfeld & Nicholson, London.
Hall, S. (2007) 'Chillingham Wild Cattle Park, Northumberland', *Landscape Archaeology and Ecology*, 6: 53–7.
Harding, P.T. and Rose, F. (1986) *Pasture-Woodlands in Lowland Britain: A Review of their Importance for Wildlife Conservation*, Institute of Terrestrial Ecology, Monks Wood Experimental Station, Huntingdon.

Jones, M. (1996) 'Deer in South Yorkshire an historical perspective', in Jones, M., Rotherham, I.D. and McCarthy, A.J. (eds) 'Deer or the New Woodlands?', *The Journal of Practical Ecology and Conservation Special Publication*, No 1, pp11–26.

Jones, M., Rotherham, I.D. and McCarthy, A.J. (eds) (1996) 'Deer or the New Woodlands?', *The Journal of Practical Ecology and Conservation, Special Publication*, No 1.

Langton, J. and Jones, G. (eds) (2005) *Forests and Chases of England and Wales c.1500–c.1850. Towards a Survey and Analysis*, St John's College Research Centre, Oxford.

Lasdun, S. (1992) *The English Park: Royal, Private and Public*, The Vendome Press, New York.

Liddiard, R. (2003) 'The deer parks of Domesday Book', *Landscapes*, 4(1): 4–23.

Mileson, S.A. (2005) 'The importance of parks in fifteenth-century society', in Clark, L. (ed.) *The Fifteenth Century V*, Boydell and Brewer, Woodbridge, pp19–37.

Peterken, G.F. (1981) *Woodland Conservation and Management*, Chapman & Hall, London

Peterken, G.F. (1996) *Natural Woodland: Ecology and Conservation in Northern Temperate Regions*, Cambridge University Press, Cambridge.

Pigott, C.D. (1993) 'The history and ecology of ancient woodlands', *Landscape Archaeology and Ecology*, 1: 1–11.

Rackham, O. (1976) *Trees and Woodland in the British Landscape*, J. M. Dent & Sons Ltd, London.

Rackham, O. (1978) 'Archaeology and land-use history', in Corke, D. (ed.) *Epping Forest – the Natural Aspect?*, Essex Nat., N.S. 2: 16–57.

Rackham, O. (1980) *Ancient Woodland: its History, Vegetation and Uses in England*, Arnold, London.

Rackham, O. (1986) *The History of the Countryside*. J. M. Dent & Sons Ltd, London.

Rackham, O. (2004) 'Pre-existing trees and woods in country-house parks', *Landscapes*, 5(2): 1–16.

Read, H. (1999) *Veteran Trees: A Guide to Good Management*, English Nature, Peterborough.

Rollins, J. (2003) *Land Marks: Impressions of England's National Nature Reserves*, English Nature, Peterborough.

Rotherham, I.D. (2007a) 'The historical ecology of medieval deer parks and the implications for conservation', in Liddiard, R. (ed.) *The Medieval Deer Park: New Perspectives*, Windgather Press, Macclesfield, pp79–96.

Rotherham, I.D. (2007b) 'The ecology and economics of medieval deer parks', *Landscape Archaeology and Ecology*, 6: 86–102.

Rotherham, I.D. (2007c) 'The implications of perceptions and cultural knowledge loss for the management of wooded landscapes: A UK case-study', *Forest Ecology and Management*, 249: 100–15.

Rotherham, I.D. (2008) 'The importance of cultural severance in landscape ecology research', in Dupont, A. and Jacobs, H. (eds) (2008) *Landscape Ecology Research Trends*. Nova Science Publishers Inc., New York, pp71–87.

Rotherham, I.D. (2009a) 'Hanging by a thread – a brief overview of the heaths and commons of the north-east midlands of England', in Rotherham, I.D. and Bradley, J. (eds) (2009) *Lowland Heaths: Ecology, History, Restoration and Management*, Wildtrack Publishing, Sheffield, pp30–47.

Rotherham, I.D. (2009b) 'Habitat fragmentation and isolation in relict urban heaths – the ecological consequences and future potential', in Rotherham, I.D. and Bradley, J. (eds) (2009) *Lowland Heaths: Ecology, History, Restoration and Management*, Wildtrack Publishing, Sheffield, pp106–15.

Rotherham, I.D. (2009c) 'Cultural severance in landscapes and the causes and consequences for lowland heaths', in Rotherham, I.D. and Bradley, J. (eds) (2009) *Lowland Heaths: Ecology, History, Restoration and Management*, Wildtrack Publishing, Sheffield, pp130–43.

Rotherham, I.D. (2010a) 'Cultural severance and the end of tradition', *Landscape Archaeology and Ecology*, 8: 178–99.

Rotherham, I.D. (2010b) *Yorkshire's Forgotten Fenlands*, Pen & Sword Books Limited, Barnsley.

Rotherham, I.D. (2011a) 'A landscape history approach to the assessment of ancient woodlands', in Wallace, E.B. (ed.) *Woodlands: Ecology, Management and Conservation*, Nova Science Publishers Inc., USA, pp161–84.

Rotherham, I.D. (2011b) 'Animals, man and treescapes – perceptions of the past in the present', in Rotherham, I.D. and Handley, C. (eds) (2011) Animals, *Man and Treescapes: The Interactions between Grazing Animals, People and Wooded Landscapes*, Wildtrack Publishing, Sheffield, pp1–32.

Rotherham, I.D. and Handley, C. (eds) (2011) Animals, *Man and Treescapes: The Interactions between Grazing Animals, People and Wooded Landscapes*, Wildtrack Publishing, Sheffield.

Rotherham, I.D., Jones, M., Smith, L. and Handley, C. (eds) (2008) *The Woodland Heritage Manual: A Guide to Investigating Wooded Landscapes*, Wildtrack Publishing, Sheffield.

Taigel, A. and Williamson, T. (1993) *Parks and Gardens*, Batsford, London.

Vera, F.W.M. (2000) *Grazing Ecology and Forest History*, CABI Publishing, Oxon, UK.

Whitaker, J. (1892) *A Descriptive List of the Deer-Parks and Paddocks of England*, Ballantyne, Hanson & Co., London.

Whitehead, G.K. (1964) *The Deer of Great Britain and Ireland*, Routledge and Kegan Paul Ltd, London.

Whitehead, G.K. (1980) *Hunting and Stalking Deer in Britain through the Ages*, Batsford, London.

8 The dynamics of pre-Neolithic European landscapes and their relevance to modern conservation

Keith J. Kirby and Ambroise Baker

Introduction

The structure and composition of a natural landscape is of interest in its own right, but to what extent is it relevant to current and future nature conservation, particularly in the highly modified cultural landscapes that characterise much of north-west Europe? This chapter explores these ideas, focusing on the English context.

When was the landscape last 'natural'?

Defining a natural landscape is itself challenging, but Peterken (1996) takes the pragmatic approach that it is a landscape where deliberate human influence is, if not absent, at least of the same order as other medium-sized mammals. The latest period when past-natural conditions may have existed over the majority of north-west Europe – and the one adopted by Vera (2000) among others – is therefore prior to the Neolithic era, which in England started about 6,000 years ago.

There were humans present in the country before that and Mesolithic communities, for example through the use of fire, probably did have at least locally significant impacts on the vegetation (e.g. Innes and Blackford, 2003: Simmons, 2003). Earlier cultures may have precipitated the extinction over large parts of Europe of mega-fauna, although more recently Allen *et al.* (2010) suggest that climate-induced vegetation change might have been sufficient. However the change to a predominantly more settled existence during the Neolithic and subsequent eras introduced substantial new elements to the landscape through new livestock (sheep, goats, horse (strictly a reintroduction)) (Yalden, 1999) and arable crops.

The pre-Neolithic landscape was itself only one stage in the evolution of the vegetation cover of Britain after the last glacial period (Godwin, 1975). Even without human intervention, the landscape would have continued to change in response to fluctuations in climate, the maturation of soils and development of blanket bog, and the continued spread of species back from post-glacial refuges. Oaks (*Quercus* spp.) were already widespread and abundant in northern Europe by 8,000 years BP (before present), lime (*Tilia* spp.) 7,000 years BP, but beech (*Fagus sylvatica* L.) was still largely confined to southern Europe as a major forest species until c.3,000 years BP (Huntley and Birks, 1983).

The landscape would not have been of uniform composition across England; variations in local climate and soils strongly determine species patterns today (e.g. Rodwell, 1991) and are likely to have been more significant in their impact before human management was imposed upon them. Peterken (1996) suggests the different types of natural woodland that might have developed on dry lowland sites, on floodplains, in the uplands, while Smith and Whitehouse (2005) draw attention to how such differences may be reflected in different sub-fossil beetle assemblages.

The landscape would also be influenced by the nature of the principal disturbance regimes or drivers that determined its structural dynamics and function, which in turn might vary over time and space. On a micro-scale this is illustrated by the range of factors that have impacted on Lady Park Wood National Nature Reserve over the last 60 years, which have included extreme droughts, disease and the depredations of mammals (Peterken and Mountford, 2005).

So there has not been one 'natural landscape' that might serve as a template for modern conservation but many, depending on what time period we choose and where in the country we look. This variation increases manyfold if our canvas is extended to Europe as a whole.

A treed but was it a wooded landscape?

The most familiar sources of information for past landscapes are pollen records (e.g. Berglund *et al.*, 1996), but other direct evidence comes from more substantial plant remains, with up to whole tree trunks found, such as are ploughed up in the Fens. Other remains include those from invertebrates such as beetles, snails or chironomid midges, fungal spores, and soil deposits etc. Indirect evidence may come from analogies with modern or historic records and extrapolations from the modern distributions and ecology of species (Russell, 1997; Wiens and Hayward, in press). Increasing use is being made of models to try to understand what the evidence of the past state of the environment might mean for how it might have functioned (e.g. Allen *et al.*, 2010).

Each approach may have its limitations; the records may be incomplete and irregularly distributed in time and space. At times the findings from different approaches may be contradictory. While there may be agreement on at least the broad parameters for past landscape structure and composition, debate as to the relative weight attached to different parts of the evidence can lead to disagreement on the overall meaning of the package.

The pollen and invertebrate records tend to indicate a shift from vegetation that was predominantly open in the immediate post-glacial period to one where broadleaved trees become more abundant in the pre-Neolithic period. Thereafter indicators of 'open' landscapes increase again. 'Open' indicators do not however disappear everywhere in the pre-Neolithic 'tree-dominated' phase, for example floodplain deposits at Bole Ings by the River Trent (Whitehouse and Smith, 2004) show a low level of pasture/dung/open ground beetles from c.8,000 years BP through to 3,000 years BP. Similarly open habitats, notably within floodplains, have been reconstructed from previous warm stages, prior to the last Ice Age, when human activities were definitely negligible (e.g. Svenning, 2002).

Much of the recent debate has been about how widespread were the open patches and what maintained them. Was the landscape a matrix of closed forest with open patches, or largely open with patches of trees (Peterken, 1996; Vera, 2000)? Separating these two hypotheses depends on whether the proxy measures (pollen records, invertebrate remains etc.) can detect mosaics of open and closed habitats. Recent studies of modern pollen/insect samples from small collecting sites (e.g. Bunting, 2002; Smith *et al.*, 2010) suggest that both can be used, with caution, to infer the local vegetation composition. On the chalk hills of Wiltshire, Allen *et al.* (2009) and French *et al.* (2007) concluded that some Neolithic monuments appear to have been established in open grassland landscapes, whereas they identified areas further down the valleys that showed evidence of extensive woodland cover comparable to that of other chalk landscapes such as the South Downs.

What factors – large herbivores, wind, flood, disease, fire etc. (assuming it was not humans) were driving the dynamics of the system and creating the open areas detected? Much debate has been around the role of large herbivores (Hodder *et al.*, 2005 and 2009), but the dominant role for them proposed by Vera (2000) does not seem to be supported by more recent studies:

- Van Vuure (2005) questions whether the aurochs, *Bos primigenius*, could have had the effect Vera proposes; Hall (2007) concludes that the aurochs in Britain was mainly a species of low-lying flat, fertile ground. The other main large grazer present on the continent at the beginning of the Holocene was the wild horse, *Equus ferus*, but Sommer *et al.* (2011) conclude that it was almost absent in the central European lowlands in the middle Holocene, and is not thought to have survived in Britain.
- Mitchell (2005) found little difference between the pollen records for *Quercus* and *Corylus* from regions where large herbivores were present and those where they were absent (specifically Ireland). Ireland does not have the same range of shade tolerant trees, notably beech, *Fagus sylvatica*, as central Europe now. However, the main spread of beech north-west from glacial refuges seems to have occurred after the Atlantic Period (Huntley and Birks, 1983), so the comparison with Ireland is valid. In addition beech has spread during historic times in the presence of intensive grazing of wood-pastures, contrary to Vera's

hypothesis, in Epping Forest (Grant, 2002) and the New Forest (Grant and Edwards, 2008).
- Struik (2001) shows that in the New Forest, a heavily grazed wood-pasture, the percentage of non-arboreal pollen found in surface samples was still more than generally found in fossil pollen records from the pre-Neolithic period.
- The experience of predator reintroduction in Yellowstone National Park suggests another mechanism by which herbivores might have been more restricted than Vera suggests in their ability to control regeneration (e.g. Ripple and Beschta, 2003 and 2010), but see also Kauffman *et al.* (2010) and Theuerkauf and Rouys (2008)).
- Dugout canoes have been recovered from various sites across Britain, for example six ranging from 4–8.3m long have come out of one recent Fenland excavation (*The Observer*, 4 December 2011: 24). The trees used for these needed to be tall and straight-trunked, suggesting that closed canopy conditions could still be found in the Bronze Age. (Even larger canoes existed – one in Dublin Museum, dated to c.4,500 years BP is 15m long.)

A critical issue has been detecting evidence for large herbivore abundance independent of their presumed effects on the vegetation record. Recently analyses of fungal spore occurrences have provided unequivocal evidence regarding the effect large-scale grazing regime changes can have on vegetation across the world (Burney *et al.*, 2003; Davis and Shafer, 2006; Gill *et al.*, 2009). There has been more limited application of these techniques in Europe but Innes and Blackford (2003) illustrate their use to suggest concentrations of animals in temporary clearances created by fire in the North York Moors (northern England). However, even where grazing was the dominant disturbance factor it might not produce the 'half-open landscape' suggested by Vera (Kirby, 2004).

Our current picture of the pre-Neolithic landscape in England is that:

- It would have looked predominantly wooded;
- It varied in both composition and structure, with significant patches of open ground, some permanent, some temporary;
- Large herbivores helped shape the nature of woodland, as deer do today, but only locally acted as the prime disturbance drivers; rather we stress that wind-throw, grazing, tree diseases, fire, flood, would all have impacted on the landscape in different combinations in different places (see for example Bradshaw *et al.*, 2003; Mitchell, 2005; Svenning, 2002; Whitehouse and Smith, 2004 and 2010) .

Elsewhere in north-west Europe there are additional tree species and large herbivores (bison, *Bison bonasus*, wild horse, *Equus ferus*) that were not in Britain in the post-glacial period. However, the evidence for large-scale open landscapes similarly appears thin; and the greater range of environmental conditions makes it less likely that a single type of disturbance factor (grazing) would predominate.

The relevance of the pre-Neolithic landscape structure to modern English conservation?

In many parts of the world the 'natural' landscape or wilderness is, at least nominally, the template for nature conservation. If the pre-Neolithic landscape were to be taken as the template for modern nature conservation in England, then whether it was predominantly open or closed woodland would become critical to the value placed on particular types of modern landscape structure.

During the 1970s and 1980s, wood-pastures and other sites with veteran trees were sometimes perceived as being more 'artificial', less valuable than closed woodland. Their value as the sites for old growth components (particularly deadwood invertebrates and lichens) (Harding and Rose, 1986; Kirby *et al.*, 1995; Read, 2000) was not so widely recognised as the value of coppice-woods for the species of young growth and open space. Part of the attraction of the temperate savannah hypothesis (Vera, 2000) for sections of the conservation movement seems to be that it gives legitimacy to wood-pasture as closer to the more natural landscape state.

However, even landscapes that appear natural to European eyes may be the product of long-established and sometimes intense human interventions (e.g. Fletcher and Thomas, 2010; Ross and Rangel, 2011). English landscapes are not anywhere near 'natural'; they are cultural (Birks *et al.*, 1988; Rackham, 1986), the product of a period of human modification as long as that between the retreat of the ice and development of Neolithic cultures. Most of the habitats that we value (Ratcliffe, 1977), meadows and heaths, wood-pastures as much as coppice-woods, are the remnants of pre-industrial farming and forestry. Indeed this diversity of landscapes and land uses is celebrated for its cultural as well as its biological heritage. Large- and medium-sized herbivore populations still play a key role in shaping the landscape, but these are largely controlled by humans; the biomass of domestic mammalian herbivores far exceeds that of wild ones (Yalden, 1999).

At least initially species diversity in England may have increased as a result of human intervention; new habitats were created by farming or at least their extent appears to have increased. Not all species benefited: there were, and continue to be, species extinctions (Natural England, 2010; Hambler *et al.*, 2011), but other species, some of them introductions, have thrived on the farmland created (e.g. brown hare, *Lepus europaeus*). Introduced species such as wild rabbits, *Oryctolagus cuniculus*, and domestic sheep have played a critical part in maintaining close-grazed turf of chalk grassland. Artificial formations such as poplar plantations and orchards are valued as strongholds for the golden oriole, *Oriolus oriolus*, and noble chafer beetle, *Gnorimus nobilis*, respectively. While many of the species we value in conservation today were present in the pre-Neolithic period, their abundances and the assemblages in which they now occur may have been very different.

Our conservation template for the last hundred years has been largely based, not on the natural landscape (whatever that might be), but on the countryside as it was experienced by the founding fathers and mothers of modern conservation. Derek Ratcliffe, in the guidelines for selecting protected sites in Britain, notes:

> An inherited knowledge of the best places to find and collect rare and local species developed during Victorian times and became part of a common fund of knowledge amongst naturalists{...}When, in 1915, Charles Rothschild compiled a list of desirable nature reserves{...}he drew on the opinions of many leading figures throughout the country. Thirty-two years later the listings of the Society's Nature Reserves Investigation Committee{...}represented the distillation of collective knowledge from a large body of informed opinion. A similar sifting was applied to the choice of SSSIs [Sites of Special Scientific Interest] when these became a statutory category.
>
> (NCC, 1989: 13)

These values as to what habitats are important, what species might be expected to occur in them, still underpin conservation today.

The vision of the *England Biodiversity Strategy* (2011) refers to a country 'where living things and their habitats are part of healthy, functioning ecosystems; where we value our natural environment, where biodiversity is embedded in policies and decisions'. However, the 'natural environment' as used in that strategy means cultural landscapes of human-modified fields, woods, heaths and moors (the priority habitats identified through the UK Biodiversity Action Plan process (DoE, 1995)), not large-scale wilderness areas where human impact is minimal.

Revisiting the cultural landscape template for conservation in England

Maintaining the conditions created by pre-industrial farming and woodland management, the nineteenth-century cultural landscape, is proving increasingly difficult:

- The social and economic systems that led to their creation have broken down and frequently can only be mimicked by public subsidy. Through agri-environment schemes Natural England pays farmers to put more stock on to abandoned commons and wood-pastures; elsewhere in the country

it pays other farmers to reduce stock numbers on upland bogs www.naturalengland.org.uk/ourwork/farming/funding/es/default.aspx).
- Biotic conditions have changed. The rabbit populations, which maintained close-cropped downland on sites even after the sheep had been removed, crashed in the 1950s (Yalden, 1999); since the 1970s, deer populations in woods have increased with major impacts on the composition of both plant and animal communities, but also making coppicing and other forms of woodland management more expensive (Fuller and Gill, 2001; Ward, 2004).
- The physical environment has changed with increased levels of nitrogen deposition (e.g. Emmett, 2002) and increasing evidence for impacts of climate change (Mitchell *et al.*, 2007). Species assemblages may change therefore even under the same grazing regimes as in the past – for example the loss of lichens from wood-pastures because of air pollution.

Whereabouts on the continuum of landscape compositions – from totally artificial to totally natural – should future conservation aim? There are already places where we conserve habitats and landscapes that have no or few analogues in the historic past because of their species diversity or rarity, for example some post-industrial sites (more artificial), or woods that have now a high forest structure (more natural (Peterken, 2000)). There are also starting to be places where we deliberately try to reduce human impact, to reduce the cultural and increase the natural components across large landscapes, an approach that in Britain has become known as re-wilding (Hughes *et al.*, 2011).

Adopting a 're-wilding' approach implies accepting change and unpredictable outcomes (Hughes *et al.*, 2011). The habitats and species assemblages may not necessarily bear much resemblance to those of traditional land-management systems, but neither may they resemble the pre-Neolithic landscape. A 're-wilded' landscape is likely to consist of a mosaic of different habitats, but we cannot predict or prescribe at what scales and what proportions the different elements might have. Species may be lost or gained; hence the conservation manager must be clear as to the priority they wish to give in a given place to 'protecting biodiversity' versus 'respect for nature's autonomy' (Ridder, 2007).

A key inspiration for re-wilding ideas in England has been what the Dutch have done at Oostvaardersplassen (Vera, 2009). Naturalistic grazing by large and medium-sized herbivores is a key part of re-wilding, because herbivores are seen as an important element of natural ecosystems in their own right, but also because of their impact on the landscape development (Hodder *et al.*, 2005). Depending on the circumstances this may mean just allowing more free ranging of domestic stock, to developing feral herds as at Oostvaardersplassen.

The advantages of using domestic stock are that they are easier to manage and the system may still just come within the scope of agri-environment schemes that provide much of the funding for conservation management. Feral herds present more difficulties in terms of how they are treated with respect, for example, to animal welfare legislation. In England, re-wilding cannot be a totally hands-off approach (Kirby, 2009), but its application in a few places is leading to the development of some novel conditions (e.g. www.wildennerdale.co.uk; www.knepp.co.uk; www.greatfen.org.uk).

Natural and cultural templates for conservation

Future tree and woodland conservation in England needs to recognise three different sets of templates:

1. 'Cultural/traditional' – where the aim is to maintain past habitats and species assemblages through mimicking past land uses such as coppice and wood-pasture management etc.;
2. 'Targeted habitat/species management' – where the aim is to conserve a particular habitat or species assemblage; these may not have any historical analogue and even if they do their conservation may not follow traditional practices;
3. 'Natural development/re-wilding' – where the aim is to allow change (within broad limits) in response to different natural or near-natural processes, without a specified endpoint.

The management of grazing animals will be an important part in all three approaches, both in terms of reducing the pressure of wild deer populations where these are preventing us from meeting our objectives; deliberately (re)introducing livestock at prescribed levels to create particular habitat mosaics; and more-or-less naturalistic grazing systems where the outcome is uncertain.

What we choose to do on each site needs to be clearly justified, perhaps as much on cultural and even spiritual grounds (Taylor, 2005) as for biodiversity reasons. Where we choose the re-wilding approach the outcome will not be the same as a pre-Neolithic landscape, but we may be able to use such areas to test hypotheses on what such landscapes were like and how they functioned.

A wider European perspective

In Britain and particularly England the dominance of the cultural landscape approach in conservation is more obvious than in much of the rest of Europe where there are much larger areas of 'more natural' landscapes. These show a range of structures, at varying scales arising from a variety of disturbance regimes. However, even areas often described as primeval, such as Białowieża Forest in Poland (Bobiec, 2012; Falinski, 1986) show evidence of past human interventions.

Deliberate management decisions were involved at Oostvaardersplassen with respect to which large herbivore species were introduced (Vera, 2000) to artificially reclaimed ground; elsewhere other conservation programmes involve reintroduction of large carnivores as well as large herbivores. Is this really that different, albeit done for different reasons, from the manipulation of landscapes and associated animal populations carried by prehistoric cultures across Europe?

The value of separating human from 'natural' impacts on the landscape is not necessarily that the latter are always 'better' in a conservation sense, but because we can and should choose to control the level of the former.

Acknowledgements

Our thanks to Frans Vera for stimulating this debate; to Kathy Hodder, Paul Buckland and James Bullock who carried out the study of naturalistic grazing for English Nature; and to English Nature/Natural England for allowing Keith Kirby the time to pursue it. We have benefited from animated discussions and borrowed ideas on this topic from many other colleagues, but the views expressed (and any errors in them) are our own.

References

Allen, J.R.M., Hickler, T., Singarayer, J.S., Sykes, M.T., Valdes, P.J. and Huntley, B. (2010) 'Last glacial vegetation of northern Eurasia', *Quaternary Science Reviews*, 29: 2604–18.

Allen, M.J., Sharples, N. and O'Connor, T. (eds) (2009) *Land and People*, Prehistoric Society Research Paper 2, Oxbow Books, Oxford.

Berglund, B.E., Birks, H.J.B., Ralka-Jasiewiczowa, M. and Write, H.E. (eds) (1996) *Palaeoecological Events during the Last 15000 Years*, Wiley, Chichester.

Birks, H.H., Birks H.J.B., Kaland, P.E. and Moe, D. (eds) (1988) *The Cultural Landscape – Past, Present and Future*, Cambridge University Press, Cambridge.

Bobiec, A. (2012) 'Białowieża Primeval Forest as a remnant of culturally modified ancient forest', *European Journal of Forest Research*, DOI 10.1007/s10342-012-0597-6.

Bradshaw, R.H.W., Hannon, G. and Lister, A. (2003) 'A long-term perspective on ungulate-vegetation interactions', *Forest Ecology and Management*, 181: 267–80.

Bunting, M. J. (2002) 'Detecting woodland remnants in cultural landscapes: Modern pollen deposition around small woodlands in northwest Scotland', *The Holocene*, 12: 291–301.

Burney, D.A., Robinson, G. and Burney, L.P. (2003) '*Sporormiella* and the late Holocene extinctions in Madagascar', *PNAS*, 100: 10800–5.

Davis, O.K. and Shafer, D.S. (2006) '*Sporormiella* fungal spores, a palynological means of detecting herbivore density', *Palaeogeography, Palaeoclimatology, Palaeoecology*, 237: 40–50.

DoE (Department of Environment) (1995) *Biodiversity Action Plan*, HMSO, London.

England Biodiversity Strategy (2011) *Biodiversity 2020*, Defra, London.

Emmett, B. (2002) 'The impact of nitrogen deposition in forest ecosystems: A review', Centre for Ecology and Hydrology contract report, Bangor.

Falinski, J. (1986) *Vegetation Dynamics in Temperate, Lowland Primeval Forest: Ecological Studies in Białowieża Forest*, Junk, The Hague.

Fletcher, M.-S. and Thomas, I. (2010) 'The origin and temporal development of an ancient cultural landscape', *Journal of Biogeography*, 37: 2183–96.

French, C., Lewis, H., Allen, M.J., Green, M., Scaife, R. and Gardiner, J. (2007) *Prehistoric Landscape Development and Human Impact in the Upper Allen Valley, Cranborne Chase, Dorset*, McDonald Institute Monographs, Cambridge.

Fuller, R.J. and Gill, R. (2001) 'Ecological impacts of deer in British woodland', *Forestry*, 74: 193–299.

Gill, J.L., Williams, J.W., Jackson, S.T., Lininger, K.B. and Robinson, G.S. (2009) 'Pleistocene megafaunal collapse, novel plant communities and enhanced fir regimes in North America', *Science*, 326: 1100–3.

Godwin, H. (1975) *History of the British Flora*, 2nd edition, Cambridge University Press, Cambridge.

Grant, M.J. (2002) 'Re-evaluating the concept of woodland continuity and change in Epping Forest: Biological and sedimentary analyses', unpublished MSc thesis, University of Reading, Reading.

Grant, M.J. and Edwards, M.E. (2008) 'Conserving idealised landscapes: Past history, public perception and future management in the New Forest (UK)', *Vegetation History and Archaeobotany*, 17: 551–62.

Hall, S.J.G. (2007) 'A comparative analysis of the habitat of the extinct aurochs and other prehistoric mammals in Britain', *Ecography*, 31: 187–90.

Hambler, C., Henderson, P.A. and Speight, M.R. (2011) 'Extinction rates, extinction-prone habitats and indicator groups in Britain and at larger scales', *Biological Conservation*, 144: 713–21.

Harding, P.T. and Rose, F. (1986) *Pasture-woodland in Lowland England*, Institute of Terrestrial Ecology, Huntingdon.

Hodder, K.H., Bullock, J.M., Buckland, P.C. and Kirby, K.J. (2005) *Large Herbivores in the Wildwood and Modern Naturalistic Grazing Systems*, English Nature (Research Report 648), Peterborough.

Hodder, K.H., Buckland, P.C., Kirby, K.J. and Bullock, J.M. (2009) 'Can the pre-Neolithic provide suitable models for re-wilding the landscape in Britain?', *British Wildlife*, 20 (supplement): 4–15.

Hughes, F.M.R, Stroh, P.A., Adams, W.M., Kirby, K.J., Mountford, O. and Warrington, S. (2011) 'Monitoring and evaluating large-scale, "open-ended" habitat creation projects: A journey rather than a destination', *Journal of Nature Conservation*, 19: 245–53.

Huntley, B. and Birks, H.J.B. (1983) *An Atlas of Past and Present Pollen Maps for Europe 0-13000 Years Ago*, Cambridge University Press, Cambridge.

Innes, J.B. and Blackford, J.J. (2003) 'The ecology of late Mesolithic woodland disturbances: Model testing with fungal spore assemblage data', *Journal of Archaeological Science*, 30: 185–94.

Kauffman, M.J., Brodie, J.F. and Jules, E.S. (2010) 'Are wolves saving Yellowstone's aspen? A landscape-level test of a behaviourally mediated trophic cascade', *Ecology*, 9: 2742–55.

Kirby, K.J. (2004) 'A model of a natural wooded landscape in Britain driven by large-herbivore activity', *Forestry*, 77: 405–20.

Kirby, K.J. (2009) 'Policy in or for the wilderness?', *British Wildlife*, 20 (supplement): 59–63.

Kirby, K.J., Thomas, R.C., Key, R.S., McLean, I.F.G. and Hodgetts, N. (1995) 'Pasture woodland and its conservation in Britain', *Biological Journal of the Linnaean Society*, 56 (supplment): 135–53.

Mitchell, F.J.G. (2005) 'How open were European primeval forests? Hypothesis testing using palaeoecological data', *Journal of Ecology*, 93: 168{}177.

Mitchell, R.J., Morecroft, M.D., Acreman, M., Crick, H.Q.P., Frost, M., Harley, M., MacLean, I.M.D., Mountford, O., Piper, J., Pontier, H., Rehfisch, M.M., Ross, L.C., Smithers, R.J., Stott, A., Walmsley, C., Watts, O. and Wilson, E. (2007) *England Biodiversity Strategy – Towards Adaptation to Climate Change*, Defra, London.

Natural England (2010) *Lost life: England's Lost and Threatened Species*, Natural England, Sheffield.

NCC (Nature Conservancy Council) (1989) *Guidelines for the Selection of Biological SSSIs*, Nature Conservancy Council, Peterborough.

Peterken, G.F. (1996) *Natural Woodland: Ecology and Conservation in Northern Temperate Regions*, Cambridge University Press, Cambridge.

Peterken, G.F. (2000) *Natural Reserves in English Woodland*, English Nature Research Report 384, Peterborough.

Peterken, G.F. and Mountford, E. (2005) 'Natural woodland reserves – 69 years of trying at Lady Park Wood Reserve', *British Wildlife*, 17: 7–16.

Rackham, O. (1986) *The History of the Countryside*, J.M. Dent, London

Ratcliffe, D.A. (1977) *A Nature Conservation Review*, Cambridge University Press, Cambridge.

Read H. (2000) *The Veteran Tree Management Handbook*, English Nature, Peterborough.

Ridder, B. (2007) 'The naturalness versus wilderness debate: Ambiguity, inconsistency and unattainable objectivity', *Restoration Ecology*, 15: 8–12.

Ripple, W.J. and Beschta, R.L. (2003) 'Wolf reintroduction, predation risk, and cottonwood recovery in Yellowstone National Park', *Forest Ecology and Management*, 184: 299–313.

Ripple, W.J. and Beschta, R.L. (2010) 'Trophic cascades in Yellowstone: The first 15 years after wolf introduction', *Biological Conservation*, 145: 205–13.

Rodwell, J.S. (1991) *British Plant Communities: I Woodlands and Scrub*, Cambridge University Press, Cambridge.

Ross, N.J. and Rangle, T.F. (2011) 'Ancient Maya agroforestry echoing through spatial relationships in the extant forest of NW Belize', *Biotropica*, 43: 141–8.

Russell, E.W.B. (1997) *People and the Land through Time: Linking Ecology and History*, Yale University Press, Yale.

Simmons, I.G. (2003) *Moorlands of England and Wales – An Environmental History 8,000 BC–AD 2,000*, Edinburgh University Press, Edinburgh.

Smith, D. and Whitehouse, N.J. (2005) 'Not seeing the woods for the trees: A palaeoentomological perspective on woodland decline', in Brinkley, M., Smith W. and Smith, D.N. (eds) *The Fertile Ground*, Oxbow Books, Oxford, pp136–61.

Smith, D., Whitehouse, N., Bunting, M.J. and Chapman, H. (2010) 'Can we characterise openness in the Holocene palaeoenvironmental record? Modern analogue studies of insect faunas and pollen spectra from Dunham Massey deer park and Epping Forest, England', *The Holocene*, 20: 1–22.

Sommer, R.S., Benecke, N., Lougas, L, Oliver, N. and Schmolcke, U. (2011) 'Holocene survival of the wild horse in Europe: A matter of open landscape?', *Journal of Quaternary Science*, 26: 805–12.

Struik, J. (2001) 'Pollen rain research within parkland landscapes', Unpublished report, Institute for Biodiversity and Ecosystem Dynamics, University of Amsterdam.

Svenning, J-C. (2002) 'A review of natural vegetation openness in north-western Europe', *Biological Conservation*, 104: 133–48.

Taylor, P. (2005) *Beyond Conservation: A Wildland Strategy*, Earthscan and BANC, London.

Theuerkauf, J. and Rouys, S. (2008) 'Habitat selection by ungulates in relation to predation risk by wolves and humans in Białowieża Forest, Poland', *Forest Ecology and Management*, 256: 1325–32.

Van Vuure, C. (2005) *Retracing the Aurochs*, Pensoft, Sofia-Moscow.

Vera, F.W.M. (2000) *Grazing Ecology and Forest History*, CABI International, Wallingford.

Vera, F.W.M. (2009) 'Large-scale nature development: The Oostvaardersplassen', *British Wildlife*, 20(5): 28–36.

Ward, A.I. (2005) 'Expanding ranges of wild and feral deer in Great Britain', *Mammal Review*, 35: 165–73.

Whitehouse, N.J. and Smith, D.N. (2004) '"Islands" in Holocene forests: Implications for forest openness, landscape clearance and culture steppe species', *Environmental Archaeology*, 9: 199–208.

Whitehouse, N.J. and Smith, D.N. (2010) 'How fragmented was the British Holocene wildwood? Perspectives on the "Vera" grazing debate from the fossil beetle record', *Quaternary Science Reviews*, 29: 539–53.

Wiens, J. and Hayward, G. (eds) (in press) *Historical Environmental Variation in Conservation and Natural Resource Management*, Oxford University Press, Oxford.

Yalden, D. (1999) *The History of British Mammals*, Poyser, London.

9 Can't see the trees for the forest

Frans Vera

Introduction

When it comes to trees, it seems that people and scientists can't see them for the forest. In science it is said that the forest returned after the last Ice Age (Roberts, 1989) and that it is the forest that regenerates (Peterken, 1996). However, it is the trees that returned after the Ice Age and it is the trees that regenerate. Trees do not only regenerate in forests. They also do so in open land. Therefore the presence of trees does not necessarily mean the presence of forest. Indeed, after the end of the last Ice Age, trees colonised in the first place the open steppe-tundra that was a legacy of the Ice Age. Whether these trees eventually formed a forest is something that has still to be proved.

Trees are only considered as being part of a forest, because in science the forest is commonly considered as the natural state of the vegetation in places where, according to climate, soil and hydrology, trees can grow (Ellenberg, 1988; Iversen, 1973; Peterken, 1996; Tansley, 1953). In the following paragraphs, I explain how this baseline has been constructed for the lowlands of Europe and what the consequences are for the preservation of biodiversity in common and for tree species in particular.

The shifting baseline syndrome

With the exception of some raised bogs, steep slopes and high tops in mountainous areas, the whole of Europe has been cultivated for thousands of years. With the loss of natural landscapes, plant and animals species disappeared from their natural distributions. Trees disappeared where arable land and pastures were established. Because they need large areas to survive, large mammals in particular, such as red deer (*Cervus elaphus*), elk (*Alces alces*), European bison (*Bison bonasus*), wolf (*Canis lupus*) and lynx (*Lynx lynx*) suffered much from the ever-growing human population and the associated conversion of nature into agricultural land and forests for wood production. Two species even became extinct: the aurochs (*Bos primigenius*), the wild ancestor of all our cattle, and the tarpan (*Equus przewalski gmelini*), the wild ancestor of our domestic horses. The last aurochs died in 1627 in an area in the neighbourhood of the village Jaktarowska, near Warsaw in Poland (Szafer, 1968) and the last tarpan in 1887 in the Zoo of Moscow (Pruski, 1963; Wrześniowski, 1878).

However, not all wildlife disappeared from the cultivated landscapes. It was these species, alongside the landscapes in which they lived, that became the baseline for nature conservation and nature management practice that arose in Europe in the nineteenth and twentieth centuries. These areas were pre-industrial landscapes, dating from before the introduction of artificial fertilisers (Figure 9.1). They are called semi- or half-natural. The diversity in species would then have been at its maximum. This maximum was supposed to have been created by man, or more specifically, by farmers. The baseline for this judgement was the idea of closed-canopy high forest as the natural vegetation in Europe (Berglund *et al.*, 1991). This forest would have been opened in Neolithic time with stone axes by the first farmers that arrived in Europe (Iversen, 1956 and 1973). As Heinz Ellenberg (1988: 1) writes: 'Central Europe would... be a monotonous wooded landscape, had man not produced a colourful mosaic of cultivated land, heath, meadow and pasture and over centuries continually cut back the forest'. All this suggests one knew how the cultivated landscape looked like in its natural state. However, the natural state is not *known*, because none of these landscapes survived in their natural state. All what was and is written about the natural state of the cultural landscape in Europe is theory. And because we do not *know*, we may suffer from 'shifting baseline syndrome' (Pauly, 1995). It is important to realise whether or not we suffer from shifting baseline syndrome, because if we make ecological models that relate to 'natural' conditions, they will have been programmed with erroneous starting points (Sheppard, 1995). What is shifting baseline syndrome?

Shifting baseline syndrome is a gradual change of the perception of what is the natural state of our environment and occurs when:

- nobody knows how nature originally looked like;
- each new generation redefines 'nature' and 'natural' according to their own experience;
- changes in the environment happen slowly and are therefore hard-to-notice.

Shifting baseline syndrome causes:

- each new generation to view the environment with the wildlife they remember from their youth as natural;
- a continuing lowering of the standard for natural to take place;
- a degraded state of nature to be accepted and considered as the normal thing.

Figure 9.1 A pre-industrial agricultural (so-called semi- or half- natural) landscape that is commonly used as a baseline for nature conservation and nature management
Source: Historic photograph

The consequences of shifting baseline syndrome are:

- the community becomes very tolerant for a creeping loss of species;
- inappropriate reference points are identified for rehabilitation and reconstruction measures for nature;
- a large educational and psychological hurdle in efforts to reset expectations and targets for nature conservation.

The starting points for reconstruction of the baseline forest for the natural vegetation

The starting point for the reconstruction of the natural vegetation in Europe was the cultural landscape. It was realised that mankind disturbed nature by cultivation. The reasoning was that if he would stop ploughing the land, cutting hay or grazing livestock, nature would rebound spontaneously to her original state. Spontaneity was considered to be synonymous with natural. What developed spontaneously on abandoned agricultural land was therefore considered as the natural conditions before cultivation. From all over Europe it was known that abandoned land developed into forest. The German forester Heinrich von Cotta summarised this in the introduction of the first print of his book *Anweisung zum Waldbau* (1816) as: 'If mankind would leave Germany, after 100 years it would be totally covered with forest' (Cotta, 1865: v). This finding was formalised in succession theory by Clements (1916) that stated that disappearance of the cause of the destruction of nature will at the same time initiate the development of a series of different types of vegetation which will eventually again result in the climax. This climax is one particular type of vegetation, which is climatologically determined. The climax is achieved when one species dominates to such an extent that it excludes the establishment of other possibly dominant species. Because the climate in Europe encourages the growth of trees, abandonment by man would result in a succession with the climax of a closed-canopy high forest (Ellenberg, 1988; Tansley, 1953). Everywhere in Europe, where the growth of trees was encouraged, the forest was defined as the imaginary natural vegetation, the so-called potential natural vegetation (PNV) (Tüxen, 1956). For Europe this was broadleaved forest (Ellenberg, 1988; Tüxen, 1956). The regeneration of this forest takes place in the gaps in the canopy created when one, several or many trees die or are blown over (Leibundgut, 1959 and 1978; Watt, 1925 and 1947) (Figure 9.2).

Figure 9.2 National Park Białowieża in eastern Poland, considered as the last remnant primeval lowland forest that would once have covered the lowland of Europe before mankind brought it under cultivation by agriculture
Source: Frans Vera

The climax forest has been constructed without taking any impact of indigenous large herbivores into account. Species such as aurochs and tarpan were absent because they were extinct. Still existing species such as red deer, roe deer, elk and European bison can, at certain densities, have an impact on the development of forest (Gill, 2006; Landolt, 1866; Moss, 1910; Tansley, 1911 and 1953). However, these were either absent or in such low densities because of human intervention that they did not prevent the establishment of forest on abandoned land. Because of the baseline forest, all European indigenous large herbivores, including the extinct aurochs and tarpan, became defined and considered as forest animals. The baseline forest and the well-known potential threat posed by large herbivores in high densities for forest regeneration lead to the conclusion that their natural densities must have been very low. In this way for red deer densities of 0.5–3.0 per 100ha, for roe deer 4.0–5.0 per 100ha and for European bison 4.0–5.0 per 1,000ha forest, are considered as natural (Anonymous, 1988 and 2002; Remmert, 1991; Wolfe and Von Berg, 1988). These densities make indigenous large herbivores ecologically non-existent in ecosystems that are considered as naturally functioning ecosystems. Because the baseline forest has been reconstructed without taking a possible impact of indigenous large herbivores on the vegetation into account, both the baseline forest and the natural densities for the indigenous large herbivores are the result of circular reasoning. It all began with the erroneous starting point of the spontaneous development of vegetation on abandoned agricultural land deprived of indigenous large herbivores.

Finally, the baseline forest has become a major educational and psychological hurdle in an effort to reset expectations and targets for nature conservation. First, because few people will be willing to have a monotonous species-poor forest in exchange for a species-rich colourful mosaic of cultivated land, heath, meadow and pasture, as Ellenberg (1986) formulated it. Second, because nature conservationists and nature managers have educated people that 'man' – and farmers will say a 'farming man' – is necessary to preserve this diversity of species by active management. Third, it is argued that the disappearance of the large indigenous wild herbivores is not a very great loss for nature. With or without them there would be forest. It is therefore legitimised to keep numbers very low by killing them. As far as cattle and horses are concerned, they are merely considered as mowing machines on four legs, and therefore only instrumental to the management of vegetation.

The wood-pasture system

Parts of the cultural landscape were the wood-pastures. They go back far in history and once covered large parts of the continent (Pott and Hüppe, 1991; Rackham, 1980 and 2003; Vera, 2000). A wood-pasture is grazed by livestock such as cattle, horses, sheep and pigs and if present, wild herbivores such as wild boar (*Sus scrofa*), red deer and roe deer. It consists of a mosaic of grassland, shrubs, thickets, solitary trees and groves, which are called forests in this chapter (see Flower, 1977; Pott and Hüppe, 1991; Rackham, 1980 and 2003) (Figure 9.3). Characteristic is a spiny, thorny, shrubby transition between grazed grassland and trees and groves. This transitional spiny, thorny vegetation is called 'mantle and fringe vegetation' (Ellenberg, 1988; Pott and Hüppe, 1991; Rackham, 1980 and 2003).

Figure 9.3 The wood pasture the Borkener Paradise in Germany; in the foreground an oak tree grows in the midst of some hawthorns; in the background is a grove surrounded by mantle and fringe vegetation of blackthorn
Source: Frans Vera

As mentioned before, at the time the natural vegetation was being theoretically reconstructed it was very well known that grazing livestock such as cattle, horse and sheep prevented the regeneration of trees in the forests in wood-pastures. This made these forests slowly change into open grassland or heath (Moss, 2010; Tansley, 1911 and 1953). This was considered as a degradation of high forest by retrogressive succession (Tansley, 1953), and therefore the wood-pasture as a whole was considered to be a degraded high forest (Ellenberg, 1988; Tansley, 1953) (Figure 9.4). The process was considered as unnatural because livestock were considered to be exotic species, introduced by man (Forbes, 1902; Moss, 1913; Tansley, 1911). With the forest as the baseline for the natural vegetation, grassland and heaths in wood-pastures were characterised as 'stolen' from the forest (Warming, 1909). The evidence for the theft was that when grazing ends, trees 'spring up' and the forest returns, as well as tree return in pieces of pasture that are fenced of from the grazing animals (Forbes, 1902; Krause, 1892; Tansley, 1911). So, as Forbes (1902: 245) formulated:

> There is little reason to doubt, therefore, what the result of leaving land entirely to Nature would be. So far as indigenous species [of trees] are concerned we have only to fence off a piece of ground from cattle, sheep, and rabbits, and quickly get a sample of indigenous forest of one or other types mentioned above.

Figure 9.4 Dyrhave, adjacent to Copenhagen in Denmark – nineteenth-century image of how the grazing of livestock such as cattle destroyed the primeval forest, because no regeneration took place in the forest and just some senile old oaks are the last witnesses of the once present forest
Source: Historic photograph

In many places in Europe, grazing livestock were removed from wood-pastures in order to restore the natural vegetation. As a result, the grasslands became colonised by shrubs and trees, and ultimately developed into a closed-canopy high forest. Long-term data show that since then there has been hardly any or no regeneration at all of the light-demanding sessile

106 *Frans Vera*

oak (*Quercus petraea*) or pedunculate oak (*Q. robur*). The diameter classes of both oak species show that oaks in the large diameter disappear because they die (Figure 9.5), while at the same time hardly any or no trees at all enter the smallest diameter class. This results in a so-called bell curve distribution that is typical for a population that is becoming extinct (Koop and Hilliger, 1987; Le Duc and Havill, 1998; Vera, 2000; Wolf, 2011 Figure 9.6). Contrary to this, shade-tolerant tree species have distribution diameter classes that reveal that many trees enter the lowest diameter class and penetrate the higher diameter classes where the old trees that disappear are replaced. These species include beech (*Fagus sylvatica*), wych elm (*Ulmus glabra*), small-leaved lime (*Tilia cordata*), hornbeam (*Carpinus betulus*) and ash (*Fraxinus excelsior*) (Bernadzki *et al.*, 1998; Emborg *et al.*, 1996; Vera, 2000; Westphal *et al.*, 2006; Wolf, 2011). This is evidenced by the structure of their diameter classes that has the shape of a 'reverse J', which indicates that there is much regeneration (Figure 9.6).

Figure 9.5 An oak that grew in openness in the former wood-pasture Sababurg in Germany – after the grazing of livestock was terminated, the park-like wood-pasture changed into a closed-canopy forest dominated by shade-tolerant species beech; light-demanding oak died because of the shade from the trees. The current forest is indicated as an 'Urwald', i.e. it is considered as a modern analogue of the primeval vegetation
Source: Frans Vera

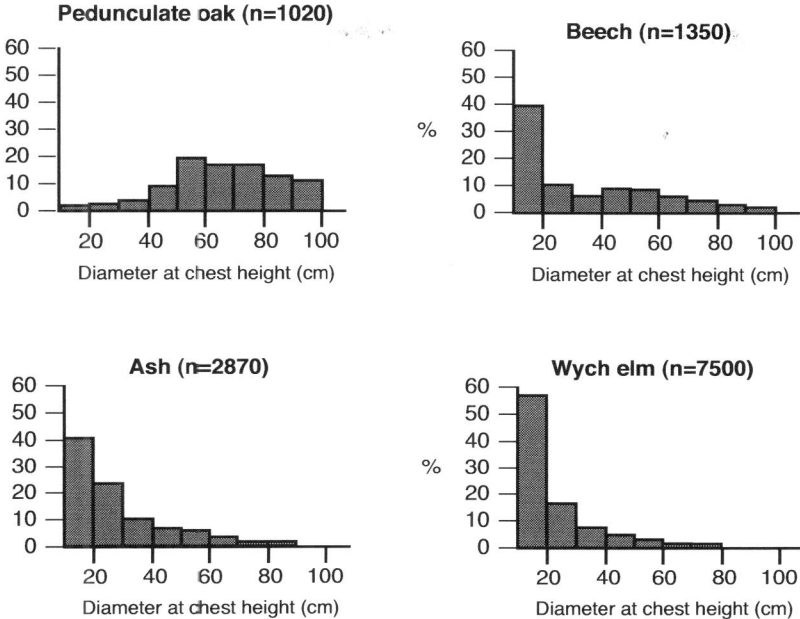

Figure 9.6 The percentage distribution per species of tree in diameter categories per species of pedunculate oak, beech, ash and wych elm in the National Park Dalby Söderkog, a former wood-pasture in southern Sweden
Note: Only trees with a trunk diameter of ≥ 10cm at chest height are included.
Source: taken from Vera, 2000 as redrawn from Malmer *et al.*, 1978: 20

This leaves us with two questions. The first is why oak could survive for hundreds of thousands of years in the presence of shade-tolerant tree species as shown by pollenanalysis, while nowadays they are outcompeted by the same in former wood-pastures, after livestock has been removed. The second is why cattle and horses are considered as 'alien species' and removed from wood-pastures in order to restore natural conditions. We know that they are indigenous species because they are domesticated descendants of the aurochs and the tarpan, which lived in Europe respectively up to AD1627 and AD1887.

The aurochs and shifting baseline syndrome

It is known from historical data that aurochs became extinct in 1627 (Szafer, 1968; Wrześniowski, 1878). The species was however unknown to science until 1827. Based on fossils founds it was scientifically described by

Bojanus as a species that became extinct by the beginning of the last Ice Age (Van Vuure, 2003; Wrześniowski, 1878). Only in 1867 did it become known from studies of historical sources that the aurochs lived in Europe until 1627 (Wrześniowski, 1878). However, only in 1927 was it further realised that from genetic research it was clear that the aurochs was the wild ancestor of domestic cattle (Van Vuure, 2003). By then the baseline for cattle had already been shifted towards them being alien species that had to be removed when natural conditions had to be restored. By then the closed-canopy high forest was already accepted in science as the baseline for the natural conditions. Consequently, the aurochs was defined as a forest species that should not have any influence on the forest; otherwise, the natural conditions would not have been a closed-canopy high forest. Later, because cattle are the domesticated descendant of the aurochs, it was accepted that cattle were indigenous. However, it was felt that cattle that created grassland in wood-pastures did not act as an ecological functional equivalent of the aurochs, because the aurochs, it was believed, lived in a closed-canopy forest. Horses were excluded from the reconstruction of the natural vegetation because the only extant species, the Przewalski horse (*Equus przewalski*) as well as its relatives like the species of zebra (*Equus* spec.) and the Asian wild asses (*Equus* spec.), all live in open grassy biotopes such as savannah and steppe (Vera, 2000). Based on the baseline of the closed canopy, horses were supposed to have left Europe after the forest returned after the last Ice Age (Sommer *et al.*, 2011). Therefore, one way or another, grasslands in wood-pastures were considered as artificial biotopes created by farmers with their livestock. All this circular reasoning did change the attitude towards the guild of the true indigenous grazers, cattle and horses in wood-pastures, in the opinion that all indigenous species of large herbivores were in natural conditions in fact functionally non-existent. It was believed that the presence of cattle and horses in wood-pastures was unnatural because they prevented the regeneration of trees in the forest and caused these forests to disappear, as happened in wood-pastures. It is true that forests in wood-pastures disappeared, but it is not true that cattle and horses prevent the regeneration of trees in wood-pasture. Trees regenerate in wood-pasture with such high densities of large herbivores, especially cattle and horse, that any regeneration of trees in the forest is impossible. To understand this, the wood-pasture has to be considered as grassland and forest together in one integral functioning system.

Regeneration of trees in wood-pasture

In wood-pastures, the regeneration of trees starts in open grassland grazed by livestock such as cattle and horses. Because the seedlings and saplings are palatable, they only survive there close to or in the direct vicinity of spiny or thorny shrubs such as blackthorn (*Prunus spinosa*), hawthorn (*Crataegus monogyna*) or herbaceous species that are defended by poisonous chemical substances (such as great yellow gentian (*Gentaina luea*)). These species are avoided by the large herbivores, and thereby defend the palatable seedlings and saplings of tree species (Bakker, 2004; Bossuyt *et al.*, 2005; Rousset and Lepart, 1999; Smit *et al.*, 2006). This phenomenon is called associational resistance (Callaway *et al.*, 2000; Ollf *et al.*, 1999; Milchunas and Noy-Meir, 2002), when protecting species nurse other species (Smit *et al.*, 2006) (Figure 9.7). In turn, the establishment of the nurse species in the grazed grassland is facilitated by swards in lawns, where tall herb species such as nettle (*Urtica dioica*) and tall grasses are less attractive to the large herbivores. These swards develop as a result of variability in the grazing intensity of the herbivores. Nurse species need this temporal protection for two to three years in order to fully develop their protective spines (Smit and Ruifrok, 2011). The regeneration of trees fails beyond the nurse species because they are eaten or trampled by the large herbivores (Bakker *et al.*, 2004; Vera, 2000; Vera *et al.*, 2006).

Besides the true grazers such as cattle and horses, the omnivorous pig may play an important role in the regeneration of trees in wood-pastures. Pigs have always been an important part of the livestock in wood-pastures.

Figure 9.7 A young oak nursed by blackthorn in the mantle and fringe vegetation of blackthorn that surrounds a grove in the wood-pasture the Borkener Paradise in Germany
Source: Frans Vera

They were fattened on the acorns of both oak species and on the fruits of wild apple, pear and cherry, called *mast* (Rackham, 1980 and 2003; Ten Cate, 1972; Vera, 2000). Their attribution to the establishment of the spiny sloe and hawthorn may exist of grubbing in the soil, creating loose bare soil. In this way, they may facilitate the establishment niche of nurse species such as blackthorn and hawthorn. The large-scale germination of blackthorn and hawthorn in abandoned arable fields (Klaudisová and Osbornová, 1990) may be an indication of this. These so-called late successional species in old fields prove to become established in the initial stage of the succession after abandonment. They are classified as later successional because it takes them years to grow above the grasses and herbs that dominate the first stage of the succession (Eglar, 1954). Birds and endozoochorous carnivores such as foxes and badgers seem the most important dispersers of the seed of blackthorn and hawthorn to swards in grazed grassland (Smit and Ruifrok, 2011). Analogues to the development of the establishment of blackthorn and hawthorn in old fields, they may become established simultaneously with the herb and grass species that become established in parts of the wood-pasture uprooted by pigs or wild boars. This activity may form a local sward that is for them a favourable establishment niche.

In mature spiny shrubs the annual shoots lack spines. They are browsed by large herbivores (Bokdam, 1987; Buttenschøn and Buttenschøn, 1978). This browsing induces a divaricate branching, which creates a canopy that is almost impenetrable for the snouts of the herbivores, which enhances the protection of the undefended palatable tree species (Bakker *et al.*, 2004). Mature shrubs of blackthorn expand clonally into open grassland by root-suckers. Tree seedlings establish in the fringes of the advancing blackthorn. In this way, trees advance into the grassland with the speed of the spreading fringes of the thorny scrub (Pott and Hüppe, 1991; Watt, 1924). Only rabbits can inhibit the expansion of blackthorn (Bakker *et al.*, 2004). Because blackthorn expands in every direction, a characteristic convex shaped group of trees develops, forming a forest called a grove (Bakker *et al.*, 2004; Vera, 2000; Vera *et al.*, 2006). This will expand rapidly and in its wave is the grove, when grazing pressure is suddenly reduced for some growing seasons, or stops altogether.

The dominance of oak in wood-pastures

A remarkable phenomenon in the wood-pasture is that compared to other tree species both oak species regenerate very well (Buttenschøn and Buttenschøn, 1985; Smith, 1980; Tansley, 1922 and 1953). They are very common in wood-pastures (Pott and Hüppe, 1991; Rackham, 1980 and 2003; Watt, 1919). This phenomenon is caused by the activity of the jay (*Garrulus glandarius*) and the wood mouse (*Apodemus sylvaticus*). Both species hoard acorns in the ground, seed by seed, at different places

(Bossema, 1979; Den Ouden et al., 2005). They are true scatter-hoarders, although wood mice will sometimes hoard several acorns in one catch (Den Ouden et al., 2005; Smit and Vermijmeren, 2011). The jay collects acorns in the oak and hoards them at a distance from the oak, from a few metres up to several kilometres, with a preference for open areas, such as large open spaces in forests, and open grasslands and fields (Bossema, 1979; Chettleburgh, 1952; Kollmann and Schill, 1996; Schuster, 1950). There they prefer a transitional area of short to long grass or brushwood, the outer edge of hedges and the fringes of thorny scrub that form mantle and fringe vegetation of forests in wood-pastures (Bossema, 1979; Chettleburgh, 1952; Rousset and Lepard, 1999; Vullmer and Hanstein, 1995). Chettleburgh (1952) observed a jay flying down into a hawthorn bush and burying an acorn at the foot of the bush. This observation explains the phenomenon of oaks, which seem to grow entwined with hawthorn in wood-pasture (Figure 9.3). In addition, jays like to bury acorns in places where the soil is loose and they can easily push the acorns into the ground (Bossema, 1979). This may be an indication of the facilitating role of pig and wild boar in wood-pastures. The distance between two hoarded acorns varies from 0.2–15m, but is generally between 0.5 and 1m (Bossema, 1979). Jays easily find the acorns they bury. The vertical structures for which jays appear to have a clear preference when they bury the acorns seem to serve as a beacon (Bossema, 1979). They dig up and eat the acorns they hide throughout the year, but do so far less in the period from April to August (Bossema, 1979). It is during this period that the seedlings appear. In June, together with their young, the jays start to look for seedlings that have grown from the acorns they buried the autumn before. When a parent bird finds a seedling, it takes hold of the stem with its beak and lifts the plant. This raises the acorn above the ground, or the soil that is brought up, and shows where it is hidden under the ground. The jay will then dig the acorn up. The jay removes the it from the seedling, peels it and feeds it to its young (Bossema, 1968 and 1979). The development of the seedlings is not hampered in their growth by the removal of the cotyledons (Bossema, 1979), because the seedlings grow in full daylight (Anderson and Frost, 1996; Sonessen, 1994). The chance that the young oak will be uprooted is small because in the full daylight immediately after germination they formed an extremely extensive root system with a long tap-root (Jarvis, 1964; Jones, 1959; Ziegenhagen and Kausch, 1995). This root system ensures that the seedling is securely anchored and not easily uprooted during the inspection of a jay. So, the disadvantage of the inspection is offset by the advantage of growing in extremely light conditions. Only very young seedlings that have grown late in the season are occasionally totally pulled out of the ground with their roots by a jay.

Wood mice transport acorns from the mother tree over a distance up to 50m (Den Ouden et al., 2005). Like the jay, they predate on the acorns they hoarded. Wood mice disperse acorns towards shrubs (Smit and Verwijmeren,

2011). At first sight, it looks like the wood mouse therefore contributes the most to the successful regeneration of oak in the spiny shrubs of wood-pastures. However, they hoard most acorns in the centre of the scrub where the chances of successful establishment of oak are less because of the very low level of daylight (Den Ouden *et al.*, 2005; Vera, 2000). To a lesser extent they hoard acorns at the outer edge of the shrubs, where oak has the best opportunity to grow successfully in wood-pastures (see Pott and Hüppe, 1991; Rackham, 1980 and 2003; Tansley, 1922 and 1953; Watt, 1919). This is where the jay prefers to hoard acorns (Bossema, 1979). This may mean that overall the jay contributes more to the successful regeneration of oak in wood pastures than wood mice (Den Ouden *et al.*, 2005).

Wood mice also hoard the seeds of beech (*Fagus sylvatica*) as does the nuthatch (*Sitta europaea*). However, nuthatch store mostly in bark crevices of trunks and thick branches and make few caches below ground (Källander, 1993; Moreno *et al.*, 1981; Perea *et al.*, 2011). This and the low density of avian seed removers in beech forests (Perea *et al.*, 2011) may explain the observed infrequent appearance of beech seedlings and the frequent appearance of oak seedlings in the grassland and thorny scrub, even adjacent to beech woods (Tansley, 1922; Watt, 1925). Shade tolerant tree species such as beech, silver fir (*Abies alba*), Norway spruce (*Picea abies*), sycamore (*Acer pseudoplatanus*), lime (*Tilia cordata* and *T. platyphyllos*), hornbeam (*Carpinus betulus*) and elm (*Ulmus glabra* and *U. leavis*) also regenerate successfully in wood-pastures in spiny shrubs and thickets by associational resistance (Rackham, 1980 and 2003; Smit *et al.*, 2006; Tansley, 1953).

Besides beechnuts, the nuthatch also hoards seeds of lime, sycamore, ash and hornbeam. Like beechnuts, these are usually hidden in bark furrows, and in the cracks of trunks and branches of trees; places that are not regeneration niches for these species (Källander, 1993; Löhrl, 1967; Matthyssen, 1998). Therefore, they are dependent on the wind in order to disperse them to nurse species sites, and this happens by accident. This may explain their subordinate appearance in wood-pastures. With the jay and the wood mouse as vectors for the distribution of the acorns to optimal generation niches, oak appears to have a huge advantage in wood-pastures above other tree species (Den Ouden *et al.*, 2005; Smit and Verwijmeren, 2011; Vera, 2000). This explains their successful regeneration and their dominance in wood-pastures in the natural distribution area of oak.

Besides the palatable tree species, palatable shrub species such as hazel (*Corylus avellana*), guelder rose (*Viburnum opulus*), bird cherry (*Prunus padus*), spindle tree (*Euonymus europeus*), elder (*Sambucus nigra*) and privet (*Ligustrum vulgare*) grow successfully by means of associational resistance. Their fleshy seeds are eaten by bird species, especially singing birds (*Passeriformes*) that defecate the seeds below the thorny shrubs where they roost (Namvar and Spethmann, 1985; Snow and Snow, 1988). This can explain why these light-demanding shrubs species are part of mantle and fringe vegetation (Ellenberg, 1988; Hondong *et al.*, 1993; Pietzarka and

Roloff, 1993; Smith, 1980; Vera, 2000). The nuthatch collects hazelnuts (Hagerup, 1942; Källander, 1993; Löhrl, 1967; Matthyssen, 1998). A pair of territorial nuthatches can deprive hazel shrubs totally from their nuts in a few days (Löhrl, 1967). Although studies show that nuthatch hides few seeds in the ground, some data suggest that from all the seeds they collect, those of hazel are the most frequently stored in the ground, 60–73 per cent compared to 16–20 per cent of beechnuts (Källander, 1993; Matthyssen, 1998). It hides the hazelnut close to hazel shrubs by pushing and hammering it into the ground and the covering the spot, as jays do for acorns (personal observation). The fact that they store seeds close to the food source (up to 40m away) (Matthyssen, 1998) and seedlings of hazel are found in open grassland and in the fringes of spiny scrub (Sanderson, 1958) makes it very plausible that the nuthatch acts as a vector for hazel, as the jay does for oak. The shrub itself can cope with heavy grazing outside the spiny scrub as a solitary shrub by forming new shoots from the roots (Bär, 1914; Jahn, 1991; Sanderson, 1958). It also spreads by underground runners (Sanderson, 1958).

Solitary trees and groves

In time, trees that are protected by associational resistance will grow above the thorny nurse species. In the case the protective shrubs do not expand clonally, like for instance hawthorn, a single tree or a few trees will grow up. Once above the protecting shrubs these trees form the large crowns that are so characteristic for open-grown trees. The shade these crowns create eventually kills the light-demanding nurse shrubs. This results in scattered, single, open-grown trees or small groups of trees in open grassland. In the case where the nurse shrubs, such as blackthorn, expand clonally by root suckers, again and again, seedlings settle in the advancing edge of the thicket. They settle only there, because within the thicket density of daylight level is too low (Vera, 2000), namely less than 2 per cent (Dierschke, 1974; Tubbs, 1988). When the trees grow above the scrub, the crowns join together, resulting in a grove. The grove advances concentrically into the grassland as the thorny blackthorn scrub advances concentrically into the grassland by means of underground suckers. In the grove, the light-demanding, spiny scrub disappears because of the shade of the closed canopy of the grove (Coops, 1988; Ekstam and Sjörgen, 1973; Watt, 1924 and 1934). The regeneration of trees is there prevented initially by the shade of the canopy but moreover by the trampling and browsing by the large herbivores (Bakker *et al.*, 2004; Mountford *et al.*, 1999; Mountford and Peterken, 2003) that enter the grove. The grazers do so through small gaps in the spiny mantle vegetation that surrounds the grove in order to look for shade and to escape from biting flies. Their presence prevents shade-tolerant species from regenerating under the canopy of oaks (Figure 9.8) even though as is known from forestry, there is sufficient light for these tolerant species (Bezaninský, 1971; Bonnemann, 1956a and 1956b; Fricke *et al.*,

1980; Vera, 2000). Regeneration of tolerant species under a canopy of oaks happened in groves in former wood-pastures after the large herbivores were removed (Vera, 2000). The animals also prevent regeneration after a gap in the canopy is formed. In this way the canopy opens up; a process that is facilitate by fungi and drought that kill more and more of the senile trees (Dobson and Crawley, 1994; Green, 1992). Grass seeds are brought in by the large ungulates in their dung and fur and as the grove becomes more open as more trees die, a grazed lawn develops (Bokdam, 2003; Mountford and Peterken, 2003) (Figure 9.9). In this way, groves change from the centre with the oldest trees, gradually into grassland (Goriup *et al.*, 1991; Mountford *et al.*, 1999; Mountford and Peterken, 2003; Peterken, 1996). When the grassland has reached a certain surface, swards will locally develop because of variability in grazing intensity of the herbivores. Spiny and thorny nurse species will establish there again and in their wake palatable shrub and tree species.

In summary, at a certain point in an area, there is grazed grassland first. Then thorny or spiny scrub or other unattractive (i.e. inedible) species of plants establish, either forming a clonally spreading scrub or remaining solitary. Then seedlings of palatable trees and shrubs establish if they are protected by these inedible nurse species. Trees grow up solitary or form-

Figure 9.8 The interior of a grove in the wood-pasture Junner Koeland, the Netherlands – the interior lacks a shrub layer as well as regeneration of trees
Source: Frans Vera

Figure 9.9 A grove that changes from the centre onwards into grassland, because of the grazing of cattle, horses and deer in Denny Wood in the New Forest, England.
Source: Frans Vera

ing groves (forests). Aged trees die and groves change into grazed grassland again. The system described is a non-linear, cyclical system. This description is a space-to-time substitution (Picket, 1987) of the well-known process of the establishment of trees in grazed grasslands by means of nurse species and the change of forests in wood-pastures by grazing animals because they prevent the regeneration of trees within the forests (Vera, 2000). In this whole process, large indigenous herbivores are steering the succession, especially the true grazers, cattle and horse (Smit and Putman, 2011; Vera, 2000).

Contrary to the closed-canopy, high-forest system the wood-pasture system enables light-demanding tree species to survive in the presence of the shade-tolerant in the context of one system. Contrary to the closed-canopy forest system, the wood-pasture system can explain the combined presence of light-demanding and shade-tolerant tree species shown by pollen diagrams from the primeval vegetation. The wood-pasture system therefore can be regarded as the closest modern analogy of the natural vegetation (Smit and Putman, 2011; Vera, 2000). Besides sessile and pedunculate oak, it concerns all light-demanding tree species. These are the wild fruit species such as wild apple (*Malus sylvestris*), wild pear (*Pyrus pyraster*) and wild cherry (*Prunus avium*), the European *Sorbus* species such as whitebeam (*S. aria*), service tree (*S. domestica*) and the chequers tree (*S.*

torminalis). On a European scale, some of these are threatened species (Kätzel *et al.*, 2011). The process also includes all indigenous shrub species (Vera, 2000). The wood-pasture system is a very diverse landscape varying from savannah-like to park-like. The mosaic of grasslands, shrubs, thickets, trees and groves – the last of these surrounded by mantle and fringe vegetation – vary in relation to each other in surface. The mosaic is shaped by a reciprocal interaction between plant and animal species. Beside a high diversity of shrub and tree species, the system is also characterised by a high diversity of animal species. This is because of the high diversity in vegetation types, vegetation structures and combinations of these (see Alexander, 1998, 2001 and 2005; Alexander *et al.*, 2006; Appelqvist *et al.*, 2001; Bossuyt *et al.*, 2005; Green, 2009; Harding and Rose, 1998; Manning *et al.*, 2006; Ranius *et al.*, 2005 and 2008; Schuffenhauer, 2011; Schulze-Hagen *et al.*, 2004; Ek and Johanesson, 2005; Vera, 2000; Vodka *et al.*, 2009).

The individualistic behaviour of tree species

Besides a high diversity of plant and animals species, the wood-pasture system is also characterised by a high diversity of tree shape. There are grove-grown and open-grown trees. The grove-grown trees are characterised by long branchless trunks and small crowns with the branches directed upwards to the light, as is known from the closed-canopy high forest. The open-grown trees have short trunks and low at the trunk are massive spreading branches that are almost perpendicular to the trunk and form huge broad crowns. They are impressive and often recognisable individually by the shape of their crown. These open-grown trees are lacking in closed-canopy forests. These trees are themselves important for many plant and animals species, in combination with the open surroundings (Antonsson and Jansson, 2001; Butler *et al.*, 2001; Green 2009; Manning *et al.*, 2006). This applies especially to both oak species to which more species are connected that any other indigenous European tree (Morris and Penning, 1974; Schuffenhauer, 2011; Ek and Johanesson, 2005; Vodka *et al.*, 2009;). Many species, like for instance the rare hermit beetle (*Osmoderma eremita*), need trees of a very old age, called the veteran tree stage. In a closed-canopy high forest, oak will never reach this stage because of the low stature that is its characteristic. This low stature is caused by the downward growth of oak starting at the age of around 300 years. The upper canopy dies off while a new canopy forms lower down the trunk (Green, 2009). This results ultimately in the characteristic short conic formed oak of an age of about a thousand years (Figure 9.10). Such oaks can be seen in Windsor Great Park in England and other wood-pasture systems (see Pater, 2010). An oak cannot develop such a shape because it will be killed before that by the shade of the trees that regenerate in the gap in the canopy that is formed by the downwards growth of the oak, or it will be killed by higher neighbouring trees (Alexander *et al.*, 2011; Vera, 2000). The killing of

Figure 9.10 An oak with its crown still intact next to an oak where the crown died off in Calke Abby Park in England and which formed a second crown low at the trunk
Note: Observe the difference in diameter between the two trunks. The smaller oak has a thicker trunk. This is an indication of its greater age.
Source: Frans Vera

veteran oaks is known from all former wood-pastures (see Pater, 2010; Rapp and Schmidt, 2006; Sperber and Thierfelder, 2008).

Both individual genetic characteristics as well as the genetic characteristics of the individual species of trees are revealed by the wood-pasture system. This individual behaviour of trees takes us back to a discussion in the early twentieth century about plant communities. As pointed out above, this was the period the high closed-canopy forest was constructed as the baseline for the natural vegetation. During this period Tansley (1916, 1935 and 1953) made a significant contribution to this construction. In his opinion the individual ecological characteristics of individual tree species were subordinate to the plant community forest. Tansley (1920: 125) formulates it thus:

> Though less like true organisms than human communities plant communities may still be regarded as quasi-organisms, or organic entities, for on the one hand they are composed of organic units, and on the other hand they are certainly entities, in the sense that they behave in many respects as wholes, and therefore have to be studied as wholes.

This is in contrast to the view of Gleason (1926). According to him the ecological characteristics of individual plant species determined the species composition of plant communities. Gleason (1926: 16 and 26) declares:

> Are we not justified in coming to the general conclusion, far removed from the prevailing opinion, that an association is not an organism, scarcely even a vegetational unit, but merely a coincidence?–In conclusion, it may be said that every species of plant is a law into itself, the distribution of which in space depends upon its individual peculiarities of migration and environmental requirements. Its disseminules migrate everywhere, and grow wherever they find favourable conditions. The species disappears from areas where the environment is no longer endurable.

Concisely, this controversy is the forest (Tansley) versus the individual tree species (Gleason), or more specifically, the forest versus the oak species. The baseline forest for the natural vegetation from before cultivation by agriculture and forestry is always presented as one unity, irrespective the ecological characteristics of the individual tree species. As many palynologists formulate it: the forest returned after the Ice Age, not the trees and the tree species. Early palynologists such as Firbas (1934 and 1935) and Godwin (1934a and 1934b) advocated the reconstruction of the history of the forest. So, all tree species, whether light demanding or shade tolerant, migrated as one organism into Europe after the Ice Age. It is remarkable that it is the pollen analysis of this migration that stresses the individual behaviour of the different tree species. So-called isopoll maps reconstruct this migration. An isopoll is a line that joins geographical localities with the same pollen percentage for a given taxon, for instance oak, at the time for which the map is drawn. These maps reveal the pattern and rates of migration of the taxa from their glacial refuge (Huntley and Birks, 1983). They show that the different taxa had different refugia and behaved individualistically through time. Plant communities are therefore impermanent assemblages of taxa (Huntley, 1990; Huntley and Webb III, 1989). Woodpastures in their recent history reveal this. When all indigenous large herbivores there became ecologically non-existent, because they were removed or culled to low densities, the park-like wood-pasture changed into a closed-canopy high forest. Both oak species, along with all other light-demanding plant species, behaved individualistically by disappearing from these areas because the environment was no longer endurable. The closed-canopy high forest as a baseline for the natural vegetation denies this individualistic behaviour. This baseline means that today we cannot see the trees for the forest. By extension, inappropriate reference points are used for identifying targets for rehabilitation measures for nature.

Bibliography and references

Alexander, K.N.A. (1998) 'The links between forest history and biodiversity: The invertebrate fauna of ancient pasture-woodlands in Britain and its conservation', in Kirby, K.J. and Watkins, C. (eds) *The Ecological History of European Forests*, CAB International Wallingford, pp73–80.

Alexander, K.N.A. (2001) 'What are veteran trees? Where are they found? Why are they important?', in Read, H., Forfang, A.S., Marciau, R. and Paltto, H., Andersson, L. and Tardy, B. (eds) *Tools for Preserving Woodland Biodiversity*, Textbook 2, Nanonex, programme September 2001, Leonardo da Vinci, Sweden, pp28–31.

Alexander, K. (2005) 'Wood decay, insects, paleoecology, and woodland conservation policy and practice – breaking the halter', *Antenna*, 29: 171–8.

Alexander, K., Butler, J. and Green, T. (2006) The value of different tree and shrub species to wildlife. *British Wildlife* October (vol. 18) 2006, 18–28.

Alexander, K., Sticker, D. and Green, T. (2011) 'Rescuing veteran trees from canopy competition', *Conservation Land Management*, Spring: 12–16.

Andersson, C. and Frost, I. (1996) 'Growth of *Quercus robur* seedlings after experimental grazing and cotyledon removal', *Acta Botanica Neerlandica*, 45: 85–94.

Anonymous (1988) *Grofwildvisie Veluwe. Ministerie van Landbouw en Visserij*, Ministerie van Landbouw en Visserij, Den Haag.

Anonymous (2002) *Strategy for the Conservation of the European Bison in the Russian Federation*, Russian Academy of Sciences, WWF Russia, Moscow.

Antonsson, K. and Jansson, N. (2001) 'Ancient trees and their fauna and flora in the agricultural landscape in the county of Östergötland', in Andersson, L., Marciau, R. and Paltto, H. (eds) *Tools for Preserving Woodland Biodiversity*, Textbook 2, Nanonex, programme September 2001, Leonardo da Vinci, Sweden, pp37–41.

Appelqvist, T., Gimdale, R. and Benston, O. (2001) 'Insects and mosaic landscapes', in Andersson, L., Marciau, R., Pallto, H., Hardy, B. and Reed, H. (eds) *Tools for Preserving Biodiversity in the Nemoral and Boreonemoral Biomes of Europe*, Nanonex, Sweden, pp14–24.

Bakker, E.S., Olff, H., Vandenberghe, C., De Maeyer, K., Smit, R., Gleichman, J.M. and Vera, F.W.M. (2004) 'Ecological anachronisms in the recruitment of temperate light-demanding tree species in wooded pastures', *Journal of Applied Ecology*, 41: 571–82.

Bär, J. (1914) 'Die Flora des Val Onsernone', *Mitteilungen aus dem botanischen Museum der Universität Zürich*, 59: 223–563.

Berglund, B.E., Malmer, N. and Persson, M. (1991) 'Landscape-ecological aspects of long-term changes in the Ystad area', in Berglund, B.E. (ed.) *The Cultural Landscape during 6000 Years in Southern Sweden– the Ystad Project*, Ecological Bulletins 41, Copenhagen, pp405–24.

Bernadzki, E. Bolibok, L. Brzeziecki, B. Zajaczkowski, J. and Zybura, H. (1998) 'Compositional dynamics of natural forests in the Białowieża National Park, northeastern Poland', *Journal of Vegetation Science*, 9: 229–38.

Bezacinský, H. (1971) ‚Das Hainbuchenproblem in der Slowakei', *Acta Facultatis Forestalis*, 8: 7–36.

Bokdam, J. (1987) Foerageergedrag van jongvee in het Junner Koeland in relatie tot het voedselaanbod', in Bie, S. de, Joenje, W. and Van Wieren, S. (eds) *Begrazing in de natuur*, Pudoc, Wageningen: 165–186.

Bokdam, J. (2003) 'Nature conservation and grazing management. Free-ranging cattle as driving force for cyclic vegetation succession', PhD thesis, Wageningen University, Wageningen.
Bonnemann, A. (1956a), Eichen-Buchen Mischbestände', *Allgemeine Forst- und Jagdzeitung*, 127: 33–42.
Bonnemann, A. (1956b), Eichen-Buchen Mischbestände', *Allgemeine Forst- und Jagdzeitung*, 127: 118–26.
Bossema, J. (1968), Recovery of acorns in the European jay (*Garrulus G. glandarius* L.)', *Proceedings Koninklijke Nedederlandse Akademie van Wetenschappen Serie C, Biological and Medical Sciences*, 71: 10–14.
Bossema, J. (1979) 'Jays and oaks: An eco-ethological study of a symbiosis', PhD thesis, Rijksuniversiteit Groningen, Groningen (also published in *Behaviour*, 70: 1–117).
Bossuyt, B., De Fré, B. and Hoffmann, M. (2005) 'Abundance and flowering success patterns in a short-term grazed grassland: early evidence of facilitation', *Journal of Ecology*, 93: 1104–14.
Butler, J.E., Rose, F. and Green, T.E. (2001) 'Ancient trees, icons of our most important wooded landscapes in Europe', in Read, H., Forfang, A.S., Marciau, R., Paltto, H., Andersson, L. and Tardy, B. (eds) *Tools for Preserving Woodland Biodiversity*, Textbook 2, Nanonex, programme September 2001, Leonardo da Vinci, Sweden, pp28–31.
Buttenschøn, J. and Buttenschøn, R.M. (1978) 'The effect of browsing by cattle and sheep on trees and bushes', *Natura Jutlandica*, 20: 79–94.
Buttenschøn, J. and Buttenschøn, R.M. (1985) 'Grazing experiments with cattle and sheep on nutrient poor, acidic grassland and heath IV: Establishment of woody species', *Natura Jutlandica*, 21: 47–140.
Callaway, R.M., Kikvidze, Z. and Kikodze, D. (2000) 'Facilitation by unpalatable weeds may conserve plant diversity in overgrazed meadows in the Caucasus Mountains', *Oikos*, 89: 275–82.
Chettleburgh, M.R. (1952) 'Observations on the collection and burial of acorns by jays in Hinault Forest', *British Birds*, 45: 359–64 (also further note in *British Birds*, 1955, 48: 183–4).
Clements, F.E. (1916) *Plant succession*, Carnegie Institute Washington, Wsahington DC.
Coops, H. (1988) 'Occurrence of blackthorn (*Prunus spinosa* L.) in the area of Mols Bjerge and the effect of cattle- and sheep-grazing on its growth', *Nature Jutlandica*, 9: 169–76.
Cotta, H. (1865) *Anweisung zum Waldbau*, Arnoldische Buchhandlung, Leipzig.
Den Ouden, J., Jansen, P.A. and Smit, R. (2005) 'Jays, mice and oaks: Predation and dispersal of *Quercus robur* and *Q. petraea* in north-western Europe', in Forget, P.-M., Lambert, J.E., Hulme, P.E. and Vander Wall, S.B. (eds) *Seed Fate: Predation, Dispersal and Seedling Establishment*, CAB International, Wallingford, pp223–39.
Dierschke, H. (1974) ‚Saumgesellschaften im Vegetations- und Standortsgefälle an Waldrändern', *Scripta Geobotanica*, (Göttingen) 6: 3–246.
Dobson, A. and Crawley, M. (1994) 'Pathogens and the structure of plant communities', *Trends in Ecology and Evolution*, 9: 303–98.
Eglar, F.E. (1954) 'Vegetation science concepts. I. Initial floristic composition, a factor in old-field vegetation development', *Vegetatio*, 4: 412–17.
Ekstam, U. and Sjörgen, E. (1973) 'Studies on past and present changes in deciduous forest vegetation on Öland', *Zoon*, Uppsala, Suppl. 1: 123–35.

Ellenberg, H. (1988) *Vegetation Ecology of Central Europe*, 4th edition, Cambridge University Press, Cambridge.
Emborg, J., Christensen, M. and Heilmann-Clausen, J. (1996) 'The structure of Suserup skov. A near-natural temperate deciduous forest in Denmark', *Forest and Landscape Research*, 1: 311–33.
Firbas, F. (1934) 'Über die Bestimmung der Walddichte und der Vegetation Waldloser Gebiete mit Hilfe der Pollenanalyse', *Planta*, 22: 109–146.
Firbas, F. (1935) 'Die Vegetationsentwicklung des Mitteleuropäischen Spätglacials', *Bibliotheca Botanica*, 112: 1–68.
Flower, N. (1977) 'A historical and ecological study of enclosed and unenclosed woods in the New Forest, Hampshire', MSc thesis, King's College, University of London.
Forbes, A.C. (1902) 'On the regeneration and formation of woods from seed naturally or artificially sown', *Transactions of the English Arboricultural Society*, 5: 239–70.
Fricke, O., Kürschner, K., and Röhrig, E. (1980) 'Unterbau in einem Stieleichenbestand', *Forstarchiv*, 51: 228–32.
Gill, R. (2006) 'The influence of large herbivores on tree recruitment and forest dynamics', in Danell, K., Duncan, P., Bergström, R. and Pastor, J. (eds) *Large Herbivore Ecology. Ecosystem Dynamics and Conservation*, Cambridge University Press, Cambridge, pp170–202.
Gleason, H.A. (1926) 'The individualistic concept of the plant association', *Torrey Botanical Club Bulletin*, 53: 7–26.
Godwin, H. (1934a) 'Pollen analysis. An outline of the problems and potentialities of the method. Part. I. Technique and interpretation', *New Phytologist*, 33: 278–305.
Godwin, H. (1934b) 'Pollen analysis. An outline of the problems and potentialities of the method. Part. II. General applications of pollen analysis', *New Phytologist*, 33: 325–58.
Goriup, P.D. Batten, L.A. and Norton, J.A. (eds) (1991) 'The conservation of lowland dry grassland birds in Europe', *Proceedings of an International Seminar held at the University of Reading, 20–22 March 1991*, Joint Nature Committee, Peterborough.
Green, T. (1992) 'The forgotten army', *British Wildlife*, 4: 85–6.
Green, T. (2009) 'Ancient trees growing downwards', www.treeworks.co.uk/downloads/Veteran_Environmental_Papers/Growing_Downwards_ancient_trees_Jan_05.pdf.
Hagerup, O. (1942) *The morphology and biology of the Corylus-fruit*. Biologiske Meddelelser, Kongelige Danske Videnskabernes Selskab 17 (6), 1–42.
Harding, P.T. and Rose, F. (1986) *Pasture-Woodlands in Lowland Britain. A Review of their Importance for Wildlife Conservation*, Natural Environment Research Council, Institute of Terrestrial Ecology, Huntingdon.
Hondong, H. Langner, S. and Coch, T. (1993) *Untersuchungen zum Naturschutz an Waldrändern*, Bristol-Schriftenreihe, Band 2, Bristol-Stiftung, Ruth und Herbert UHL - Forschungsstelle für Natur- und Umweltschutz.
Huntley, B. (1990) 'European vegetation history: Palaeovegetation maps from pollen data – 13.000 yr BP to present', *Journal of Quaternary Science*, 5: 103–22.
Huntley, B. and Birks, H.J.B. (1983) *An Atlas of Past and Present Pollen Maps of Europe: 0–13000 Years Ago*, Cambridge University Press, Cambridge.

Huntley, B. and Webb III, T. (1989) 'Migration: Species' response to climatic variations caused by changes in the earth's orbit', *Journal of Biography*, 16: 5–19.
Iversen, J. (1956) 'Forest clearance in the Stone Age', *Scientific American*, 194(3): 36–41.
Iversen, J. (1973) 'The development of Denmark's nature since the last Glacial', *Danmarks Geologiske Undersøgelse*, V. Raekke nr. 7-c (*Geological Survey of Denmark. V. Series No. 7-c*).
Jahn, G. (1991) 'Temperate deciduous forests of Europe', in Röhrig, R. and Ulrich, B. (eds) *Temperate Deciduous Forests. Ecosystems of the World*, 7, Elsevier, Amsterdam: 377–503.
Jarvis, P.G. (1964) 'The adaptability to light intensity of seedlings of *Quercus petraea* (Matt.) Liebl', *Journal of Ecology*, 52: 545–71.
Jones, E.W. (1959) 'Biological flora of the British Isles', *Quercus* L. *Journal of Ecology*, 47: 169–222.
Källander, H. (1993) 'Food caching in the European nuthatch *Sitta europaea*', *Ornis Svecica*, 3: 49–58.
Kätzel, R., Schulze, T., Becker, F., Schröder, J., Riederer, J., Kamp, T., Wurm, A. and Huber, G. (2011) 'Seltene Baumarten in Deutschland – Erfassung und Erhaltung', *AFZ-DerWald*, 19: 39–41.
Klaudisová, A. and Osbornová, J. (1990) 'Abandoned fields in the region', in Osbornová, J., Kovářová, M., Lepš, J. and Prach, K. (eds) *Succession in Abandoned Fields. Studies in Central Bohemia, Czechoslovakia*, Geobotany 15, Kluwer Academic Publishers, Dordrecht, pp3–21.
Kollmann, J. and Schill, H.-P. (1996) 'Spatial patterns of dispersal, seed predation and germination during colonization of abandoned grassland by *Quercus petraea* and *Corylus avellana*', *Vegetatio*, 125, 193–205.
Koop, H. and Hilgen, P. (1987) 'Forest dynamics and regeneration mosaic shifts in unexploited beech (*Fagus sylvatica*) stands at Fontainebleau (France)', *Forest Ecology and Management*, 20: 135–50.
Kraus, H.L. (1892) ,Die Heide. Beitrag zur Geschichte des Pflanzenwuchses in Nordwesteuropa', *Engleis Bot. Jahrb.*, XIV: 517–39.
Landolt, E. (1866) *Der Wald. Seine Verjüngung, Pflege und Benutzung. Bearbeitet für das Schweizervolk*, Friedrich Und Schulthe Verlag, Zürich.
Le Duc, M.G. and Havill, D.C. (1998) 'Competition between *Quercus petraea* and *Carpinus betulus* in an ancient wood in England: Seedling survivorship', *Journal of Vegetation Science*, 9: 873–80.
Leibundgut, H. (1959) 'Über Zweck und Methodik der Struktur und Zuwachsanalyse von Urwäldern', *Schweizerische Zeitschrift für Forstwesen*, 110: 111–24.
Leibundgut, H. (1978) ,Über die Dynamik europäischer Urwälder', *Allgemeine Forstzeitschrift*, 33: 686–90.
Löhrl, H. (1967) *Die Kleiber Europas*, Die Neue Brehm-Bücherei, Ziemsen Verlag, Wittenberg Lutherstadt.
Malmer, N., Lindgren, K. & Persson, S. (1978) Vegetational succession in a south-Swedish deciduous wood. *Vegetatio* 36, 17–29.
Manning, A.D., Fischer, J. and Lindenmayer, D.B. (2006) 'Scattered trees are keystone structures: Implications for conservation', *Biological Conservation*, Doi:10.1016/j.biocon.2006.04.023.
Matthyssen, E. (1998) *The Nuthatches*, T. & A.D. Poyser, London.
Milchunas, D.G. and Noy-Meir, I. (2002) 'Grazing refuges, external avoidance of herbivory and plant diversity', *Oikos*, 99: 113–30.

Moreno, J., Lundberg, A. and Carlson, A. (1981) 'Hoarding of individual nuthatches *Sitta europaea* and marsh tits *Parus palustris*', *Ecography*, 4: 263–9.

Morris, M.G. and Perring, F.H. (eds) (1974) *The British Oak. Its History and Natural History*, The Botanical Society of the British Isles, E.W. Classey, Berkshire.

Moss, C.E. (1910) 'The fundamental units of vegetation: Historical development of the concepts of the plant association and the plant formation', *New Phytologist*, 9: 18–53.

Moss, C.E. (1913) *Vegetation of the Peak District*, The University Press, Cambridge.

Mountford, E.P. and Peterken, G. (2003) 'Long-term change and implications for the management of wood-pastures: Experience over 40 years from Denny Wood, New Forest', *Forestry*, 76: 19–43.

Mountford, E.P., Peterken, G.F., Edwards, P.J. and Manners, J.G. (1999) 'Long-term change in growth, mortality and regeneration of trees in Denny Wood, an old-growth wood-pasture in the New Forest (UK)', *Perspectives in Plant Ecology, Evolution and Systematics*, 2: 223–72.

Namvar, K. and Spethmann, W. (1985) 'Waldbaumarten aus der gattung *Ulmus* (Ulme, Rüster)', *Allgemeine Forstzeitschrift*, 40: 1220–5.

Ollf, H., Vera, F.W.M., Bokdam, J., Bakker, E.S., Gleichman, J.M., Maeyer, K. de and Smit, R. (1999) 'Shifting mosaics in grazed woodlands driven by the alternation of plant facilitation and competition', *Plant Biology*, 1: 127–37.

Pater, J. (2010) *Europas alte Bäume. Ihre Geschichten, ihre Geheimnisse*, Kosmos, Stuttgart.

Pauly, D. (1995) 'Anecdotes and the shifting baseline syndrome', *Trends in Ecology & Environment*, 10: 430.

Perea, R., San Miguel, A. and Gil, L. (2011) 'Flying vs. climbing: Factors controlling arboreal seed removal in oak-beech forests', *Forest Ecology and Management*, 262: 1251–7.

Peterken, G.F. (1996) *Natural Woodland. Ecology and Conservation in Northern Temperate Regions*, Cambridge University Press, Cambridge.

Picket, S.T.A. (1987) 'Space-for-time substitution as an alternative to long-term studies', in Likens, G.E. (ed.) *Long-Term Studies in Ecology. Approaches and Alternatives*, Springer Verlag, New York, pp110–35.

Pietzarka, U. and Roloff, A. (1993) 'Dynamische Waldrandgestaltung. Ein Modell zur Strukturverbesserung von Waldaussenrändern', *Natur und Landschaft*, 68: 555–60.

Pott, R. and Hüppe, J. (1991) *Die Hudenlandschaften Nordwestdeutschlands*, Westfälisches Museum für Naturkunde, Landschafsverband Westfalen-Lippe, Veröffentlichung der Arbeitsgemeinschaft für Biol.-ökol, Landesforschung, *ABÖL*, nr. 89, Münster.

Pruski, E. (1963) 'Ein regenationsversuch des Tarpans in Polen', *Zeitschrift für Tierzüchtung und Tierzüchtungsbiologie*, 79: 1–30.

Rackham, O. (1980) *Ancient Woodland. Its History, Vegetation and Uses in England*, Edward Arnold, London.

Rackham, O. (2003) *Ancient Woodland. Its History, Vegetation and Uses in England*, New Edition, Castlepoint Press, Kirkcudbrightshire.

Ranius, T., Aguado, L.O., Antonson, K., Audisio, P., Ballerio, A., Carpaneto, G.M., Chobot, K., Gjurašin, B., Hanssen, O., Huijbregts, H., Lakatos, F., Martin, O., Necualiseanu, Z., Nikitsky, N.B., Paill, W., Pirnat, A., Rizun, V., Tamutis, V., Telnov, D., Tsinkevich, V., Versteirt, V. Vignon, V., Vögeli, M. and Zach, P. (2005) 'Osmoderma eremite (Coleoptera, Scarabaeidea, Cetoniitae) in Europe',

Animal Biodiversity and Conservation, 28: 1–44.
Ranius, T., Eliasson, P. and Johansson, P. (2008) 'Large-scale occurrence patterns of red-listed lichens and fungi on old oaks are influenced both by current and historical habitat density', *Biodiversity Conservation*, 17: 2371–81.
Rapp, H.-J. and Schmidt, M. (Hg) (2006) *Baumriesen und Adlerfarn. Der Urwald Sababurg' Im Reinhardswald*, Euregioverlag, Kassel.
Remmert, H. (1991) 'The mosaic-cycle concept of ecosystems. An overview', in Remmert, H. (ed.) *The Mosaic-Cycle Concept of Ecosystems*, Springer, Berlin, pp11–21.
Roberts, N. (1989) *The Holocene. An Environmental History*, Basil Blackwell Oxford, New York.
Rousset, O. and Lepart, J. (1999) 'Shrub facilitation of *Quercus humilis* regeneration in succession on calcareous grasslands', *Journal of Vegetation Science*, 10: 493–502.
Sanderson, J.L. (1958) 'The autecology of *Corylus avellana* (L.) in the neighbourhood of Sheffield with special reference to its regeneration', PhD thesis, University of Sheffield, Sheffield.
Schuffenhauer, F. (2011) 'Einheimische Eichenwälder als Orte der Biodiversität im Wald', *AFZ-DerWald*, 19: 32–5.
Schulze-Hagen, K. (2004) 'Allmenden und ihr Vogelreichtum – Wandel von Landschaft, Landwirtschaft und Avifauna in den letzten 250 Jahren', *Charadrius*, 40: 97–121.
Schuster, L. (1950) 'Über den Sammeltrieb des Eichelhähers (*Garrulus glandarius*)', *Vogelwelt*, 71: 9–17.
Sheppard, C. (1995) 'The shifting baseline syndrome', *Marine Pollution Bulletin*, 30: 766–7.
Smit, C., and Putman, R. (2011) 'Large herbivores as "environmental engineers"', in Putman, R., Appolonia, M. and Andersen, R. (eds) *Ungulate Management in Europe*, Cambridge University Press, Cambridge, pp260–83.
Smit, C. and Ruifrok, J.L. (2011) 'From protégé to nurse plant: Establishment of thorny shrubs in grazed temperate woodlands', *Journal of Vegetation Science*, 22: 377–86.
Smit, C. and Vermijmeren, M. (2011) 'Tree-shrub associations in grazed woodlands: First rodents, then cattle', *Plant Ecology*, 212: 483–93.
Smit, C., Den Ouden, J. and Müller-Schärer, H. (2006) 'Unpalatable plants facilitate tree saplings survival in wooded pastures', *Journal of Applied Ecology*, 43: 305–12.
Smith, C.J. (1980) *Ecology of the English chalk*, Academic Press, London.
Snow, B. and Snow, D. (1988) *Birds and Berries. A Study of an Ecological Interaction*, T. & A.D. Poyser, Calton.
Sommer, R.S., Benecke, N., Lõngas, L., Nelle, O. and Schmölcke, U. (2011) 'Holocene survival of the wild horse in Europe: A matter of open landscape?', *Journal of Quaternary Science*, 26: 805–12.
Sonnesson, L.K. (1994) 'Growth and survival after cotyledon removal in *Quercus robur* seedlings, grown in different natural soil types', *Oikos*, 69: 65–70.
Sperber, G. and Thierfelder, S. (2008) *Urwälder Deutschlands. Nationalparks, Naturwaldreservate und andere Schutzgebiete*, BLV Buchverlag, München.
Szafer, W. (1968) 'The ure-oxe, extinct in Europe since the seventeenth century: An early attempt at conservation that failed', *Biological Conservation*, 1: 45–7.
Tansley, A.G. (ed.) (1911) *Types of British Vegetation*, Cambridge University Press, Cambridge.
Tansley, A.G. (1916) 'The development of vegetation. Review of Clements' "Plant succession", 1916, *Journal of Ecology*, 4: 198–204.

Tansley, A.G. (1920) 'The classification of vegetation and the concept development', *Journal of Ecology*, 8: 118–49.
Tansley, A.G. (1922) 'Studies on the vegetation of the English Chalk II. Early stages of redevelopment of woody vegetation on chalk grassland', *Journal of Ecology*, 10: 168–77.
Tansley, A.G. (1935) 'The use and abuse of vegetational concepts and terms. *Ecology* 16, 284–307', in Real, L. and Brown, J.H. (eds) (1991) *Foundations of Ecology. Classic papers with commentaries*, The University of Chicago Press, Chicago, London, pp319–41.
Tansley, A.G. (1953) *The British Islands and their Vegetation*, Vols 1 and 2, 3rd edition, Cambridge University Press, Cambridge.
Ten Cate, C.L. (1972) *'Wan god mast gift....'. Bilder aus der Geschichte der Schweinezucht im Walde*, Pudoc, Wageningen.
Tubbs, C.R. (1988) *The New Forest. A Natural history*, The New Naturalist, Collins, London.
Tüxen, R. von (1956) 'Die heutige potentielle natürliche Vegetation als Gegenstand der Vegetationskartierung', *Angewandte Pflanzensoziologie*, 13: 1–42.
Van Ek, T. and Johannesson, J. (eds) (2005) 'Multi-purpose management of oak habitats. Examples of best practice from the county of Östergötland, Sweden', County administration of Östergötland, report 2005, 16.
Van Vuure, C. (2003) 'De Oeros. Het spoor terug', Wageningen UR, Rapport 186.
Vera, F.W.M. (2000) *Grazing Ecology and Forest History*, CABI Publishers, Wallingford.
Vera, F.W.M, Bakker, E. and Olff, H. (2006) 'The influence of large herbivores on tree recruitment and forest dynamics', in Danell, K., Duncan, P., Bergström, R. and Pastor, J. (eds) *Large Herbivore Ecology, Ecosystem Dynamics and Conservation*, Cambridge University Press, Cambridge, pp203–231.
Vodka, S., Konvika, M. and Cizek, L. (2009) 'Habitat preferences of oak-feeding xylophagous beetles in a temperate woodland: Implications for forest history and management', *J. Insect Conserv.*, 13: 553–62.
Vullmer, H. and Hanstein, U. (1995) 'Der Beitrag des Eichelhähers zur Eichenverjüngung in einem naturnah bewirtschafteten Wald in der Lüneburger Heide', *Forst und Holz*, 50: 643–6.
Warming, E. assisted by Vahl, M. (1909) *Oecology of Plants: an Introduction to the Study of Plant-Communities*. Claredon Press, Oxford.
Watt, A.S. (1919) 'On the causes of failure of natural regeneration in British oakwoods', *Journal of Ecology*, 7: 173–203.
Watt, A.S. (1924) 'On the ecology of British beech woods with special reference to their regeneration. Part II. The development and structure of beech communities on the Sussex downs', *Journal of Ecology*, 12: 145[-]204.
Watt, A.S. (1925) 'On the ecology of British beech woods with special reference to their regeneration. Part II, sections II & III. The Development and structure of beech communities', *Journal of Ecology*, 13: 27–73.
Watt, A.S. (1934) 'The vegetation of the Chiltern Hills with special reference to the beechwoods and their seral relationship, Part II', *Journal of Ecology*, 22: 445–507.
Watt, A.S. (1947) 'Pattern and process in the plant community, *Journal of Ecology*, 35: 1–22.
Westphal, J., Tremer, N., Von Oheimb, G., Hansen, J., Von Gadow, K. and Härdtle, W. (2006) 'Is the reverse J-shaped diameter distribution universally applicable in European virgin beech forests?', *Forest Ecology and Management*, 223: 75–83.

Wolf, A. (2011) 'Determining the rate of change in a mixed deciduous forest monitored for 50 years', *Annals of Forest Science*, 68: 485–95.

Wolfe, M.L. and von Berg, F.C. (1988) 'Deer and forestry in Germany. Half a century after Aldo Leopold', *Journal of Forestry*, 86: 25–31.

Wrześniowski, A. (1878) 'Studien zur Geschichte des polnischen Tur (Ur, Urus) *Bos primigenius Bojanus*', *Zeitschrift für wissenschaftliche Zoologie*, 30, Supplement 45: 493–555.

Ziegenhagen, B. and Kausch, W. (1995) 'Productivity of young shaded oaks (*Quercus robur* L.) as corresponding to shoot morphology and leaf anatomy', *Forest Ecology and Management*, 72: 97–108.

10 Ancient trees and wood-pastures

Observations on recent progress

Ted Green

Introduction

'Science doesn't stand still' is a phrase I have used frequently for many years. It is certainly very true in the world of the natural environment and its ecology. Frans Vera's studies condensed ideas from his long-term research and his book *Grazing Ecology and Forest History* in 2000 opened so many doors and windows for open-minded naturalists. To this end, rather than wasting time defending Vera with the sceptics, I want to discuss some of the questions raised and the potential answers proposed. His study has become a springboard for inspiring positive research into many aspects of his work and related fields. A good example is perhaps that our oaks are pivotal in developing his work and were in fact one of the main reasons for his original studies. In explaining that oak and pine are very light-demanding species, which would subsequently lead to their eventual demise in woodlands when in competition with shade-tolerant species such as beech and hornbeam, has led to the answering of so many questions in ecology.

The great Frances Rose described Vera's work as landmark. For Frances himself, as early as the 1960s, was beginning to question the concept of continuous cover and dense forest of light-demanding tree species (Rose, 1992 and 1993). His doubts as a field-man and ecologist had been aroused by his encyclopaedic knowledge of botany and especially his studies of lichens where many of these specialist species only occurred on old, open-grown trees. Oliver Rackham (1986) too was having his doubts and used the word 'savanna' to describe the north-west European landscape. Keith Alexander (1998), also working on parkland and old trees, found that his decaying wood beetles also required sunlight and warmth. They again are associated with old, open-grown trees. For myself, there were questions about the co-evolutionary links with jays preferring to bury acorns in the open and of course, this was also brought into focus by Mick Crawley's study of oak regeneration establishing in open ground after two significant crashes in rabbit populations due to outbreaks of myxomatosis (Dobson and Crawley, 1994).

Therefore we could say that some key words are open-grown trees, wood-pasture, scrub, and with a major driver, grazing and browsing animals. So having received such a wealth of information, and such stimulation especially in recent times, an inevitable question is where actually are we today, a decade on from Vera publishing his work?

The importance of open-grown trees

The now-annual conferences at Sheffield have been very successful in bringing many people together from a large array of disciplines to discuss these issues. These began in 1992 with the meeting on *'Ancient Woodlands: Their Archaeology and Ecology – a Coincidence of Interest'* (Beswick and Rotherham, 1993). This pivotal meeting helped to trigger the emergence of the Ancient Tree Forum, and the major follow-up event in 2003, *'Working and Walking in the Footsteps of Ghosts'* (Rotherham and Handley, 2003), helped to bring together key individuals to discuss and debate the ideas of Frans Vera on the influence of large grazing herbivores in the European landscape. From these meetings and the consequent collaborations, the relationship between grazing mammals and the survival of large, open-grown trees was beginning to be recognised. Further events dedicated to these ideas took place in 2005 (Rotherham, 2005) and 2007 (Rotherham, 2007), and with *The End of Tradition?* in 2010 (Rotherham *et al.*, 2010a and b) and *Animals, Man and Treescapes* (Rotherham and Handley, 2011) in 2011, the platform was set for this present book. The National Lowland Heaths Conference, held in Nottingham in 2002, with a keynote address by Frans Vera also proved a hugely influential meeting (Rotherham and Bradley, 2011).

In the many ensuing discussions, for my part, the open-grown tree plays a significant role and to quote Darwin 'You can't have Science without speculation'. I would argue that by quoting Darwin I imply how after Vera some of us have increased our understanding of some of the natural processes that include human activities that surround us today. For many of us, again the open-grown tree plays a significant role in our speculation and expansion of knowledge. For example, gathering data on the value of trees and shrubs for biodiversity research it soon became apparent that open-grown, light-demanding trees have a far greater and diverse assemblage of biodiversity both in numbers of species and mass than shade-tolerant tree species growing in dense dark woodland. However, the situation changes dramatically in woodland when the value of decaying wood and dead standing trees are considered in relation to biodiversity. While in woodland a light-demanding tree species is usually at a distinct disadvantage in competition with shade-tolerant tree species, it seems that shade-tolerant trees frequently grow successfully as open-grown trees in open conditions. However, at times of stress such as prolonged drought, they appear to be far less able to cope with the conditions than open-growing trees that are light demanding. Usually, open-grown trees with no interference from other competing trees are able to develop a large open-grown crown with a subsequent far greater leaf area and big supporting limbs. This is in comparison with the reduced crowns and few side limbs of woodland trees.

Presumably as early farmers, our ancestors discovered the value of open-grown trees when growing fruit trees and shrubs. Of course, the same applies to growing cabbages. With all such planting, you do not plant to close to each other; and of course, you never have dense woodland of fruit trees either! In fact, to date, observations have led me to the conclusion that perhaps only holly, ivy and yew are the only fruit-bearing trees and shrubs capable of producing some flowers and fruit in reduced quantities in low-level light, such as in dense shady woodland. In other words, perhaps all our other trees and shrubs require varying but significant amounts of sunlight to produce flowers and the subsequent fruit. Even shade-tolerant trees will produce the bulk of their flowers and fruit in the canopy in woodlands with their reduced crown areas. Reduced flowers produced by trees growing in woodlands obviously means less pollen production in comparison with pollen produced by trees growing in the open. In addition, pollen dispersal over any distance will be hampered greatly by surrounding vegetation and consequent reduced wind strengths. In terms of theoretical reconstruction of past landscapes, this begs the question: have these observations in the past been considered when interpreting the evidence from pollen cores?

Biodiversity impacts

Interestingly, the value to biodiversity of shade-tolerant woodland trees increases very substantially as they begin to produce non-living wood (decaying wood) either through the death of limbs or the whole tree due to competition for resources and light from more vigorous neighbours or death through old age. The decay of non-living hardwood (found in oak, sweet chestnut and yew) and ageing ripewood (found in other tree species such as beech, ash and Celtic maple) is a perfectly natural process and one can say that it occurs through the co-evolutionary complex relationships between the tree, fungi, bacteria and a host of other micro-organisms. All trees, whether growing in the open or dense woodland, if allowed to age naturally will eventually hollow. However, deadwood, from leaves, twigs and limbs to completely dead trees, can often behave quite differently when produced from an open-grown tree in comparison to the dense woodland tree. When constantly exposed to sunlight and wind, the conditions can lead to desiccation with virtually no or very slow breakdown through decay. Consequently, there is a whole spectrum of decay rates depending on the degree of moisture conditions found in woodlands. Presumably, moisture content is fundamental for fungi and other micro-organisms in recycling nonliving wood. Invertebrates such as wood-boring beetles (Figure 10.1) and ants, perhaps having a greater role in facilitating decay in partially desiccated wood, behave in a similar fashion to termites in warmer climes.

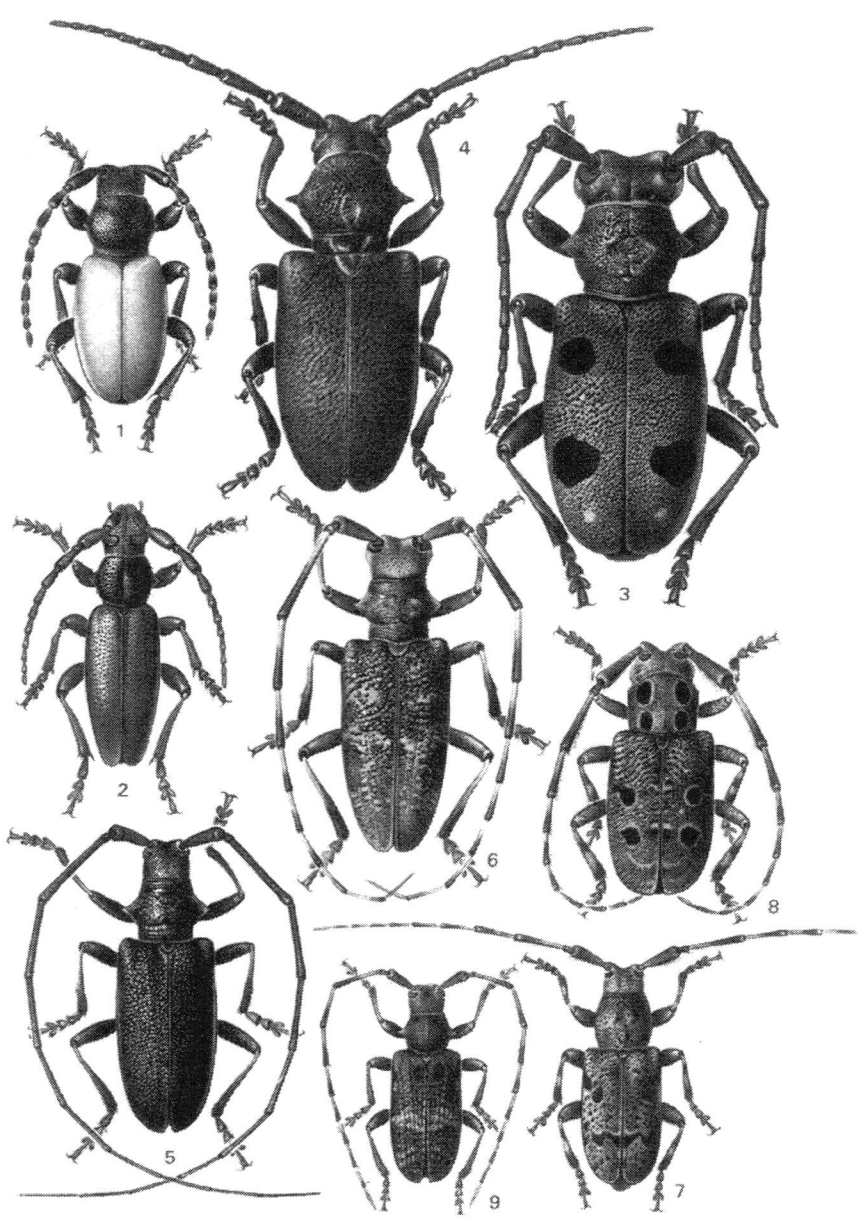

Figure 10.1 Deadwood beetles
Source: Ian D. Rotherham

Today it is widely accepted that the roles of people in the natural world could be described as 'culture and history' and these disciplines have been represented at the conferences in Sheffield. Keith Alexander and I are calling our next article in preparation, '*Ancient trees: Islands in history, islands in biodiversity*'. Obviously, we have called them islands because the oak, especially, does not survive to any great age in woodlands; it prefers to be isolated, open-grown, ideally away from any competition. Because of their age, some over, a thousand years old, we can assume they have become reservoirs of biodiversity and we hope, will provide an essential biological continuity for future generations. From the evidence of the organisms found living with them on the outside, the inside and especially below ground, our conclusion is that this has been the situation for several millennia. We believe this to be the case especially for the soil-inhabiting micro-organisms. They also provide us with a great insight into our past, for example, when William the Conqueror stood on a hill at Windsor and declared 'This is where I will build my home', he was possibly surrounded by open-grown trees and scrub in parkland, savannah-like landscape stretching many miles in all directions and at least to Winchester! Small fragments like Windsor Park remain today. William was an ardent hunter and loved 'The Chase' on horseback. Therefore, it is quite reasonable to assume that this vast area was a parkland landscape because it is virtually impossible to ride a horse at any speed in woodland, especially dense woodland.

Nowadays we have methods to determine the age of these ancient oaks and an ever-increasing knowledge of how trees grow and develop. Especially the large limbs growing low down indicate that these trees have spent their whole lifetime growing in the open away from other trees, especially from concentrations of other trees. Most of the remaining trees are now in parklands (Figure 10.2); with some like Windsor, which dates back to Saxon times. While other notable individual trees scattered across the countryside were perhaps meeting places for individuals and groups for a whole host of reasons. Several trees were hanging trees and perhaps others were places of worship.

The survival of ancient trees and the tradition of pollarding

Compared with the rest of northern Europe, Britain has relatively large numbers of old trees. There are many reasons for this; and we can certainly thank the aristocracy for the continuity they have provided in the parklands on their estates. The arrival of coal and the railways no doubt drastically reduced the demand for wood fuel for domestic and industrial use. Equally, there have been no wars in the UK for 400 years. Wars mean large armies and refugees moving across the land and the associated need for timber and fuel. It is very unlikely that any army would ever carry wood and water any distance; they obtained them wherever they went. The chances are that in previous times, armies, refugees, travellers and especially drovers, who

Figure 10.2 Parkland oaks
Source: Ian D. Rotherham

cut trees for use en route, did not cut trees to the ground. They either cut new pollards or shreds there and then as they went, or used existing pollard trees started in the past and cut in the traditional way of that period. It is quite possible and logical that travellers would never damage or destroy a watering place, and so behaved likewise with a working pollard or a fruit tree. Pollards are working trees providing people with a multitude of different materials (Figure 10.3). Usually with age the trunks are hollow living columns of sapwood, so again would provide little or no immediate use if cut down. Without doubt, the cutting of working trees has been a part of European rural life for millennia; in fact preserved pollard oaks have been found recently in aggregate extractions from the River Trent. These have been carbon dated at 3,400 years old and shredded oaks in the River Meuse, Belgium were dated as 1,800 years old.

Perhaps right up until recent times, in rural communities across Europe, cut working trees were seen as essential for the host of products they could produce. Unfortunately, the recognition of tree archaeology, especially with trees that are still living, is still very much in its infancy. On reflection, it is perhaps the case throughout archaeology. There are good examples too with art historians where paintings by Brueghel, for example, have been studied in detail with every individual in the painting pored over

Figure 10.3 Re-grown pollard
Source: Ian D. Rotherham

down to their buttons and the houses studied too. However, nobody has looked at the trees! In fact, when we do examine the trees it is most informative — they have all been cut and there are several different types of pollard including shreds featured throughout these paintings!

As our interest and knowledge of working trees increases, it is becoming apparent that we have underestimated their importance to society in the past. Again, another good example is the reference to the 10,000 oaks required to build the ship, *HMS Victory*. This would represent an incredibly vast area of individual trees. However, questions need to be asked such as were these large oaks, small oaks or timber just from small limbs? Oliver Rackham points out that in those days the shipwright's wages were paid with timber. Therefore, it is quite likely the trees went into the shipyard at one end and the chances are 95 per cent that they went out the other end to build London! So yes, *HMS Victory* might have taken pieces from 10,000 trees to build it but never 10,000 whole trees. In those days too, it is very likely that were times in ship and house building when curved timbers were perhaps more important than straight timber. Now there appears to be only a single reference to a large spreading curved limb being cut from a standing living tree. In this case it is described how brushwood was placed underneath to cushion the fall and stop the timber from breaking. To me this is a mystery! Because when felling a large tree with several substantial side limbs, on hitting the ground, perhaps 75 per cent of the limbs will snap, break or shatter in the wrong places on impact; making them useless. Therefore, the value of the whole tree is greatly reduced.

Cutting working trees, pollards and shreds for tree fodder (tree hay) is another area where there appears to be scant reference or attention to their importance by historians. From what we do know we can assume that tree fodder was important especially to the owner of a small number of animals in long hard winters or in long hot dry summers. Yorkshire and Lancashire have several place names beginning with Hol or Hollins, which apparently referred to an orchard of holly pollards or even a single pollard. Holly leaves and twigs were used as tree fodder to feed the animals in periods of prolonged frost and snow. Apparently, this gradually died out with the arrival of the alternative food of the turnip. Holly wood also became exceptionally valuable in making the moving parts for the spinning and weaving industries of Lancashire and Yorkshire.

There are still scattered groups and individual holly pollards in the UK. For example at the Stiper Stones in Shropshire it appears that only the tips of the branches were cut, that is only up to 20cm was ever cut, presumably the reasoning being that fodder bundles (fagots) had to be carried long distances down the hill and valley. Holly being slow growing would mean that if larger limbs were cut, it would take several years before they could be cut again. New Forest ponies too can still manage to find some holly for themselves but the extensive re-pollarding recently carried out on ancient hollies has been a tragic disaster, with a large proportion not surviving or now in very poor condition. This was presumably through ignorance and lack of research into pollard restoration techniques. Sadly, this is yet another example of bad conservation affecting part of our unique northern European cultural history and biodiversity. In days gone by, you could have a hand cut off for destroying venison and vert. Vert is holly and ivy so perhaps we should reinstate this ancient law!

There are examples of pollards being cut on a commercial scale. One example is the hornbeam pollard orchards that can still be found on the outskirts of London. It was said that the bakers of London prized the hornbeam fagots for making the best bread. Actually French researchers have concluded that hornbeam has the highest calorific value of all broadleaved wood. Another example is the thousands of black poplars in the Teruel region of northern Spain at Calamocha. The cut and seasoned pollarded limbs at 5m in length and 15–20cms in diameter were used to provide rafters for the houses of north-east Spain and Barcelona. When weight and length are considered, these pollard poles are reputed to be the strongest broadleaved wood for rafters.

Also in the Basque region, in response to the increasing lack of curved timbers for shipbuilding, trees were grown specifically for this purpose. Young oaks had their main stem cut when they were very young at about 2m and the topside limbs tied down and trained to make a curved limb. These can still be seen today in the remaining mature trees. While these trees were maturing, the technology for bending wood under steam, which gave a uniform ideal bend, was developing. Therefore, the limbs were cut at about 4m out from the main stem and the cut wood was used to make the charcoal to make the steam to bend the wood (i.e. two uses). Later the technology to make iron and steel ships developed, and so the trees were cut in the same area again to make the charcoal to make the iron and steel (three uses). The Ancient Tree Forum went to the Basque Country to research these trees and pointed out their importance to Europe's cultural heritage (four uses) and their value to biodiversity (five uses!).

Now at the beginning of the twenty-first century, the cutting of trees for fodder and the starting of new pollards has been begun again in the UK. We are relearning an old practice and obviously making some mistakes but gaining experience. While cutting tree fodder is still carried out, usually in the poorer and remoter parts of Europe, it is by an ageing population. Much of the knowledge of the methodology could soon be lost with the demise of these aged practitioners. We hope now to have followed their age-old practices and to have safeguarded the cultural knowledge to be passed to future generations. In general, trees are cut on a rotation of one to four years depending on growth and performance. This work is done in the period at the end of June to early July before the trees starts to withdraw the nutrients and micronutrients contained in the leaves, in a similar fashion to hay. The cut small branches or twigs, up to 2m in length are air-dried by hanging them above ground and the leaves remain on the twigs. When the leaves are dried, they are stored in a dry shelter so that the leaves remain on the twigs, and later fed to animals in the winter. Another method is to cut the branches and twigs in high summer in the Mediterranean region, and the green leaves and twigs are fed to the animals. It appears that this practice was used particularly during periods of prolonged drought and high temperatures when the improved grasslands are 'burned off' and yellow. Cutting leaf fodder is perhaps the only available 'greenery' to feed to animals. Sadly this type of use has been lost in the UK. It is suggested that the minerals, nutrients, micronutrients and trace elements trapped in the trees' leaves provide animals with elements that may never occur in present, modern-day diets.

There is also evidence that many plants contain medicinal properties that are potential alternatives to manufactured industrial pharmaceuticals and that the animals may instinctively seek out plants to self-medicate. For example, plants with tannin content have been found to reduce nematode (worm) infestations in sheep. Interestingly at three sites in the UK we have found old redundant stone compounds for keeping animals and these have ancient sallow trees growing beside the walls. Sallow contains tannin and aspirin too. So what can those old shepherds and herdsman tell us? In addition, have their knowledge and talents been lost through so-called modern improvements?

Ancient trees, young trees, ancient soils

They say an oak tree grows for 300 years, rests for 300 years and spends some 300 years gracefully expiring. Like the 'tree watcher' before me, who made this observation, the oak fascinates me. I call our oaks the 'Rolls Royce' of the tree world because like the Rolls Royce, all the rest are but mere imitations. The oak tree has an incredible longevity, few other trees can equal it and even fewer surpass it. However, the part that fungi play throughout its life, both without and within, is fundamental and aids the tree in its longevity (Figure 10.4). For example, the endophytic fungi that colonise the nonliving tissue will defend their territory against perceived invaders, whether friend or foe, and therefore act in an 'antibiotic' manner. The mycorrhizal fungi that colonise and develop with the tree's roots provide an essential, extended food-gathering system and defend their territory against invaders such as the pathogenic honey fungus and the array of other pathogens that we find especially in our degraded soils. So every tree becomes a unique and individually dynamic support system for fungi. These endophytic fungi and likewise the tree's mycorrhizal partners that colonise the tree's roots may well have already been centuries old when they colonised them as an acorn or seedling. They have passed from tree to tree in a continuous process since woodland and individual tree creation. This is best described as providing unbroken biological continuity in the tree both above and below the soil.

Ancient trees and wood-pastures 137

Figure 10.4 Fungi
Source: Ian D. Rotherham, collection of antiquarian prints

Today entomologists, mycologists and other scientists view the oak as some vast skyscraper: a living pulsating ecosystem in its own right. No other tree in England has such an enormous biodiversity of associated and dependent organisms; and some are extremely rare and not only on the tree but also inside the tree and in the surrounding soil.

The fascination with oaks has turned to concern when travelling around the European continent in search of ancient oaks: where are they? Without doubt, Britain is now the major custodian of perhaps the largest majority of Europe's ancients and veterans north of the Mediterranean, as well as being an integral part of our nation's treescape primarily in parklands, on commons and in hedgerows. However, those ancient trees that still do exist in continental Europe are usually recognised for their importance and uniqueness and recorded as part of their nation's heritage, often marked on maps and with an information board.

By contrast, in the UK, apart from some individual landowners and organisations, we take old oaks and ancient trees for granted. Since the most recent Dutch elm disease outbreak, with the tragic loss of tens of thousands of elms that graced the countryside, and then the horrendous hurricanes of 1987 and 1989, we have begun to realise the fragility and frailty of this resource. This is a uniquely British landscape portrayed in the past by painters such as Turner and Constable and of huge cultural value (Figure 10.5). So there is even more reason to conserve what remains of our treescape. After all, no one would question the preservation of old buildings, art treasures and our ancient monuments, burial mounds and other parts of a nation's heritage. In fact, our ancient trees are actually our living heritage. Therefore, why should we not have individual trees designated as sites of special scientific interest? One has to ask, do we want future generations simply to look at their old paintings and photographs? They might then ask questions. Why did they not look after their ancient trees like works of art? Nature's sculptures, biological storehouses so full of life, gene banks of Europe's future trees; the list of questions would be endless.

Figure 10.4 Ancient park oak
Source: Ian D. Rotherham, collection of antiquarian prints

However, it now appears that in the last few years voices of concern, such as Oliver Rackham's in the 1980s and the Ancient Tree Forum's from the mid-1990s onwards, have begun to be heard. Together with the launch in the 1990s of English Nature's Veteran Tree Initiative, we may have turned the corner and never again will these vast reservoirs of biodiversity, which have graced the countryside for centuries, be under such direct threat. Of course, this has not been possible without the constant and significant contribution of the Tree Council and its members down the years. Of the dozens of campaigns carried out by the Tree Council, the Green Monuments Campaign for the protection of our heritage trees, the publications of British and Scottish Heritage Trees, the Hedgerow Campaign, National Tree Week and of course the involvement by members, including in the Ancient Tree Hunt, have all been hugely influential. The impacts of all this activity have been recognised throughout the tree world. I have only mentioned coppicing in passing, because generally, the history of this practice is already well documented; but in recent years, the importance of veteran coppice trees has also begun to be recognised.

Conclusions

Interaction between diseases of grazing animals and tree generation in open areas

There are examples of diseases and parasites affecting populations of grazing animals. By reducing the number of animals this in turn would allow the generation of trees and shrubs. In the UK when the rabbit (*Oryctolagus cuniculus*) population virtually disappeared in the mid-1950s, due to the virus myxomatosis, oak became established from acorns. These seeds were in open areas and buried by jays and wood mice in the un-grazed meadows that resulted from the cessation of grazing. The seedlings were found up to 75m from the nearest seed source. Several other tree species appeared including elm (*Ulmus procera*) suckers that continued to extend annually until the return of the rabbit. However, further tree establishment ceased in subsequent years when the meadows became too dense and rank. Similarly, in East Africa, the different age classes of the groves of *Acacia tortilis* can be directly linked to the periods of reduction in populations of grazing animals through outbreaks of disease (Dobson and Crawley, 1994). There is also an account in *The Drove Roads of Scotland* (Haldane, 2006) of drovers bringing cattle to markets in the south from as far afield as the Scottish Isles and losing all their animals to disease before they had reached their resting destinations in Norfolk and Suffolk before sale:

> There, the cattle disease which raged that autumn throughout England attacked them, and his drove died first in scores and then in hundreds{...}our conditions are such that several drovers have run their beasts and left them dying in the lanes and highways and nobody to own them.

Sir Arthur Oliver, principal of the Royal (Dick) Veterinary College wrote: 'A serious outbreak of cattle plague (rinderpest) occurred in Great Britain in 1745 following its extension throughout the greater part of the

Continent'. The outbreak lasted until 1757, though cattle plague existed in some parts up to 1770, and there is no record of any other major outbreak of cattle plague in this country again until 1865. Mortality was exceedingly high, probably running to 90 per cent in badly affected herds. It is very unlikely that the cause of the cattle's death was specific and therefore quite possible that a large proportion of other grazing and browsing animals, whether domestic or wild, could have succumbed over a large area. Obviously, a reduction in animal numbers would have positive implications for tree and shrub colonisation for several years. Perhaps it is worth posing a question about the impact the decrease in the human population had on tree cover after the Black Death. Therefore, there can be situations in the forest where trees die, assisted by diseases, and glades/savannah can develop. These areas are grazed by herbivores, perhaps later assisted by people, but when animal numbers are reduced by diseases – often exacerbated by periods of prolonged drought – there are opportunities for the forest to regenerate to some extent, generally around the margins of surviving isolated trees, groves and edges of woodland.

Oaks and jays

Evidence for tree and shrub generation is found in the fascinating evolutionary relationship between the oak and the jay (*Garrulus glandarius*) and other members of the crow family. Other relationships include: azure-winged magpies (*Cyanopica cyanus*) and the evergreen oaks (*Quercus ilex* in Spain and Portugal); the nutcracker (*Nucifraga caryocatactes*) and the arolla pine (*Pinus cembra*) and other tree seeds in European mountain ranges; and the rook (*Corvus frugilegus*) and the walnut (*Juglans ssp*). It is also very likely – but as yet not widely recognised – that small mammals such as wood mice and other rodents can play an equally significant role in tree seed dispersal (Crawley and Willoughby, *pers. comm.*). In most cases, trees that are successfully established will be of either an isolated open-grown form or in small groves; these are created from a rodent seed cache after the animal dies and so abandons its hoard, and the seedlings are protected by thorns. Thrushes and several mammal species are perhaps the key species in distributing berries and other soft fruit tree and shrub seed with excreta playing a significant role. In other words, again in support of Vera's ideas, an acorn buried in an open space by a jay or a small mammal and left to germinate can become a tree. This, when able to grow unrestricted without competition from other trees, can develop into what is termed an 'open-grown' form, optimising its light-gathering power by developing a shape similar to a sphere. The tree thus produces copious volumes of pollen. In later years, if the browsing animals reappear, the lower limbs within reach can be browsed and the light levels increased beneath the canopy. In 1904, William Menzies speculated about the oaks at Windsor:

It is interesting to conjecture how these ancient trees, both in the Park and the Forest, sprung up, and survived, unprotected as they were, against cattle and deer. It is probable that many of them must have come up in the thickets of thorns and bushes, and by the sides of roads and division fences.

Menzies also quotes Arthur Standish, writing in 1613, who speaks of bushes as 'the Mother and Nurse of Trees', and that 'but for them there would be no timber in the common land'. He goes on to say 'There is an old forest proverb: the Thorn bush is the mother of the Oak'.

It is remarkable to think that even by 1613 this proverb was considered old! In the UK, ancient trees, especially open-grown, have an incredible wealth of associated biodiversity – visible and invisible – and their great age provides biological links and continuity with past generations of trees. These trees are also significant features in those landscapes, with perhaps the closest resemblance to the savannah that graced the European landscape.

Since 10,000 oaks of a hundred years old are no substitute for a five hundred-year-old oak (Rackham, 1986), we need to consider their future conservation to be a very serious matter. We may well ask, are we simply going to stand by and watch these ancient and veterans merely fade away? I hope that with the annual conferences at Sheffield organised by Ian Rotherham, there is now a great opportunity to arrest any further drain of our past 'cultural history and the natural world' through the coming together of like-minds from many disciplines.

Bibliography and references

Alexander, K.N.A. (1998) 'The links between forest history and biodiversity: The invertebrate fauna of ancient pasture woodland in the British Isles and its conservation', in Kirby, K. and Watkins, C. (eds) *The Ecological History of European Forests*, CABI International, Oxon, Wallingford, pp.73–80.

Alexander, K.N.A. (1999) 'The invertebrates of Britain's wood-pastures', *British Wildlife*, 11(2): 108–117.

Beswick, P. and Rotherham, I.D. (eds) (1993) 'Ancient woodlands: Their archaeology and ecology – a coincidence of interest', *Landscape Archaeology and Ecology*, 1.

Brasier, C. (1999) *Phytophthora Pathogens of Trees; Their Rising Profile in Europe*, Forestry Information Note, Forestry Commission, October 1999, Edinburgh.

Dobson, A. and Crawley, M. (1994) 'Pathogens and structure of plant communities', *Tree*, 9(10): 393–8.

Edlin, H.L. (1956) *Trees Woods and Man*, Collins New Naturalist, London.

Green, T. (1992) 'The forgotten army', *British Wildlife*, 4(2): 85–6.

Green, T. (1996) 'Pollarding: Origins and some practical hints', *British Wildlife*, 18(2): 100–05.

Green, T. (1998) 'Fungi first', in *Proceedings of Bring back the Bison*, Farnborough, Hants.

Green, T. (2000) *Growing downwards*, Tree Line, UK and Ireland.

Green, T. (2001) 'Comment: Should ancient trees be designated as sites of special scientific interest', *British Wildlife*, 12: 164–6.
Green, T. (2002) *The role of invisible biodiversity in pasture landscapes*, in *Pasture Landscapes and Nature Conservation*, B. Redecker, P. Finck, W. Hardtle, U. Riecken, and E. Schroder (eds), Springer, Netherlands.
Green, T. (2010) 'Natural origin of the commons: People, animals and invisible biodiversity', *Landscape Archaeology and Ecology*, 8: 57–62.
Haldane, A.R.B. (2006) *The Drove Roads of Scotland*, Birlinn Ltd, Edinburgh.
Menzies, W. (1904) *Windsor Park and Forest*, London.
Merryweather, J.W. (2001) 'Comment: Meet the glomales – the ecology of mycorrhiza', *British Wildlife*, 13(2): 86–93.
Merryweather, J.W. (2006) 'Secrets of the soil', *Resurgence*, 235.
Merryweather, J.W. (2007) 'Comment: planting trees or woodland? An ecologist's perspective', *British Wildlife*, 18(4): 250–8.
Rackham, O. (1986) *The History of the Countryside*, J. Dent and Sons, London.
Rose, F. (1992) 'Temperate forest management: its effects on bryophytes and lichen floras and habitats', in Bates, J.W. and Farmer, A.M. (eds) *Bryophytes and Lichens in a Changing Environment*, Clarendon Press, Oxford, pp211–33.
Rose, F. (1993) 'Ancient British woodlands and their epiphytes', *British Wildlife*, 5: 83–93.
Rotherham, I.D. (ed.) (2005) 'Crisis and continuum in the shaping of landscapes', *Landscape Archaeology and Ecology*, 5.
Rotherham, I.D. (ed.) (2007) *The History, Ecology and Archaeology of Medieval Parks and Parklands*, Wildtrack Publishing, Sheffield.
Rotherham, I.D. and Bradley, J. (eds) (2011) *Lowland Heaths: Ecology, History, Restoration and Management*, Wildtrack Publishing, Sheffield.
Rotherham, I.D. and Handley, C. (eds) (2003) 'Working and walking in the footsteps of ghosts. The pre-published proceedings and notes of the international conference. Sheffield, May 29–June 1 2003', *Peak District Journal of Natural History and Archaeology Special Publication*, 2, Wildtrack Publishing, Sheffield, pp1–2.
Rotherham, I.D. and Handley, C. (eds) (2011) *Animals, Man and Treescapes*, Wildtrack Publishing, Sheffield.
Rotherham, I.D, Agnoletti, M. and Handley, C. (2010a) *The End of Tradition? Aspects of Commons and Cultural Severance in the Landscape*, Volume 1, Wildtrack Publishing, Sheffield.
Rotherham, I.D, Agnoletti, M. and Handley, C. (2010b) *The End of Tradition? Aspects of Commons and Cultural Severance in the Landscape*, Volume 2, Wildtrack Publishing, Sheffield.
Vera, F.W.M. (2000) *Grazing Ecology and Forest History*, CABI Publishing, Wallingford.

11 Grazed wood-pasture versus browsed high forests

Impact of ungulates on forest landscapes from the perspective of the Białowieża Primeval Forest

Tomasz Samojlik and Dries Kuijper

Introduction

After the retreat of the last glacier and significant warming of the climate c.10,000 years BP, Europe's steppe and tundra were gradually replaced by forest. By 8,000 BP most of the continent was covered by forest (Peterken, 1996; Williams, 2006). The questions how this forest landscape looked and, more specifically, how open the primeval forests of Europe were, have been discussed in recent decades by several authors. Using palaeoecological evidence, the *high forest hypothesis* was proposed. It states that lowland Europe was originally covered by dense, closed-canopy primeval forests and it was only the subsequent waves of large-scale human impact from c.6,000 BP onwards that opened the forest and later decreased its area (Birks, 2005; Mitchell, 2005). The high forest hypothesis was widely accepted by forest ecologists, yet it was raised for discussion after publication of Frans Vera's competing *wood-pasture hypothesis* (Vera, 2000). According to this hypothesis the impact of large herbivores on post-glacial woodland vegetation was strong enough to create and maintain a mosaic landscape of grasslands with shrubs and small forested patches. This idea received much attention and had a large influence on current forest management (Hampicke and Plachter, 2010; Kirby, 2004; Kirby *et al.*, 1995; Putman, 1986; Van Wieren, 1995) and nature conservation management in general.

In view of Vera's hypothesis, wooded pastures were a natural component of the landscape in prehistoric times, when now-extinct larger grazers such as tarpan (*Equus ferus ferus*) and aurochs (*Bos primigenius*) used to roam large parts of Europe (Vera, 2000). Livestock grazing is therefore nowadays often used in these areas as a substitute for the lost role of the large herbivores. For example, internationally famous nature reserves such as the New Forest (UK) and the Oostvaardersplassen area (the Netherlands) are managed in compliance with the ideas of the wood-pasture hypothesis. Despite the wide application of these ideas in nature management, there is still much scientific debate on how large the impact of these extinct herbivores really was on the landscape; did they create a half-open parkland, or were their densities and impact too low to maintain open areas and instead closed forest dominated (e.g. Birks, 2005; Mitchell, 2005)?

In this chapter, we aim to present a view of primeval forest landscapes from the perspective of the Białowieża Primeval Forest (BPF). Interestingly, the example of BPF has already been used in the discussion both by supporters of high forest and wood-pasture hypotheses. In spite of this, we aim to offer a fresh look based on historical and ecological data from current research projects dealing with the environmental history of BPF and the ungulate impact on the regeneration of its tree stands. We discuss what role the community of large ungulates had and has in shaping the forest landscape. We argue that it is unlikely that large ungulates were able to create a half-open landscape everywhere, emphasising the need to broaden our view on management of forest landscapes.

Grazed wooded pastures

According to the wood-pasture hypothesis (Vera, 2000), large predominantly grazing herbivores created a part-open landscape in prehistoric Europe composed of short grazed vegetation alternated by patches of shrubs and old-growth tree stands. This landscape was created by the now extinct ancestors of our domestic cattle – aurochs (*Bos primigenius*) and horse – tarpan (*Equus ferus*). As both these species occurred in herds and are classified as typical grazers (Hofmann, 1989) they exerted strong impact on the vegetation. On the most intensively grazed patches of herbaceous vegetation, they prevented the establishment of shrubs and trees, and hence maintained these areas open. Establishment of trees is supposed to be only possible when 'safe' regeneration sites are being created by thorny bushes, which prevent grazing. This can be compared to the effect of putting a small fence up to protect young trees in grassland. According to Vera (2000) these natural large herbivores created this part-open mosaic landscape; humans who started to settle in the landscape were guided by this. They were most likely to settle in the open places or at the border of forest and grassland. When the aurochs and tarpan became extinct, in 1627 (Van Vuure, 2005) and 1887 (Bennett and Hoffman, 1999) respectively, their roles were taken over by their domesticated relatives. These included horses and cows, together with other domesticated livestock such as sheep, goats and pigs. Of course other factors also contributed to the maintenance of the part-open landscape, such as human-induced fires and wood exploitation. With increasing human populations their impact on the transformation of the landscape also dramatically increased. However, in several nature reserves throughout Europe, a glimpse of the historic low-intensity use of humans and their livestock of forest landscapes can still be observed. A famous example is the New Forest in the UK (Tubbs, 1986). Similar landscapes that are currently maintained by livestock-grazing can be found throughout Europe such as the Borkener Paradies (Germany), Mols Bjerge (Denmark) and the Junner Koeland (Netherlands) (Vera, 2000). In Germany these landscapes are referred to as 'Hudewald', literally meaning a forest used for herding of livestock, a forest-pasture. These landscapes are nowadays much appreciated as they have high aesthetic value and, due to their heterogeneity, contain high biodiversity.

Following the development of the wooded pasture hypothesis, based on an impressive review of existing knowledge, by Vera (2000), several studies have revealed the mechanisms behind the role of herbivores in these landscapes. The crucial process by which palatable plant species, such as oak (*Quercus robur*) (Figure 11.1) can regenerate in these grazed landscapes is by means of 'associational resistance' (Olff *et al.*, 1999). This means that unpalatable plants offer protection against grazing and allow the establishment and growth of the palatable species when they grow in close vicinity. Unpalatable plant species are not being eaten by herbivores (or are less preferred) because they are protected by physical (e.g. thorns, hairs) or chemical defences (low digestibility or toxicity). Because they are avoided, they can get established in short grazed grasslands and form dense thickets that become impenetrable to grazing herbivores. As a result, inside the thickets there are herbivore-free places in which palatable tree species such as oak can establish themselves (Bakker *et al.*, 2004) (Figure 11.1). The thorny shrubs, such as blackthorn or sloe (*Prunus spinosa*) and common hawthorn (*Crataegus monogyna*), seem to play an important role in facilitating the establishment of oak in temperate grazed woodlands as indicated by studies from a variety of systems across Europe (Bakker *et al.*, 2004; Burrichter *et al.*, 1980; Coops, 1988; Tansley, 1922; Vera, 2000; Watt, 1919). The tall-growing oaks will eventually outgrow and out-shade the thickets of thorny shrubs, which in the long run will disappear. This process results in patches or clumps of tall oaks in short grazed vegetation (Figure 11.1). As young oak trees cannot establish under the closed canopy of tall oaks because of lack of light and also through constant grazing, the stands of old oaks will eventually collapse and be transformed once again to grassland. The process of thorny shrub establishment followed by oak regeneration, transformed to grassland after the oaks have died, occurs as a continuous cycle (Figure 11.1). This results in a pattern of shifting mosaics in which open grasslands, thorny thickets and large solitary oak trees alternate each other in space and time (Olff *et al.*, 1999).

The process of shifting mosaics can nowadays be observed in many nature reserves that are being managed by means of livestock grazing, especially with cows and horses (Bakker *et al.*, 2004). These types of herbivores have in common that they are predominantly grazers and relatively unselective ones. Small and highly selective herbivores, such as rabbits, can easily pick out the palatable species, even though they are growing in close association with unpalatable species (Bakker *et al.*, 2004). Even smaller species such as rodents are important seed predators in these systems. Also

Figure 11.1 Tree regeneration in grazed woodlands in Borkener Paradies, Germany, according to the grazed woodland hypothesis: thorny bushes establish in grazed vegetation (1) and offer protection for palatable species (e.g. oak) to become established (2); grazing and the lack of light lead to solitary oaks in the landscape (3); after collapse of old oaks, the process starts again with short grazed vegetation (4)
Source: D. Kuijper

thorny shrubs offer little protection against them (Rousset and Lepart 2000; Smit *et al.*, 2006). Hence, protection by unpalatable species will only work with relative large and unselective herbivores (Bakker *et al.*, 2004; Olff *et al.*, 1999).

An alternative view: The browsed high forest

The internationally well-known Białowieża Primeval Forest (52°45 N, 23°50 E) is located in eastern Poland (600km^2) and western Belarus (850km^2). It offers one of the best preserved examples of temperate lowland forest composed of multi-species tree stands. In 1921, the best preserved central part of the BPF was proclaimed as the Strict Reserve, which later became Białowieża National Park (BNP). At present, the BNP consists of a 47.5km^2 area of strictly protected old-growth forest in which no human intervention has been allowed since 1921 and tourist access is only permitted with a guide. Before 1921, human impact on tree-stand structure and composition was minimal (Jędrzejewska *et al.*, 1997; Samojlik, 2007). The area outside the BNP is managed for forestry purposes, and hence wood exploitation and regulation of ungulate numbers is taking place. As neither of those activities takes place inside the BNP, this area is unique in providing the opportunity to study the role of large herbivores, together with their natural predators wolf (*Canis lupus*) and lynx (*Lynx lynx*), without human interference. Different forest types can be found in this area, with coniferous forest dominating the driest, nutrient-poor soils and deciduous forest occurring in productive soils rich in organic matter (Faliński, 1986).

In contrast to the wooded pastures described above, this high forest system has an ungulate community that is dominated by browsers. A unique aspect of BPF is that it is one of the very few European lowland forests that is still host to the complete native extant ungulate assemblage. Five ungulate species occur throughout the forest system. The ungulate community is dominated in terms of numbers by the red deer (*Cervus elaphus*) (6.0 individuals per km^2 in 2008). Even though Hofmann (1989) classified red deer as an 'intermediate feeder', the diet of red deer in our study area is dominated by woody species (49–96 per cent) year-round (Dzięciołowski, 1970; Gębczyńska, 1980). The second most abundant ungulate is wild boar (*Sus scrofa*) (5.4/km^2 in 2008), whose diet is mainly composed of plant material (Genov, 1981) but cannot be regarded as a typical grazer. Besides, typical 'concentrate selectors' (browsers) such as roe deer (*Capreolus capreolus*) (2.4/km^2 in 2008) and moose (*Alces alces*) (0.08/km^{-2} in 2008) are present in the forest. Both these species have a high proportion of woody species in their diet (Gębczyńska, 1980; Morow, 1976). Typical 'grass eaters' (*sensu* Hofmann, 1989) such as the European bison (*Bison bonasus*), comprise only a small proportion (0.49/km^{-2}) of the community. Besides, its diet is also composed to a large extent of woody material ranging from 11–13 per cent in summer (Gębczyńska *et al.*, 1991) up to 65 per cent in winter for individuals that do not receive supplementary feeding, such as those occurring in our part of the study area (Kowalczyk *et al.*, 2010). Because browsers, generally speaking, are more selective at the individual plant level than pure grazers such as livestock (Searle and Shipley, 2008), regeneration of trees may work in a different way in comparison to the well-studied wooded pastures.

Vera (2000) attributed a large role to the extinct aurochs and tarpan in creating a part-open landscape in the historic BPF. According to his hypothesis, when both species became extinct, their role was replaced by livestock (cattle) that were herded into the forest. The wood-pastures created by the extinct species were the preferred grazing grounds for their domestic counterparts. In his view, Białowieża's wood-pastures started to be overgrown only after ungulate species underwent a significant decline in numbers (or were driven into extinction, like the European bison), and cattle pasturing in the forest was prohibited. The majestic old oaks that are nowadays present in the BPF originate from the time when the forest was open and provided favourable growing conditions for this species.

From the historical point of view, there is little evidence that aurochs and tarpan were ever present in BPF in ecologically meaningful numbers. What is more, the presence of aurochs and tarpan in the forest has never been confirmed, not by written sources (available from the fifteenth century onwards) nor by archaeological finds. The group of ungulates in BPF in historical times (covered by the written sources, i.e. starting from fifteenth century) was composed of European bison (which went extinct in 1919 and was reintroduced to the wild in early 1950s), red deer (extinct during the Little Ice Age in the second half of the seventeenth century and reintroduced in 1865), moose (eradicated from BPF during World War I but that re-colonised the forest after 1945), roe deer (continuously present in BPF) and wild boar (permanently inhabiting BPF) (Jędrzejewska et al., 1997). Cattle pasturing in the forest, present in past centuries (see later for details), was prohibited in the Polish part in 1973 (Jędrzejewska et al., 1997). So how does tree regeneration work only with wild browsing ungulates?

Tree regeneration without grazers but with wild ungulate browsers

Recent studies have revealed much about the role that the ungulate community is playing in shaping BPF. This area differs from the grazed pastures in several ways, but one obvious way is the dense cover of trees. There are very few open spaces in the National Park (see Figure 11.5). In fact, only 0.8 per cent of the area consists of open grassland (Michalczuk, 2001) and the rest is mainly a closed forest. There are open spaces inside the forest, but they are relatively small and usually originate from fallen single old trees creating gaps in the tree canopy. On average the canopy cover by large trees is more than 70 per cent (Kuijper et al., 2010b). The small-scale forest gaps form the crucial factor allowing tree regeneration.

After a canopy tree has fallen down, more light penetrates to the soil surface (Figure 11.2). The herbaceous vegetation directly responds to this and increases in cover. Likewise, tree saplings can be found in higher densities in these canopy gaps (Bobiec, 2007). Both factors increase forage availability for the herbivores that visit these small forest gaps much more intensively than the surrounding forest. When all ungulate species are combined, more than a two-fold higher visitation frequency occurs inside forest gaps, and especially red deer visit these places more than seven times longer than patches of closed forest (Kuijper et al., 2009). As a result, tree saplings growing in these forest gaps experience a higher level of browsing (Cromsigt and Kuijper, 2011; Kuijper et al., 2009). Hence, on the one hand increased light allows them to grow, but on the other hand increased herbivore browsing retards their growth. The constant browsing by ungulates in forest gaps leads to the development of a high density of short tree saplings. Trees are typically 'cut' at a height of 50–100cm (Cromsigt and Kuijper, 2011), as if they were a nicely maintained hedge (Figure 11.2). This height fits very well with the preferred foraging height of the dominant herbivore in the system, red deer (Renaud *et al.*, 2003). Next to keeping these regenerating trees at a certain height, ungulate browsing influences the species composition. Although many tree species occur inside these regeneration gaps, there is one species clearly dominant: hornbeam (*Carpinus betulus*) (Bobiec, 2007; Cromsigt and Kuijper, 2011). An easy explanation for this would be that the ungulates do not like to forage on this species. However, the contrary is true. Hornbeam constitutes an important food plant for most of the ungulates in the system (Kuijper, 2011). The reason why hornbeam is dominating in these forest gaps is because it is highly tolerant to browsing. Whereas many other tree species cannot cope with continuous browsing, hornbeam will continuously re-sprout and develop into a hedge-like structure (Cromsigt and Kuijper, 2011). This seems to be its key to success in this forest.

Figure 11.2 Natural and anthropogenic hedges: ungulate browsing inside forest gaps in the Białowieża primeval forest leads to the development of hedge-like structures mainly of hornbeam as in the Strict Reserve (1); this process resembles the constant cutting of hornbeam into a hedge as in this 25 year-old cut hedge at a farm in Varsen, Netherlands (2), demonstrating hornbeam's high tolerance to browsing
Source: D. Kuijper

Long-term exclosure studies carried out since 2001 in the BNP demonstrate that hornbeam is more tolerant to browsing than other species. Inside exclosures, which exclude ungulates by a large fence, a variety of tree species is present. Outside the exclosures, hornbeam is clearly the dominant species (Kuijper *et al.*, 2010a). It seems that this species is actually profiting from ungulate browsing, mainly because it removes the other competing tree species that are less resistant to browsing. This is demonstrated by another study that compared how tree-species composition is related to fluctuations in ungulate numbers. During the last 70 years, large fluctuations in the number of ungulates in BPF occurred (Jędrzejewska *et al.*, 1997). These fluctuations clearly affected the species composition of regenerating trees, as monitored on long-term permanent transects (Bernadzki *et al.*, 1998). With growing ungulate numbers, more browsing-tolerant tree species regenerated, such as hornbeam, lime (*Tilia cordata*) and elm (*Ulmus glabra*). Again, hornbeam was the species most strongly positively correlated with ungulate density (Kuijper *et al.*, 2010b).

There are clear effects of ungulate browsing both on species composition as well as the number of regenerating trees. Inside the above-mentioned exclosures, tree density was more than three times higher than in the presence of ungulates (Kuijper *et al.*, 2010a). However, trees are regenerating with ungulates present, even during peaks in their population density (Kuijper *et al.*, 2010a and b). Despite a high density of five different ungulate species and their concentration in forest gaps, these gaps are closed in by tall trees within around two to three decades (Figure 11.3). The 'hedges' mentioned earlier are not being maintained so intensively as to prevent the growth of trees. Often in the centres of these hedges, trees escape from the browsing of ungulates (Cromsigt and Kuijper, 2011). Once trees are taller than c.200 cm, ungulates do not nibble at their tops and they can grow into the canopy without problem. Self-thinning prevents all trees from becoming canopy trees; only the fast growers will make it to the top. Hence, ungulates in this forest can only retard the regeneration of trees but seem to be unable to maintain patches of open landscape.

Besides herbivores, humans shaped BPF

Besides large herbivores, humans have played an important role in shaping the composition of BPF. Therefore one should not look at this area as a wilderness area where human influences can be ignored. The first factor that is connected with human activity is the openness of the forest. According to data by Mitchell and Cole (1998), until the end of the eighteenth century the percentage of pollen characteristic for open areas (i.e. the fraction of non-arboreal pollen (NAP)) fluctuated around 10 per cent but never exceeded 12 per cent. As the limit value of this index for dense forest is 40 per cent (Birks, 2005), BPF makes a perfect example of closed-canopy dense forest. At the same time, it does not mean that there were no gaps in BPF cover during recent centuries. There were, partially created by drivers described above and in part by human activities. From studies of the environmental history of BPF, there are several examples of settlements in patches inside the forest in historical times. Archaeological excavation evidence shows that at least since the first century BC, small settlements were located in the forest, probably in natural or anthropogenic glades. So far, traces of two ancient (first century BC to first century AD), two early mediaeval (ninth to eleventh centuries) and one sixteenth-century settlements were discovered (Krasnodębski *et al.*, 2005, 2008 and 2011; Samojlik, 2007). Nevertheless, none of the discovered settlements lasted more than one or two centuries. Following the abandonment of these settlements in

Figure 11.3 Tree regeneration in browsed high forest of the Strict Reserve of BPF: collapse of old trees creates gaps in the canopy (1) and offers regeneration gaps for trees; intensive visitation by ungulates to forest gaps creates shortly browsed trees (2); only browsing-tolerant tree species (hornbeam) can sustain chronic browsing and will eventually escape out of reach of the ungulates (3); gaps are closed within several decades and turned into closed forest (4), which only open when a tree falls down
Source: D. Kuijper

combination with the cessation of anthropogenic factors opening the forest, all these glades returned into dense forest. There is only one exception to this general pattern. One glade in which a village existed c.2,000 years ago is now still devoid of trees because of its current hunting function (dating back to the nineteenth century) and the annual mowing of the entire glade by foresters (Samojlik, 2007). This illustrates that the present large herbivores alone cannot maintain patches in an open grassland state without human activities to support them.

In the modern period, dating from the first mention of Białowieża in the *Polish Chronicle* by Długosz (by the year 1409), the traditional utilisation of the forest was the main factor opening up the forest. Riverside meadows were scythed on a regular basis and forest glades, especially close to the forest border, were used for cattle pasturing. Furthermore, traditional beekeeping, which involved carving beehives in trees and attracting bees and then collecting honey and wax, was the cause of frequent, small-scale forest fires. In the seventeenth century this fire activity was strengthened by the introduction of new ways of utilisation: potash, wood-tar, birch-tar and charcoal burning (Samojlik, 2005 and 2010; Samojlik and Jędrzejewska, 2004). All these activities combined, meant that in the period from the end of the sixteenth century to the end of the eighteenth century, small fires occurred in BPF every 7–13 years. This was confirmed by a recent dendrological study (Niklasson et al., 2010). These different traditional activities most probably created several open glades with single old trees – in such places BPF would then resemble Western European forests such as the New Forest (Figure 11.4).

Livestock pasturing was also a part of the traditional utilisation of the forest in this area. Data on the number of livestock and the extent of the area was used for pasturing in BPF are fragmented, as mentioned by Jędrzejewska et al. (1997). Still, the available information gives indications of periods in which domestic ungulates could have played a significant role in the forest. The first – 'royal forest' period – lasted from the fourteenth to the end of the eighteenth century, when the forest served as hunting ground of kings. From detailed descriptions of human activities inside BPF from 1559 (VAK, 1867) and 1639 (VAK, 1870), listing several types of traditional forest uses allowed by the king's access rights in the specific parts of the forest, one can learn much of the forest uses. The most widespread form of allowed utilisation was haymaking in forest meadows and river valleys (27 out of 44 so-called access areas). Cattle are mentioned for only four of those access areas, in which their users were granted another privilege of feeding and keeping cattle during winter on those meadows, where hay is scythed (Samojlik, 2010; VAK, 1867). Livestock grazing during the vegetative season was not mentioned as being allowed in the inner part of the forest in the royal period. What supports this conclusion is the distance between access areas inside the forest and the villages that had the right to utilise them, which ranged between 5km and 55km (on average 27km) (Samojlik and Jędrzejewska, 2004). This distance makes it unlikely that cattle were walking back and forth to the village, but instead were kept in winter only at some designated places in the forest. Furthermore, royal commissioners sent to the forest every few decades officially announced that 'in the forest separated for game animals and in the backwoods no one is allowed to pasture their livestock, especially with dogs' (CAHR, 1700). What was present, though, was cattle pasturing on the edge of the forest in the vicinity of the settlements. Despite limited evidence that livestock pasturing took place in the forest in this period, Harnak (1764) observed that cattle was being driven for a mile or more into the forest and they destroyed shoots and shrubs. His proposal to restrain cattle pasturing to a quarter of a mile (1.8km) into forest interior (Harnak, 1764) suggests that at the end of this period there was a growing use of the forest for livestock pasturing and a growing concern about the impact of this. The second period started after the Russian administration of BPF in 1795 and a law prohibiting livestock pasturing inside the forest was issued. Whether it was obeyed is a matter of discussion. Official reports and articles on the forest that appeared in ministerial periodicals all sounded very similar: '[beaters and riflemen] have plots and pastures measured near their settlements, and they are not allowed to cross their borders' (Ronke, 1830) or 'cattle pasturing is strongly limited and allowed only in places devoid of European bison' (Bobrovski, 1863). Most probably livestock grazing was present only in the pastures near settlements, but the new villages inside BPF (dating back to the second half of the eighteenth century), as well as growing human population dwelling on the edge of the forest, caused a significant pressure of domestic animals on the forest. In 1875, according to Genko (1902–3) and Karcov (1903), livestock grazing posed such a threat to European bison population (both by food competition and transmission of diseases) that wooden fences were built to separate the inner, 'bison'-part of the forest with the outer parts used for cattle pasturing. Statistics cited by Jędrzejewska et al. (1997) estimate the total number of cattle pastured in BPF at the end of nineteenth century as up to 8,300, concentrated in the clearings around settlements.

In the third period, starting from 1915, two World Wars, persecution of local inhabitants, transportation of entire families to the east and decline in human population resulted in substantial decline of livestock pasturing in BPF (Jędrzejewska et al., 1997). In the Polish part of the forest, livestock pasturing was practiced until 1973, whereas, in the Belarussian part, a small number of livestock is still allowed to graze in open meadows alongside rivers.

Summing all of the above, in the first period (fourteenth–eighteenth centuries) almost no livestock were present inside the forest (with the exception of winter feeding on four access areas). Only by the end of this period did pasturing on the edge of the forest rise to an extent observed as destructive for the forest and game. The second period (1795–1915) was characterised by a large increase in the total population of livestock in the area and likely in the area too. Officially, livestock pasturing occupied 38 per cent of the area of the BPF (Jędrzejewska et al., 1997) and most likely caused a substantial problem for forest regeneration. In the third period (1915–present), livestock pasturing rapidly declined in the area and eventually disappeared from the Polish part of BPF.

From the above, it is clear that at least in the second period, when peak numbers of livestock recorded in the area were regarded by the administration as destructive for the forest, livestock grazing likely contributed to other human activities that created glades in the forest. This period coincides with similar concerns about problems with livestock pasturing observed in other parts of Europe, for example in Dutch woodlands (Dirkx, 1998).

The conclusion that it was mainly the anthropogenic activities that were responsible for opening the forest and maintaining its openness is supported by historical and current observations of areas deprived of human disturbance. In the second half of the seventeenth century, a series of wars and uprisings lasting until 1667 caused a drastic decline in the population of villages located on the border of the forest. A forest inventory from 1696 revealed that 26 per cent of all arable land was abandoned and was already overgrown with young forest, illustrating the speed at which forest returns when human activities are ceased. Similar patterns can be observed in the riverside meadows inside the forest. In the Polish part of the forest, the riverside meadows have not been scythed since 1970s–1980s, in contrast to the Belarussian part where utilisation of meadows persists until today. A contemporary Landsat-based map of BPF shows that there is a striking difference between Polish river valleys, which are almost completely closed by willow bushes and reed, whereas the Belarussian ones are still open meadows (DANCEE, 2002). Satellite images also show that glades created for game keeping and hunting purposes in the nineteenth century in the area of the contemporary Strict Reserve of BNP have all been quickly overgrown since their abandonment in 1915. These examples illustrate once again that when human interference is gone, large herbivores are not able to keep the forest open.

Implications for management

Grazed woodlands versus browsed high forest

The case study of the BPF as discussed above clearly demonstrates that humans have played a large role in shaping this forest system. The lack of evidence of the presence of extinct large grazing herbivores – aurochs and tarpan – raises the interesting question of whether these species where ever present here, or present in high enough densities to have a considerable impact on the landscape? The ongoing studies in this area show that the ungulate community, which is composed of mainly browsing herbivore species, does have important impacts on the forest systems. They influence the tree-regeneration process by reducing the amount of regenerating trees and influencing the species composition. Despite the occurrence of five large ungulate species they are not able to prevent tree regeneration altogether and maintain open glades. Hence, without human influences, open areas get overgrown by trees within a period of several decades. The example of BPF does not stand by itself. Several studies have indicated that also in prehistoric times, large herbivores have not been able to create a half-open landscape everywhere (Birks, 2005; Mitchell, 2005).

Throughout Europe, grazed woodlands can be found that have a long history of human use for the exploitation of wood and pastures for their livestock. These wooded pastures may have been a natural component of the landscape in prehistoric times when presence of tarpan and aurochs kept these half-open woodlands and promoted structural diversity in the vegetation (Vera, 2000). Contemporary livestock grazing is used as a substitute for the lost role of these large herbivores. The example of BPF and the evidence that large, now extinct, herbivore species were not present everywhere in our opinion asks for a broader view on management of woodlands. Their densities may have been high enough to have profound effects on plant communities or vegetation dynamics in some parts of Europe, such as their supposed preferred habitat –fertile riverine flatlands (Hall, 2008). However, they may have never occurred or only in low densities in other parts of Europe and there is general consensus that their role has been marginal in large parts of Europe where closed forests prevailed (Birks, 2005; Mitchell, 2005). Hence, a one-sided focus on the introduction of substitute-grazing species in nature management across Europe may be undesirable.

The studies from BPF demonstrate that the processes guiding the tree-regeneration are different in grazed woodlands versus browsed high forest (Figures 11.1 and 11.3). One of the underlying reasons seems to be the differences in herbivore community between these two types of systems. Whereas grazers such as cattle, horses and sheep dominate the grazed woodlands, browsers (red deer, roe deer and moose) and intermediate feeders (red deer, European bison) dominate in our studied forest system. Due to differences in behaviour and diet selection between these types of herbivores, their effects on the landscape also differ (see also Bobiec *et al.*, 2011a). Grazers typically occur in groups and intensively forage in open grasslands. In this way they can prevent tree regeneration. Browsers and intermediate feeders typically forage alone or in small groups and are generally more selective than grazers (Searle and Shipley, 2008). They often select their food on the level of individual plants. These differences may explain in large part why ungulates are unable to keep patches open inside our forest system. In addition one should consider the differences in density. Whereas livestock density typically ranges between 0.4 and 1.9 individuals/ha^{-1} in grazed woodland (Bakker *et al.*, 2004), peak density of all ungulates combined in our system is c.0.14 individuals/ha^{-1} (Jędrzejewska *et al.*, 1997 and 2002).

Grazed woodlands and browsed high forest; where and when?

Livestock grazing is commonly used in management in areas across Europe. The resulting landscapes from this management approach often have high aesthetic value and are appreciated by the general public. Open grasslands alternate with patches of solitary oaks and mosaics of shrubs (Figure 11.4). Also due to this high heterogeneity, often at small scale, these landscapes have high biodiversity. High diversity of insects, plants, birds and small mammal species find suitable foraging or reproduction habitats.

The typical landscape of a browsed high forest contrasts greatly with those of grazed woodlands. In general browsed high forests consist of rather closed forest with only small patches of open habitats (forest gaps).

Figure 11.4 Watercolour by Jan Henryk Müntz *Białowieża Forest – the Bear Hunt* (1784); note the dense forest just outside the border of the anthropogenic glade here
Source: Prints Office of the Library of Warsaw University, Inw. G. R.829, k. 36

Also these habitats offer opportunities for light-demanding species such as oak to regenerate (Bobiec *et al.*, 2011b; Kuijper *et al.*, 2010b). However, due to the differences in underlying mechanisms, the oaks growing in these areas look entirely different. Whereas, relatively low oaks with large crowns are typical for grazed woodlands (Figure 11.4), tall oaks with narrow crowns are characteristic for browsed high forest (Figure 11.5). Also the branching pattern differs. Whereas oak in grazed woodlands have low branches (Figures 11.4 and 11.5), oaks in high forest have high branches, indicating that they regenerated in a more closed habitat.

Of course the aesthetic value or beauty of a landscape is open for debate. Not in favour of the high forest is the finding that people often prefer relatively orderly landscapes above more natural, wild landscapes (Gundersen and Frivold, 2008; Hull and Revell, 1989; Schroeder and Daniel, 1981). There is a general tendency to rate the landscapes more highly that have a more pronounced human influence. However, the aesthetic value of a landscape should not be the only argument for choosing for a certain management regime. Regarding the conservation value, it is hard to rank which landscape is of higher value, grazed woodland or browsed high forest. Both have their own set of typical species associated with the characteristic habitats. Regarding our example of the high forest of BPF, its fauna harbours over 12,000 species, many of which cannot be found elsewhere (Jaroszewicz, 2004).

Figure 11.5 Oaks in grazed woodland and browsed high forest: (1) example of a woodland grazed by livestock (horses and cattle), the Borkener Paradies in Germany, with typical large oak trees with low branches and wide crowns; these landscapes often have high aesthetic value as well as harbouring high biodiversity; (2) example of a browsed high forest with wild ungulates only (red deer, roe deer, wild boar, moose and European bison) in BPF in Poland; typical are the tall oak trees with long stem, high branches and narrow crowns – these landscapes have their own beauty and offer hotspots of biodiversity

Source: Tomasz Samojlik and Dries Kuijper

In our opinion we should not get trapped in a discussion on how the prehistoric landscape in Europe looked; was it dominated by a part-open landscape or was it rather a closed forest? Of course it is highly relevant to think about the proper reference system that should guide management actions. However, it is not that only one view is correct and we should superimpose this view everywhere. Most likely both views are correct. The prehistoric landscape in Europe most likely consisted of large stretches of closed high forest dominated by browsing ungulates (Birks, 2005; Mitchell, 2005) interspersed by open or part-open landscapes dominated by grazing herbivores (Vera, 2000). That grazing large herbivores were not everywhere present, or clearly had habitats where they concentrated, has been shown by the locations of bones and fossil remains (Hall, 2008). Hence, it is not about choosing which view is right; it is all a matter of scale. In some areas a wooded pasture may be the best reference; in others it is a high forest.

Areas where the herbivore community was naturally dominated by grazers are likely in the low-elevation areas, such as river valleys and deltas, salt marshes (Hall, 2008) and areas with low rainfall that prevent trees from growing, such as steppes. Management practices should in these areas also focus on using livestock as a substitute of grazing herbivore species to mimic best the natural processes. However, in the higher and drier areas, and those far away from rivers or coast, a closed high forest seems to be the most natural reference. In these areas livestock should not be introduced but one should strive to get herbivore communities dominated by browsers.

Nowadays, there is a pronounced focus on the use of livestock in nature management with the idea of introducing 'missing' components to make the system more complete. We argue that in many areas this may indeed be true; in others we may create a less natural situation. In the case of BPF, it shows that wild ungulates can also play an important role in shaping certain landscapes. It shows that we should not regard livestock grazing as a necessary component to mimic natural processes. In many areas one would better mimic the natural processes by not introducing large herbivores and instead by relying on wild (mainly browsing) ungulates only. In conclusion, in our opinion a one-sided focus on introducing livestock grazing in woodlands is unhelpful. We argue that there is more to gain by diversifying our management practices and aiming for part-open landscapes maintained by grazers in some areas, while aiming at closed old-growth forest maintained by browsers elsewhere.

References

Bakker, E.S., Olff, H., Vandenberghe, C., De Maeyer, K., Smit, R., Gleichman, J.M. and Vera, F.W.M. (2004) 'Ecological anachronisms in the recruitment of temperate light-demanding tree species in wooded pastures', *Journal of Applied Ecology*, 41: 571–82.

Bennett, D. and Hoffman, R.S. (1999) 'Equus caballus', *Mammalian Species*, 628: 1–14.

Bernadzki, E., Bolibok, L., Brzeziecki, B., Zajączkowski, J. and Żybura, H. (1998) 'Compositional dynamics of natural forests in the Białowieża National Park, northeastern Poland', *Journal of Vegetation Science*, 9: 229–38.

Birks, H.J.B. (2005) 'Mind the gap: How open were European primeval forests?', *Trends in Ecology and Evolution*, 20(4): 154–6.

Bobiec, A. (2007) 'The influence of gaps on tree regeneration: A case study of the mixed lime-hornbeam (*Tilio-Carpinetum* Tracz. 1962) communities in the Białowieża Primeval Forest', *Polish Journal of Ecology*, 55: 441–55.

Bobiec, A., van der Burgt, H., Meijer, K., Zuyderduyn, C., Haga, J. and Vlaanderen, B. (2000) 'Rich deciduous forests in Białowieża as a dynamic mosaic of developmental phases: premises for nature conservation and restoration management', *Forest Ecology and Management*, 130: 159–75.

Bobiec, A., Kuijper, D.P.J., Niklasson, M., Romankiewicz, A. and Solecka, K. (2011a) 'Oak (*Quercus robur* L.) regeneration in early successional woodlands grazed by wild ungulates in the absence of livestock', *Forest Ecology and Management*, 262: 780–90.

Bobiec, A., Jaszcz, E. and Wojtunik, K. (2011b) 'Oak (*Quercus robur* L.) regeneration as a response to natural dynamics of stands in European hemiboreal zone', *European Journal of Forest Research*, 130: 785–97.

Bobrovski, P.O. (1863) *Materiały dla geografii i statistiki Rossii, sobrannye ofitserami generalnaho shtaba. Grodnenskaya Gubernya. Chast pervaya*, St. Petersburg, Drukarnya Departamenta Gheneralnoho Shtaba, pp404–59 (in Russian).

Burrichter, E., Pott, R., Raus, T. and Wittig, R. (1980) 'Die Hudelandschaft *Borkener Paradies* im Emstal bei Meppen', *Abhandlungen aus dem Landesmuseum für Naturkunde zu Münster in Westfalen*, 42: 1–69.

CAHR (Central Archives of Historical Records) (1700) *Commission in Białowieża, Sokółka, and Nowy Dwór Forests during the glorious reign of His Majesty King August II, hereditary Prince and Elector of Soxony (...) in Anno praesenti 1700 finished* (1700), Archiwum Kameralne I–12, mf 44338, Central Archives of Historical Records, Warsaw (in Polish).

Coops, H. (1988) 'The occurrence of blackthorn (*Prunus spinosa* L.) in the area of Mols Bjerge and the effect of cattle and sheep grazing on its growth', *Natura Jutlandica*, 15: 169–76.

Cromsigt, J.P.G.M. and Kuijper, D.P.J. (2011) 'Revisiting the browsing lawn concept: Evolutionary interactions or pruning herbivores?', *Perspectives in Plant Ecology, Evolution and Systematics*, 13: 207–15.

DANCEE (Danish Cooperation for Environment in Eastern Europe) (2002) *Białowieża Forest. Forest classification based upon Landsat ETM, August 2002*, Danish Cooperation for Environment in Eastern Europe, Ministry of Environment, Prins Engineering (map).

Dirkx, G.H.P. (1998) 'Wood-pasture in Dutch common woodlands and the deforestation of the Dutch landscape', in K.J. Kirby and C. Watkins (eds) *The Ecological History of European Forests*, CAB International, Wallingford, pp53–62

Dzięciołowski, R. (1970) 'Foods of the red deer as determined by rumen content analyses', *Acta Theriologica*, 15(6): 89–110.

Faliński, J.B. (1986) *Vegetation Dynamics in Temperate Lowland Primeval Forests: Ecological Studies in Białowieża forest*, Dr. W. Junk Publishers, Dordrecht, Netherlands.

Gębczyńska, Z. (1980) 'Food of the roe deer and red deer in the Białowieża Primeval Forest', *Acta Theriologica*, 40: 487–500.

Gębczyńska, Z., Gębczyńska, M. and Martynowicz, E. (1991) 'Food eaten by free-living European bison in Białowieża Forest', *Acta Theriologica*, 36: 307–13.

Genko, N. (1902–3), 'Charakteristika Belovezhskoi Pushchi i istoricheskiya o nei dannyya', *Lesnoi Zhurnal*, 22: 1014–56; 22: 1269–302; 23: 22–56 (in Russian).

Genov, P. (1981) 'Food composition of wild boar in North-eastern and Western Poland', *Acta Theriologica*, 10: 185–205.

Gundersen, V.S. and Frivold, L.H. (2008) 'Public preferences for forest structures: A review of quantitative surveys from Finland, Norway and Sweden', *Urban Forestry and Urban Greening*, 7: 241–58.

Hall, S.J.G. (2008) 'A comparative analysis of the habitat of the extinct aurochs and other prehistoric mammals in Britain', *Ecography*, 31: 187–90.

Hampicke, U. and Plachter, H. (2010) 'Livestock grazing and nature conservation objectives in Europe', in U. Hampicke and H. Plachter (eds) *Large-scale Livestock Grazing. A Management Tool for Nature Conservation*, Springer-Verlag, Berlin-Heidelberg, pp3–25.

Harnak, G. (1764) *Summariusz z Podatków Łowieckich* [*The Summary of the Hunting Taxes*], Lithuanian State Historical Archives in Vilnius, inventory no. SA 11575.

Hedemann, O. (1939) *L'histoire de la foret de Białowieża (jusqu'a 1798)*, Instytut Badawczy Lasów Państwowych, Rozprawy i Sprawozdania Seria A, Nr 1, Warsaw (in Polish with French summary).

Hofmann, R.R. (1989) 'Evolutionary steps of ecophysiological adaptation and diversification of ruminants: A comparative view of their digestive system', *Oecologia*, 78: 443–57.

Hull R.B. and Revell, G.R.B. (1989) 'Cross-cultural comparison of landscape scenic beauty evaluations: A case study in Bali', *Journal of Environmental Psychology*, 9: 177–91.

Jaroszewicz, B. (2004) 'Białowieża Primeval Forest – a treasure and a challenge', in B. Jędrzejewska and J.M. Wójcik (eds) *Essays on Mammals of Białowieża Forest*, Mammal Research Institute, Polish Academy of Sciences, Białowieża, pp3–12.

Jędrzejewska, B., Jędrzejewski, W., Bunevich, A.N., Miłkowski, L. and Krasiński, Z.A. (1997) 'Factors shaping population densities and increase rates of ungulates in Białowieża Primeval Forest (Poland and Belarus) in the 19th and 20th century', *Acta Theriologica*, 42: 399–451.

Jędrzejewski, W., Schmidt, K., Theuerkauf, J., Jędrzejewska, B., Selva, N., Zub, K. and Szymura, L. (2002) 'Kill rates and predation by wolves on ungulate populations in Białowieża Primeval Forest (Poland)', *Ecology*, 83: 1341–56.

Karcov, G. (1903) *Belovezhskaya Pushcha. Eya istoricheskii ocherk, sovremennoe okhotniche khozaistvo i Vysochaishie okhoty v Puchche*, A. Marks, St Petersburg (in Russian).

Kirby, K.J. (2004) 'A model of a natural wooded landscape in Britain as influenced by large herbivore activity', *Forestry*, 77(5): 405–20.

Kirby, K.J., Thomas, R.C., Key, R.S., McLean, I.F.G. and Hodgetts, N. (1995) 'Pasture woodland and its conservation in Britain', *Biological Journal of the Linnaean Society*, 56 (supply): 135–53.

Kowalczyk, R., Taberlet, P., Coissac, E., Valentini, A., Miquel, C., Kamiński, T. and Wójcik, J. M. (2010) 'Influence of management practices on large herbivore diet – case of European bison in Białowieża Primeval Forest (Poland)', *Forest Ecology and Management*, 261: 821–28.

Krasnodębski, D., Samojlik, T., Olczak, H. and Jędrzejewska, B. (2005) 'Early mediaeval cemetery in the Zamczysko range, Białowieża Primeval Forest', *Sprawozdania Archeologiczne*, 57: 555–83.

Krasnodębski, D., Dulinicz, M., Samojlik, T., Olczak, H. and Jędrzejewska, B. (2008) 'Cmentarzysko ciałopalne kultury wielbarskiej w Uroczysku Wielka Kletna (Białowieski Park Narodowy, woj. podlaskie)' ['A cremation cemetery of the Wielbark culture in Kletna range (Białowieża National Park, Podlasie province)'], *Wiadomości Archeologiczne*, 60: 361–76 (in Polish with English summary).

Krasnodębski, D., Olczak, H. and Samojlik, T. (2011) 'Wczesnośredniowieczne cmentarzyska Puszczy Białowieskiej' [Early mediaeval cemeteries in Białowieża Forest], in S. Cygan, M. Glinianowicz and P. Kotowicz (eds) *In silvis, campis... et urbe średniowieczny obrządek pogrzebowy na pograniczu polsko-ruskim*, Instytut Archeologii Uniwersytetu Rzeszowskiego, Rzeszów-Sanok, pp144–74 (in Polish with English summary).

Kuijper, D.P.J. (2011) 'Lack of natural control mechanisms increases wildlife-forestry conflict in managed temperate European forest systems', *European Journal of Forest Research*, 130: 895–909.

Kuijper, D.P.J., Cromsigt, J.P.M.G., Churski, M., Adam, B., Jędrzejewska, B. and Jędrzejewska, W. (2009) 'Do ungulates preferentially feed in forest gaps in European temperate forests?', *Forest Ecology and Management*, 258: 1528–35.

Kuijper, D.P.J., Cromsigt, J.P.G.M., Jędrzejewska, B., Miścicki, S., Churski, M., Jędrzejewska, W. and Kweczlich, I. (2010a) 'Bottom-up versus top-down control

of tree regeneration in the Białowieża Primeval Forest, Poland', *Journal of Ecology*, 98: 888–99.
Kuijper, D.P.J., Jędrzejewska, B., Brzeziecki, B., Churski, M., Jędrzejewski, W. and Żybura, H. (2010b) 'Fluctuating ungulate density shapes tree recruitment in natural stands of the Białowieża Primeval Forest, Poland', *Journal of Vegetation Science*, 21: 1082–98.
Mitchell, F.J.G. (2005) 'How open were European primeval forests? Hypothesis testing using palaeoecological data', *Journal of Ecology*, 93:168–77.
Mitchell, F.J.G. and Cole, E. (1998) 'Reconstruction of long-term successional dynamics of temperate woodland in Białowieża Forest, Poland', *Journal of Ecology*, 86: 1042–59.
Morow, K. (1976) 'Food habits of moose from Augustów Forest', *Acta Theriologica*, 21: 101–16.
Niklasson, M., Zin, E., Zielonka, T., Feijen, M., Korczyk, A.F., Churski, M., Samojlik, T., Jędrzejewska, B., Gutowski, J.M. and Brzeziecki, B. (2010) 'A 350-year tree-ring fire record from Białowieża Primeval Forest, Poland: Implications for Central European lowland fire history', *Journal of Ecology*, 98: 1319–29.
Olff, H., Vera, F.W.M., Bokdam, J., Bakker, E.S., Gleichman, J.M., De Maeyer, K. and Smit, R. (1999) 'Shifting mosaics in grazed woodlands driven by the alteration of plant facilitation and competition', *Plant Biology*, 1: 127–37.
Peterken, G.F. (1996) *Natural Woodland. Ecology and Conservation in Northern Temperate Regions*, Cambridge University Press, Cambridge.
Putman, R.J. (1986) *Grazing in Temperate Ecosystems: Large Herbivores and the Ecology of the New Forest*, Croom Helm, London.
Renaud, P.C., Verheyden-Tixiera, H. and Dumont, B. (2003) 'Damage to saplings by red deer (*Cervus elaphus*): Effect of foliage height and structure', *Forest Ecology and Management*, 181: 31–7.
Ronke de, E. (1830) 'Do Wielmożnego Jarockiego, profesora Królewskiego Warszawskiego Uniwersytetu. Niektóre uwagi względem Puszczy Białowieskiej' ['To the Honourable Mr Jarocki, professor of the Royal University of Warsaw. Some remarks concerning Białowieża Forest'], *Dziennik Powszechny Krajowy*: 84–5 (in Polish).
Rousset, O. and Lepart, J. (2000) 'Positive and negative interactions at different life stages of a colonizing species (*Quercus humilis*)', *Journal of Ecology*, 88: 401–12.
Samojlik, T. (ed.) (2005) *Conservation and Hunting. Białowieża Forest in the Time of Kings*, Mammal Research Institute PAS, Białowieża, Poland.
Samojlik, T. (2007) 'Antropogenne przemiany środowiska Puszczy Białowieskiej do końca XVIII wieku' [Anthropogenic changes of the environment of Białowieża Primeval Forest until the end of 18th century], unpublished PhD thesis, Mammal Research Institute PAS, Białowieża, Poland (in Polish).
Samojlik, T. (2010) 'Traditional utilisation of Białowieża Primeval Forest (Poland) in the 15th to 18th centuries', *Landscape Archaeology and Ecology*, 8: 150–64.
Samojlik, T. and Jędrzejewska, B. (2004) 'Użytkowanie Puszczy Białowieskiej w czasach Jagiellonów i jego ślady we współczesnym środowisku leśnym' [Utilization of Białowieża Forest in the times of Jagiellonian dynasty and its traces in the contemporary forest environment], *Sylwan*, 148(11): 37–50 (in Polish with English summary).
Schroeder, H.W. and Daniel, T.C. (1981) 'Progress in predicting the perceived scenic beauty of forest landscapes', *Forest Science*, 27: 71–80.

Searle, K.R. and Shipley, L.A. (2008) 'The comparative feeding behaviour of large browsing and grazing herbivores', in I.J. Gordon, H.H.T. Prins (eds) *The Ecology of Grazing and Browsing*, Ecological Studies 195, Springer, Berlin, pp201–16.

Smit, C., Gusberti, M. and Muller-Scharer, H. (2006) 'Safe for saplings; safe for seeds?', *Forest Ecology and Management*, 237: 471–7.

Tansley, A.G. (1922) 'Studies on the vegetation of the English Chalk II. Early stages in the redevelopment of woody vegetation on chalk grassland', *Journal of Ecology*, 10: 168–77.

Tubbs, C.R. (1986) *The New Forest*, Collins, London.

VAK (Vilenskaya Arkheogeograficheskaya Kimmissieya) (1867) *Inspection of forests and animal passages in the former Grand Duchy of Lithuania (...) of the 1559* (1867), Vilenskaya Arkheogeograficheskaya Kimmissieya, Vilnius (in Russian).

VAK (Vilenskaya Arkheogeograficheskaya Kimmissieya) (1870) *Ordinatia of the King's Białowieża and Kamieniec forestry in October 1639* (1870), Vilenskaya Arkheogeograficheskaya Kimmissieya, Vilnius (in Polish and Russian).

Van Vuure, C. (2005) *Retracing the Aurochs. History, Morphology and Ecology of an Extinct Wild Ox*, Pensoft, Sofia-Moscow.

Van Wieren, S.E. (1995) 'The potential role of large herbivores in nature conservation and extensive land use in Europe', *Biological Journal of the Linnaean Society*, 56: 11–23.

Vera, F.W.M. (2000) *Grazing Ecology and Forest History*, CAB International, Wallingford.

Watt, A.S. (1919) 'On the causes of failure of natural regeneration in British oakwoods', *Journal of Ecology*, 7: 173–203.

Williams, M. (2006) *Deforesting the Earth. From Prehistory to Global Crisis*, University of Chicago Press, Chicago.

12 The influence of grazing animals on tree regeneration and woodland dynamics in the New Forest, England

Adrian C. Newton, Elena Cantarello, Alexander Lovegrove, Dina Appiah and Lorretta Perrella

Summary

Vera's theory proposes that intense browsing pressure exerted by populations of large herbivores has historically had a major influence on the composition and structure of wooded vegetation in Europe. In addition, according to Vera (2000), high herbivore pressure leads to cyclical vegetation dynamics, in which tree regeneration is dependent on facilitation by spiny shrubs. Although the theory has had a significant impact on environmental policy and management, it has not been adequately tested. Here we present data from three field surveys undertaken in the New Forest, southern England, which examine tree regeneration in 1) woodland, 2) shrubland and 3) heathland. The implications of herbivory for woodland dynamics are also explored using a spatially explicit modelling approach. Results indicate that as hypothesised by Vera (2000), facilitation can be observed on woodland peripheries. However, tree regeneration was also widely observed in woodland and in heathland, where it is not necessarily dependent on the presence of spiny shrubs. Model outputs projected woodland expansion even in the presence of high herbivore pressure and without facilitation, a finding that is consistent with field observations at this site. These results highlight the need to identify with greater precision the situations under which Vera's theory is most likely to apply, and suggest the need for caution in using the theory to inform conservation management.

The theory developed by Vera (2000) has led to a major reappraisal of the role of large herbivores in woodland dynamics, reflecting the potential impacts of such herbivores on the process of tree regeneration. Described simply, the theory has two main elements. First, Vera (2000) proposed that for much of the post-glacial period, the intense browsing pressure exerted by large populations of herbivores could have maintained extensive areas largely free of tree cover in north-western and central Europe. The herbivores included *Bos primigenius* (aurochs), *Equus przewalski gmelini* (tarpan), *Bison bonasus* (European bison), *Cervus elaphus* (red deer), *Alces alces* (elk), *Capreolus capreolus* (roe deer), *Castor fiber* (beaver) and *Sus scrofa* (wild boar). This contradicts traditional successional theory, which suggested that this region was largely covered by dense forest (Clements, 1916; Ellenberg, 1988; Iversen, 1973; Tansley, 1935). Second, Vera (2000) proposed that under conditions of high herbivore pressure, vegetation dynamics would be cyclic, in which tree species would primarily regenerate outside woodland under the canopy of spiny shrubs such as *Prunus spinosa* (blackthorn), where they would be protected from herbivory. Groves of trees would therefore become established within shrub vegetation, which would mature over time and eventually collapse, owing to the prevention of tree regeneration under a forest canopy by herbivory. The groves of trees would therefore be replaced by grassland, which would subsequently be colonised by shrubs, reinitiating the vegetation cycle. The result would be a park-like mosaic of woodland and grassland, rather than closed woodland; a situation that would be driven and maintained by populations of large wild herbivores (Vera, 2000).

If Vera's theory is correct, it has major implications for conservation policy and management in Europe, particularly where the objectives are to manage or restore forest close to its natural condition prior to human impact (Mitchell, 2005). For example, the theory has supported development of the concept of 're-wilding', which involves the (re)introduction of populations of large herbivores that are allowed to roam freely over extensive areas to provide 'naturalistic grazing' (Hodder and Bullock, 2009). Examples of large-scale naturalistic grazing initiatives include the Oostvaardersplassen (Vera, 2009) and Veluwezoom National Park (Hodder *et al.*, 2005) in the Netherlands. Such efforts have inspired similar approaches in the UK (Newton *et al.*, 2009; Taylor, 2009), where the issue of re-wilding and its consequences is now recognised as having high policy relevance (Sutherland *et al.*, 2006). Similarly, the concept of 'Pleistocene re-wilding' is currently being explored in both North and South America (Donlan *et al.*, 2006; Galetti, 2004; Martin, 2005; Rubenstein *et al.*, 2006).

Vera's theory has stimulated a great deal of debate among communities of both researchers and management practitioners (Hodder *et al.*, 2005; Kirby, 2004; Mitchell, 2005; Rackham, 2003). Much of this debate has focused on interpretation of palaeoecological evidence, which has so far provided limited support for Vera's theory (Birks, 2005; Bradshaw *et al.*, 2003; Mitchell, 2005; Svenning, 2002; Whitehouse and Smith, 2010). However, relatively little research has been conducted on the impacts of herbivory on contemporary vegetation, with the explicit objective of testing Vera's theory (e.g. see Bakker *et al.*, 2004; Olff *et al.*, 1999).

This chapter describes some recent research undertaken in the New Forest, UK, which aimed to provide a partial test of Vera's theory. Specifically, the research involved a series of field surveys to examine patterns of tree regeneration in relation to herbivore pressure, and to evaluate the extent to which regeneration is dependent on facilitation, involving the protection of young trees from herbivory by spiny shrubs. The implications of herbivory for woodland dynamics were also explored, using a spatially explicit modelling approach.

Study area: The New Forest

The New Forest National Park, situated on the south coast of England, is the largest area of semi-natural vegetation in lowland Britain (Tubbs, 2001). Covering 57,100ha, it contains extensive areas of three habitats that are now the focus of conservation concern in lowland Europe, namely ancient woodland, heathland and mire wetlands (Tubbs, 2001). Because of its history as a 'Royal Forest' and the continuation of common rights of pasture, the New Forest has largely escaped the agricultural intensification that has transformed the wider landscape in much of Europe. As a medieval relict, the Forest is one of the few remaining pastoral landscapes in Europe; and owing to its size, high biodiversity value and the maintenance of traditional land-use practices it is one of the most significant. The area is cited by Vera (2000) as one of the principal examples that supports his theory.

The Forest has a long history of human settlement that has directly influenced its modern-day appearance. The most significant event in the Forest's history was when William the Conqueror designated the area as a 'Royal Forest', in about 1079 (Tubbs, 1968). The primary function of a Royal Forest was the preservation of deer and the vegetation they depended on; fallow deer (*Dama dama*) were probably introduced to the Forest at this time (Tubbs, 1968). At the same time, people living within the Forest boundary continued their traditional land-use practices, which revolved around five rights of common: those of pasture, mast, turbary, estovers and marl (Newton, 2011). Pasture is the grazing of livestock on common land; mast the use of pigs in autumn (also known as pannage) to consume acorns and beech mast; turbary the cutting of peat for fuel; estovers the collection of timber for firewood; and marl the removal of clay for building work (Tubbs, 2001). One of the most striking features of the New Forest is the long-term maintenance of such medieval land-use patterns, although modern commoning practices are rather more limited than in the past, and primarily involve the depasture of livestock (New Forest ponies and cattle). In recent years, about 550 commoners have typically depastured around 6,000–7,500 animals, whereas deer numbers have been maintained at around 2,000 (Newton, 2010a). Pannage still occurs to a limited extent. Much of the traditional cutting of gorse and heather has been replaced by modern habitat management, which uses different techniques.

The New Forest is of exceptional importance for biodiversity, as reflected in its many designations; for example, it is recognised as internationally important under the European Union Habitats Directive for the presence of nine habitats (Newton, 2010b). The species richness of many groups is high; for example, more than two thirds of the British species of reptiles and amphibians, butterflies and moths, fish, bats, dragonflies and damselflies are found in the New Forest (Newton, 2010b). The current high biodiversity value of the New Forest is attributable, at least in part, to the survival of traditional patterns of land use and the activities of large herbivores (Newton, 2012). The Forest can usefully be divided into the Open Forest, over which livestock and deer are free to roam, and the Silvicultural Inclosures, which are areas designated for timber production from which large herbivores may be excluded. Although some inclosures are fenced, this is not necessarily effective in excluding animals, and in recent years, there has been a trend towards removal of fences from many inclosures (Smith and Burke, 2010).

Field surveys

A series of three field surveys was conducted over the period 2005–10 to examine the impact of large herbivores on the vegetation of the New Forest. These included surveys of tree regeneration in: 1) woodland; 2) shrubland, in areas where woodland expansion is occurring; and 3) heathland, in chrono-sequences of recovery following burning. Results of each of these surveys are summarised below.

Survey 1: Woodland

Woodland structure and species composition were assessed in all the woodlands of the New Forest (Newton *et al.*, 2012), focusing on the woodland units (WUs) defined by Natural England for monitoring habitat condition. The survey therefore included both ancient pasture woodlands and plantations of exotic and native tree species, within both the Open Forest and the Silvicultural Inclosures. A total of 173 WUs was surveyed, using 50 x 50m plots, located randomly. Within each plot, the number of individuals of each tree species was counted, and the diameter at breast height (dbh) of each tree >10cm dbh was measured using a diameter tape. The total number of saplings in the plot was recorded (defined as trees <10cm dbh but >1.3m height), together with the total number of tree seedlings (i.e. individuals <1.3m height) of each species, which were recorded in a 10 x 10m subplot. In each plot, a series of ten variables was assessed by visual observation to provide a measure of browsing pressure, following Reimoser *et al.* (1999). The variables were bark stripping, shrub layer, presence of browse line, tree seedling and sapling projection in relation to ground vegetation height, evidence of trampling, evidence of browsing on seedlings, presence of patches of bare soil, dung, moss and ground vegetation height. The data obtained were divided into two groups, 'inclosed' and 'non-inclosed', based on whether or not the survey plots were located within Silvicultural Inclosures.

A total of 38 tree species was recorded as adult trees, of which 20 were classified as non-native. This reflects the widespread establishment of tree plantations in the New Forest, particularly in the inclosures. As adult trees

(>10cm dbh), nine exotic conifers were encountered only in the non-inclosed plots; the native broadleaved species *Carpinus betulus* (hornbeam) was similarly only encountered in non-inclosed plots. Nine tree species were found to occur at significantly higher densities as adults in the non-inclosed plots than in the inclosures, namely *Crataegus monogyna, Fagus sylvatica, Frangula alnus, Fraxinus excelsior, Ilex aquifolium, Malus sylvestris, Salix cinerea, Sorbus torminalis* and *Taxus baccata* (P<0.05 in each case, Mann-Whitney tests).

Saplings were recorded of for a total of 39 species, of which 19 were classified as non-native. This highlights the fact that a wide diversity of tree species are regenerating within the New Forest woodlands, including the majority of species present as adult trees. Mean sapling densities varied markedly between species, reaching maximum values of >2000 individuals/ ha^{-1} in the case of holly (*Ilex aquifolium*) and hawthorn (*Crataegus monogyna*) (Table 12.1). Both oak (*Quercus robur*) and beech (*Fagus sylvatica*), which are canopy dominants in most of the native woodlands of the New Forest, were both widespread as saplings in both inclosed and non-inclosed stands (Table 12.1). However, sapling densities of oak were higher in non-inclosed than inclosed sites. Five other species displayed a similar pattern to oak, namely *Crataegus monogyna, Frangula alnus, Ilex aquifolium, Prunus spinosa* and *Salix cinerea*. Conversely, five non-native species displayed higher sapling densities in inclosed sites, namely *Castanea sativa, Picea abies, Pinus sylvestris, Pseudotsuga menziesii* and *Tsuga heterophylla* (Table 12.1).

Results of the browsing survey indicated that all of the inclosed plots were characterised by at least moderate browsing pressure (total score 8–10), whereas all of the non-inclosed plots were associated with at least heavy browsing pressure (score ≥11). Median browsing score was significantly higher in the non-inclosed than in the inclosed plots (values of 18 and 14 respectively; $P<0.001$, Mann-Whitney U test). Overall the incidence of bark stripping was found to be significantly higher in the non-inclosed plots ($P=0.001$, Mann Whitney test), along with a less well developed shrub later ($P=0.004$), greater incidence of bare soil ($P=0.046$) and higher dung abundance ($P<0.001$). When sapling density was correlated with browsing impact score, significant relationships were found in *I. aquifolium* ($r=0.31$, $P<0.001$), *B. pendula* ($r=-0.25$, $P=0.001$), *Frangula alnus* ($r=0.18$, $P=0.019$), *P. sylvestris* ($r=-0.21$, $P=0.005$) and *Prunus spinosa* ($r=0.21$, $P=0.004$). In the case of seedling density, significant relationships were found only in *I. aquifolium* ($r=-0.40$, $P<0.001$), *B. pendula* ($r=-0.15$, $P=0.045$), *P. sylvestris* ($r=-0.21$, $P=0.006$) and *T. baccata* ($r=0.23$, $P=0.002$).

Survey 2: Shrubland adjacent to woodland

The study was conducted in June and July 2006, in three areas of the New Forest, as described by Appiah (2006). The study areas were Redrise Hill in the south-west of the New Forest, and Cadman Pool and Eyeworth Pool,

Table 12.1 Sapling densities (n ha⁻¹) of each tree species, within both inclosed and non-inclosed stands, assessed in the woodlands of the New Forest (Survey 1)

	Species name	Non-inclosed			Inclosed			P value (Mann-Whitney test)
		Mean	Standard error (SE)	Max.	Mean	SE	Max.	
*	Abies grandis	0	0	0	2	1	150	0.150
	Acer campestre	0	0	25	0	0	0	0.084
*	Acer pseudoplatanus	0	0	25	0	0	4	0.782
	Alnus glutinosa	0	0	4	0	0	20	0.366
	Betula pendula	8	3	175	40	13	825	0.583
	Carpinus betulus	0	0	0	0	0	12	0.408
*	Castanea sativa	0	0	0	2	1	75	**0.040**
*	Chamaecyparis lawsoniana	0	0	0	0	0	25	0.241
*	Corylus avellana	1	1	50	7	7	750	0.574
	Crataegus monogyna	126	39	2078	35	13	1070	**0.016**
	Fagus sylvatica	20	8	525	17	5	384	0.892
	Frangula alnus	2	1	50	0	0	25	**0.000**
	Fraxinus excelsior	2	1	50	1	1	56	0.183
	Ilex aquifolium	248	63	3600	106	31	2025	**0.000**
*	Larix decidua	0	0	0	1	1	100	0.241
	Malus sylvestris	1	1	25	0	0	25	0.100
*	Picea abies	0	0	0	3	2	200	**0.040**
*	Picea sitchensis	0	0	0	3	2	175	0.095
*	Pinus nigra	0	0	0	1	1	60	0.061
*	Pinus sylvestris	1	1	44	13	7	600	**0.025**
*	Populus alba	0	0	25	0	0	0	0.224
	Prunus spinosa	6	5	325	0	0	28	**0.024**
*	Pseudotsuga menziesii	0	0	0	13	5	350	**0.000**
	Quercus robur	4	1	50	2	1	25	**0.049**
*	Quercus rubra	0	0	0	0	0	4	0.408
	Salix cinerea	1	1	25	0	0	4	**0.028**
*	Sorbus aria	0	0	4	0	0	25	0.782
	Sorbus aucuparia	1	1	44	0	0	8	0.183
	Taxus baccata	0	0	4	0	0	0	0.224
*	Tsuga heterophylla	0	0	0	2	2	152	**0.040**
	Ulex europea	0	0	4	0	0	0	0.224
	Viburnum opulus	0	0	8	0	0	0	0.241

Note: The final column presents the results of Mann-Whitney tests, for the difference in median values of sapling density between inclosed and non-inclosed plots. Values in bold are significant at $P \leq 0.05$. *indicates species not considered to be native to the study area.

which are both located in the north-west. The sites were selected using a stratified random sampling approach, as areas where trees were actively colonising areas of grassland or heathland adjoining a woodland edge. Candidate areas were identified using aerial photographs supported by field reconnaissance. Sample plots of 100 x 100m were each aligned along a woodland edge, and laid out perpendicular to the edge. Within each sample plot, all patches of thorny scrub were measured as well as all individual adult trees, saplings and seedlings. Individuals were classified as seedlings if they were <1.5m tall; saplings if they were >1.5m tall and <10cm dbh and as adult trees if they were ≥10cm dbh and >1.5m tall.

Patches of thorny scrub were systematically surveyed by identifying all the plant species present and recording the number of tree seedlings and saplings encountered. The relative abundance of each thorny shrub species present in the patches was also estimated by recording the canopy cover of each species present in the patches. The sizes of all patches were obtained by measuring the lengths and breadths of each patch.

Stem densities of tree species varied between the three plots, with values of 781, 3344 and 2091 individuals/ha^{-1} recorded in Plots 1 (Cadnam's Pool), 2 (Eyeworth Pool) and 3 (Redrise Hill) respectively. Corresponding densities of spiny shrub species were 345, 286 and 359 individuals/ha^{-1}. Overall, alder buckthorn (*Frangula alnus*) was the most abundant tree species present, followed by oak (*Quercus robur*) and silver birch (*Betula pendula*) (Table 12.2). More tree individuals were associated with holly (*Ilex aquifolium*) and bramble (*Rubus fruticosus* agg.) than the other thorny species present, with totals of 1594 and 1063 respectively, although the relative importance of different shrub species in terms of their role in facilitation differed among sites.

One aspect that has received little attention in previous research is the spatial characteristics of patches of thorny shrubs. As noted by Vera (2000), patches of such shrubs could potentially expand with time, for example through the clonal spread of species such as blackthorn (*Prunus spinosa*). The size and shape of a patch of shrubs may influence the degree of accessibility to a browsing herbivore, and consequently determine its effectiveness in terms of facilitation. Potentially, there may be critical thresholds of patch size above which facilitation is more likely to occur. In the current investigation, mean patch size was found to vary both within and among species (Table 12.3), varying from 14.8m^2 in hawthorn (*Crataegus monogyna*) and holly, to more than 25m^2 in the case of gorse (*Ulex europea*). Conversely, holly was by far the most numerous of the thorny shrubs in the areas surveyed, in terms of the total number of patches recorded. Shrub species differed substantially in the number of tree stems recorded per patch, values ranging from 2.4 in rose (*Rosa* spp.) to 51 in the case of crab apple. However, when analysed by linear regression, the relationship between mean patch size and mean number of stems per patch was not found to be significant ($P > 0.05$).

Table 12.2 Density of regeneration of woody species associated with thorny shrubs, in areas of woodland expansion

(a) Plot 1 (Cadman's Pool)

Thorny shrub species	Tree species (number of individuals)				
	Alder Buckthorn	Beech	Oak	Silver Birch	Yew
Blackthorn	2	9	81	0	0
Bramble	2	2	34	1	0
Crab apple	0	7	76	0	1
Gorse	2	1	22	0	0
Hawthorn	0	3	50	0	1
Holly	4	26	131	2	1
Rose	2	0	4	0	0

(b) Plot 2 (Eyeworth Pool)

Thorny shrub species	Tree species (number of individuals)							
	Alder Buckthorn	Beech	Oak	Rowan	Scots Pine	Silver Birch	White-beam	Yew
Bramble	637	12	55	80	1	22	30	32
Crab apple	209	0	1	0	0	2	0	0
Gorse	9	0	3	0	0	0	0	0
Hawthorn	72	0	0	0	0	0	0	0
Holly	821	14	98	93	3	41	31	33

(c) Plot 3 (Redrise Hill)

Thorny shrub species	Tree species (number of individuals)					
	Alder Buckthorn	Beech	Oak	Rowan	Scots Pine	Silver Birch
Bramble	14	2	42	28	2	67
Crab apple	0	2	4	0	0	5
Gorse	15	2	54	31	0	134
Hawthorn	0	2	8	0	0	5
Holly	21	2	82	33	3	155
Rose	4	0	19	0	0	4

Survey 3: Heathland subjected to rotational burning

A further survey was undertaken of tree regeneration, focusing on heathland (Perrella, 2009). Three sites were selected that contain a mosaic of different ages since the occurrence of controlled burns, which are currently undertaken as part of heathland management in the New Forest. The sites were Bolderwood, Hasley Inclosure and Holm Hill, and were

Table 12.3 Characteristics of patches of thorny shrubs in relation to the regeneration of tree species, assessed in Survey 2 (data pooled across the three survey plots)

Thorny shrub species	Characteristics of patches of thorny shrubs				
	Mean patch size (m^2) ($\pm$ SD)	Patch size range (m^2)	Total number of patches recorded	Total number of tree stems recorded	Mean number of tree stems per patch
Blackthorn	18.1± 14.9	1.60–42.6	12	92	7.7
Bramble	17.5±15.8	0.3–75.7	61	1,063	17
Crab apple	23.6±12.5	11.0–39.7	6	307	51
Gorse	25.5±39.9	2.7–192.2	23	273	12
Hawthorn	14.8± 14.2	0.4–42.6	19	141	7.4
Holly	14.8± 21.7	0.4–192.2	99	1,594	16
Rose	22.3± 20.1	2.5–75.7	14	33	2.4

identified through inspection of aerial photographs in a geographic information system together with maps of burned areas obtained from the Forestry Commission. Within each study site, four treatments were sampled based on the age of the controlled burn patches, giving three replicate chrono-sequences. These were surveyed in July and August 2008 using randomly located plots, which were each 2.5 x 2.5m in size. Twenty-five plots were randomly located within each treatment, giving a total of 300 plots. Areas of mire were avoided, as was a 10m buffer zone around main footpaths and tracks, to minimise disturbance from human recreation. The following variables were recorded during the field survey: 1) percentage cover of principal plant species, 2) growth phase of heather (*Calluna vulgaris*), 3) mean height of heather, 4) percentage cover of spiny shrub species, 5) browsing pressure, 6) presence/absence of individual tree species, and 7) the number of seedlings of each tree species.

The growth phases of heather were differentiated as pioneer, building, mature or degenerate phases, following JNCC (2004). Browsing pressure was estimated on a scale of 1 (very high) – 5 (no evidence) by recording the abundance of pony, deer and cattle faeces, damage to vegetation (loss of leaf tissue or damage to stem) and evidence of animal tracks or hoofprints. Saplings were defined as trees <10cm dbh but >1.3m height, and seedlings as individuals <1.3m height.

Browsing pressure was found to be significantly higher in treatment 1 (0 years since burn) than in the other treatments (P <0.001, Kruskal-Wallis test). A total of 153 tree seedlings and two saplings were recorded across all of the plots, but their distribution was uneven, 60 per cent being recorded from one of the three study areas (Holm Hill). Overall, seedlings or saplings were found in 23.3 per cent of plots, with the number per plot ranging from 0–26. Scots pine (*Pinus sylvestris*) was the most common (51 per cent), followed by birch (*Betula* spp.) (28.4 per cent) and oak (*Quercus robur*) (14.3 per cent). Other species recorded at relatively low abundance included *Sorbus aucuparia* (three seedlings), *Frangula alnus* (five seedlings) and one *Ilex aquifolium* regenerating from a burnt stump. Young trees were present in all treatments, but were most abundant in sites that had been burned ten years previously (48 per cent of observations), with fewest (3 per cent) in sites burned the previous year. Similarly, tree species were found associated with all growth stages of heather, but most (65 per cent) were found in the plots containing the building phase, an association that was statistically significant ($P = 0.001$, Kruskal-Wallis test).

The only spiny scrub species encountered in the areas surveyed was gorse (*Ulex europaeus*), which was present in 18.3 per cent of plots. A chi square test of association was performed, which showed that the association of tree seedlings and gorse was not statistically significant ($x^2 = 1.384$, d.f = 1, $P = 0.239$). Similar results were obtained when the different tree species were analysed individually. These results therefore provide no evidence that facilitation is supporting the colonisation of heathland by tree species. It should be noted that these survey plots differed from those in Survey 2, which were adjacent to woodlands, by being relatively isolated from potential seed sources. The distance of each plot from the nearest adult tree ranged from 0–204m, with an overall mean (±SE) of 54 (±3.2)m. Evidence of browsing damage was observed on 21 per cent of seedlings, but the results differed between species. Only 7.6 per cent of *Pinus sylvestris* seedlings had been browsed, compared to 31.8 per cent of *Betula* spp. and 59.1 per cent of *Quercus robur* seedlings.

Modelling of vegetation dynamics

One of the main challenges to fully testing Vera's theory is the need to monitor vegetation dynamics over long timescales, to determine whether cyclical dynamics take place as hypothesised. An alternative technique is to use some form of modelling approach, which would enable long-term forest dynamics to be explored. A wide variety of different process-based modelling approaches has been applied to the study of forests (Bugmann, 2001; Liu and Ashton, 1995; Newton, 2007; Porte and Bartelink, 2002). However, few modelling investigations have explicitly considered the potential role of large herbivores in the dynamics of forest landscapes.

Newton *et al.* (2012) describe the application of a spatially explicit model, LANDIS II, to explore the dynamics of a wooded landscape under high herbivore pressure. LANDIS II is designed to simulate the dynamics of forested landscapes through the incorporation of a variety of ecological processes, including succession, disturbance, competition and seed dispersal (Mladenoff, 2004; Scheller *et al.*, 2007). The LANDIS group of models employs an object-oriented modelling approach operating on raster maps, with each cell containing species, environment, disturbance and harvesting information. LANDIS models have been used to explore the dynamics of a wide variety of forested landscapes in different parts of the world (e.g. Cantarello et al., 2011; Newton et al., 2011; Scheller et al., 2007).

In the New Forest, the LANDIS II model was parameterised, calibrated and tested as described by Newton et al. (2012), then used to produce a series of modelled scenarios. Simulations were conducted for 300 years. Two forms of disturbance were explored in the scenarios: the mortality of young trees caused by the activities of large mammals ('browsing') and the effects of burning ('fire'), reflecting the use of fire in current heathland management. The time steps were set at ten years for tree succession, ten years for fire disturbance and one year for browsing. The scenarios were defined as follows: Scenario 1, no disturbance (neither fire nor browsing); Scenario 2, browsing only; Scenario 3, fire only; Scenario 4, fire plus browsing; Scenario 5, browsing, fire and protection from herbivory by presence of spiny shrubs.

Modelling results provided only limited support for Vera's theory (Newton et al., 2012). According to Vera (2000), regeneration of tree species should be limited to the periphery of woodlands under conditions of high browsing pressure, whereas both model outputs and field survey data (Survey 1) indicated that tree regeneration is widespread within woodlands. As a result, there was little tendency of woodland stands to break up with maturity and be replaced by grassland, as required for woodland dynamics to be cyclic. The model also enabled the role of facilitation by spiny shrubs in the pattern of woodland expansion to be explored. Vera (2000) hypothesised that woodland would expand concentrically around existing woodland patches. Although such a pattern was projected by the LANDIS II model, this was obtained in scenarios both with and without facilitation. The protection of tree regeneration by spiny shrubs was therefore found to have little effect on the pattern of woodland spread, although it did substantially increase the rate of expansion (Newton et al., 2012). For example, simulated woodland extent after 300 years was projected to be 30 per cent higher with facilitation than without (Scenarios 5 versus 4) (Table 12.4). Woodland expansion was found to occur in all of the scenarios explored (Table 12.4), even in the presence of high herbivore pressure and without facilitation, a result that is consistent with the results of the field surveys reported here.

Discussion

Vera's theory has been of undoubted value in challenging the prevailing orthodoxy relating to forest succession, and in stimulating debate regarding the ecological role of large herbivores in forest landscapes. The theory has also had a major influence on conservation policy and practice, supporting the widespread (re)introduction of large herbivores as a conservation management approach (Hodder et al., 2005). For example,

Table 12.4 Projected extent of occurrence of principal tree species under different disturbance regimes

Species	Initial value	Scenario				
		1	2	3	4	5
Betula pendula	7,126	1,148	698	1,741	577	3,389
Fagus sylvatica	6,658	2,6612	15,038	18,638	12,314	17,378
Ilex aquifolium	5,909	7,079	3,049	4,057	2,099	4,891
Pinus sylvestris	5,051	2,086	5,812	1,683	3,204	3,116
Quercus robur	8,566	13,505	9,117	11,642	8,776	12,059
Combined	10,326	28,370	22,678	21,248	17,671	22,909

Note: Values presented are the areas (ha) occupied by each species, represented by the occurrence of individuals ≥ 10 years old. The initial values are those at the onset of model scenarios; the values given under each scenario are those projected to occur after 300 years following the simulations described in the text. 'Combined' refers to the total area where one or more of these five species was present. Scenario 1, no disturbance (neither fire nor browsing); Scenario 2, browsing only; Scenario 3, fire only; Scenario 4, fire plus browsing; Scenario 5, browsing, fire and protection from herbivory by presence of spiny shrubs.

grazing animals are now being widely reintroduced to lowland heathland in the UK, with the aim of improving habitat condition and preventing succession to woodland (Newton *et al.*, 2009). Vera's theory has been used to inform the current management plans of many protected areas, including the New Forest itself (Newton, 2010b). However, Vera's theory has not been widely tested and therefore it should be applied with caution. The need for such caution is illustrated by the case of lowland heathland in the UK. Here there is very little robust evidence that introduction of herbivores will have the positive impacts that are anticipated (Newton *et al.*, 2009).

The results of the research presented here demonstrate that woodland expansion is likely to occur in the New Forest even under high herbivore pressure and in the absence of protection of young trees by spiny shrubs. The current results are supported by analyses of historic maps, which indicate that native woodland area increased substantially in the New Forest during much of the twentieth century, despite high herbivore numbers (Small and Haggett, 1972). Woodland expansion has continued in recent decades, as indicated by analyses of aerial photographs (C. Bradley, Forestry Commission, pers. comm.), despite coinciding with browsing pressures that are historically at a very high value (Newton, 2011). This raises questions regarding the ability of large herbivores to maintain areas free of trees in partially wooded landscapes, as hypothesised by Vera (2000). The current results are supported by palaeoecological evidence, which has consistently suggested that the original natural landscape of north-west Europe was predominantly closed Forest, with open areas likely to have been restricted or very localised (Birks, 2005; Bradshaw *et al.*, 2003; Mitchell, 2005; Svenning, 2002; Whitehouse and Smith, 2010).

One of the main uncertainties in Vera's theory is the role of herbivory in woodland collapse and the conversion of woodlands to grassland, which is required for cyclic dynamics to occur. The results of Survey 1, presented here, highlight the fact that tree regeneration is widespread in the New Forest woodlands, despite high herbivore pressure. This may be attributed, at least in part, to the protection of young trees within protective microsites, such as fallen branches or trunks and spiny shrubs occurring within woodlands, as documented by Morgan (1991). This suggests that the facilitation by spiny shrubs hypothesised by Vera (2000) is not limited to woodland peripheries, but can occur under a woodland canopy, reducing the likelihood of cyclical dynamics. However, there are areas of the New Forest where break-up of beech stands can be observed (Newton et al., 2010), such as in Denny Wood, as documented by Mountford and Peterken (2003). The role of herbivory in stand break-up therefore requires further analysis; evidence suggests that soil characteristics and climate change might also be influential (Newton et al., 2010).

The results of Survey 2 indicate an association between young trees and spiny shrubs at woodland margins, as hypothesised by Vera (2000). However, these results highlight the diversity of species that can be associated with facilitation processes in an area such as the New Forest; a total of eight tree species was observed in association with a total of seven spiny shrub species. While the emphasis in Vera's original account was primarily on the regeneration of oak in blackthorn scrub (Vera, 2000), these results highlight the potential complexity of facilitation processes operating at the community scale, which could usefully be the focus of future research. Further research is also required on the spatial dynamics of shrub cover. In particular, it would be useful to address how this varies in relation to site characteristics and browsing pressure. Information is needed on the relationship between the characteristics of shrub patches and their effectiveness in terms of supporting tree regeneration through facilitation. It may be that particular tree species are more likely to be associated with particular shrub species, or particular patch characteristics, such as size or shape.

The results of Survey 3 show that while facilitation can be observed in some areas of the New Forest, such as woodland margins, it is not necessarily occurring everywhere. Young trees, particularly of Scots pine, birch and oak, are widespread colonists of heathland, even in the absence of spiny shrubs. The current results highlight the preference of large herbivores for heathland areas immediately after a burn, which is often associated with an increase in abundance of relatively palatable grasses such as *Molinea caerulea* (Tubbs, 2001). Over time, because of competitive interactions with grass species, less palatable ericaceous shrubs will increasingly tend to predominate. The relatively low browsing pressure associated with relatively mature heathland may account for the higher abundance of young trees on such sites, especially of relatively unpalatable species, such as Scots pine. These results highlight the potential complexity of the facilitation process, with both browsing pressure and the distribution of shrubs varying both in time and space. Further research is required to understand fully such dynamics and the role of soil characteristics in determining the locations where facilitation is more likely to occur. For example, the spatial association of oak with blackthorn reported by Bakker *et al.* (2004) at a floodplain site in the New Forest may be limited to the relatively rich soils characteristic of such sites.

The current research highlights the potential value of integrating field survey data with ecological modelling approaches, for exploring the potential influence of herbivory in the dynamics of wooded landscapes. As noted by Weisberg *et al.* (2006), modelling approaches can be of particular value for exploring the many complex interacting factors influencing herbivore-vegetation interactions at the landscape scale. Few previous attempts have been made to apply modelling techniques specifically to explore Vera's theory. One example is provided by WOODPAM (Gillet, 2008), which employed a mosaic compartment modelling approach to examine the impacts of cattle grazing. In application of this model to a site in Switzerland, Gillet (2008) found little evidence of cyclical vegetation dynamics, in common with the results of the current investigation. Potentially, such models could be used to refine Vera's theory, by identifying situations under which facilitation and cyclical dynamics are more likely to occur. Such conditions might prevail under higher herbivore densities than those currently being experienced in the New Forest, although these are historically at a high value (Newton, 2011; Newton *et al.*, 2010) and are higher than other European sites where Vera's theory is believed to apply (Bakker *et al.*, 2004). Modelling techniques may also be of value for informing habitat-management approaches involving introduction of large herbivores by enabling the potential impacts of herbivory to be projected.

Acknowledgements

Thanks to Andrew Brown, Natalia Tejedor and Gillian Myers for assistance with the woodland survey, and Richard Reeves (New Forest Centre), Ed Mountford (JNCC), Jonathan Spencer, Berry Stone and Simon Weymouth (Forestry Commission) for assistance with accessing information about the New Forest.

References

Appiah, D.A.K. (2006) 'Comparing changes in land use on species and habitat aanagement in the New Forest using a geographical information system (GIS)', unpublished MSc dissertation, Bournemouth University, Bournemouth.

Bakker, E.S., Olff, H., Vandeneberghe, C., De Maeyer, K., Smit, R., Gleichman, J.M. and Vera, F.W.M. (2004) 'Ecological anachronisms in the recruitment of temperate light-demanding tree species in wooded pastures', *Journal of Ecology*, 41: 571–82.

Birks, H.J. (2005) 'Mind the gap: How open were European primeval forests?', *Trends in Ecology and Evolution*, 20: 154–6.
Bradshaw, R.H.W., Hannon, G.E. and Lister, A.M. (2003) 'A long-term perspective on ungulate-vegetation interactions', *Forest Ecology and Management*, 181: 267–80.
Bugmann, H. (2001) 'A review of forest gap models', *Climatic Change*, 51: 259–305.
Cantarello, E., Newton, A.C., Hill, R.A., Tejedor-Garavito, N., Williams-Linera, G., López-Barrera, F., Manson, R.H. and Golicher, D.J. (2011) 'Simulating the potential for ecological restoration of dryland forests in Mexico under different disturbance regimes', *Ecological Modelling*, 222: 1112–28.
Clements, F.E. (1916) *Plant Succession. An Analysis of the Development of Vegetation*, Publication No. 242, Carnegie Institution, Washington DC.
Donlan, C.J., Berger, J., Bock, C.E., Bock, J.H., Burney, D.A., Estes, J.A., Foreman, D., Martin, P.S., Roemer, G.W., Smith, F.A., Soulé, M.E. and Greene, H.W. (2006) 'Pleistocene rewilding: An optimistic agenda for twenty-first century conservation', *The American Naturalist*, 168: 1–22.
Ellenberg, H. (1988) *Vegetation Ecology of Central Europe*, Cambridge University Press, Cambridge.
Galetti, M. (2004) 'Parks of the Pleistocene: Recreating the cerrado and the Pantanal with megafauna', *Natureza e Conservação*, 2(1): 93–100.
Gillet, F. (2008) 'Modelling vegetation dynamics in heterogeneous pasture-woodland landscapes', *Ecological Modelling*, 2: 171–218.
Hodder, K.H. and Bullock, J.M. (2009) 'Really wild? Naturalistic grazing in modern landscapes', *British Wildlife*, June: 37–43.
Hodder, K.H., Bullock, J.M., Buckland, P.C. and Kirby, K.J. (eds) (2005) *Large Herbivores in the Wildwood and Modern Naturalistic Grazing Systems*, English Nature Research Reports No 648, English Nature, Peterborough.
Iversen, J. (1973) 'The development of Denmark's nature since the last glacial', *Danmarks Geologiske Undersøgelse*, 7: 1–126.
JNCC (Joint Nature Conservation Committee) (2004) *Common Standards Monitoring Guidance for Lowland heath, Version August 2004*, Joint Nature Conservation Committee, Peterborough.
Kirby, K.J. (2004) 'A model of a natural wooded landscape in Britain driven by large-herbivore activity', *Forestry*, 77: 405–20.
Liu, J.G. and Ashton, P.S. (1995) 'Individual-based simulation-models for forest succession and management', *Forest Ecology and Management*, 73: 157–75.
Martin, P.S. (2005) *Twilight of the Mammoth: Ice Age Extinction and the Rewilding of America*, University of California Press, Berkley.
Mitchell, F.J.G. (2005) 'How open were European primeval forests? Hypothesis testing using palaeoecological data', *Journal of Ecology*, 93: 168–177.
Mladenoff, D.J. (2004) 'LANDIS and forest landscape models', *Ecological Modelling*, 180: 7–19.
Morgan, R.K. (1991) 'The role of protective understorey in the regeneration system of a heavily browsed woodland', *Vegetatio*, 92: 119–132.
Mountford, E.P. and Peterken, G.F. (2003) 'Long term change and implications for the management of woodpastures: Experience over 40 years from Denny Wood, New Forest', *Forestry*, 76: 19–43.
Newton, A.C. (2007) *Forest Ecology and Conservation. A Handbook of Techniques*, Oxford University Press, Oxford.

Newton, A.C. (ed.) (2010a) *Biodiversity in the New Forest*, Pisces Publications, Newbury, Hampshire.

Newton, A.C. (2010b) 'Synthesis: status and trends of biodiversity in the New Forest', in A.C. Newton (ed.) *Biodiversity in the New Forest*, Pisces Publications, Newbury, Hampshire, pp218–28.

Newton, A.C. (2011) 'Social-ecological resilience and biodiversity conservation in a 900-year-old protected area', *Ecology and Society*, 16: 13.

Newton, A.C. (2012) 'Biodiversity conservation and the traditional management of common land: the case of the New Forest', in I. Rotherham (ed.) *Common Land: Cultural Severance and the Environment*, Springer Verlag, Berlin (in press).

Newton, A.C., Stewart, G.B., Myers, G., Diaz, A., Lake, S., Bullock, J.M. and Pullin, A.S. (2009) 'Impacts of grazing on lowland heathland: a systematic review of the evidence', *Biological Conservation*, 142: 935–47.

Newton, A.C., Cantarello, E., Myers, G., Douglas, S. and Tejedor, N. (2010) 'The condition and dynamics of New Forest woodlands', in A.C. Newton (ed.) *Biodiversity in the New Forest*, Pisces Publications, Newbury, Hampshire, pp132–47.

Newton, A.C., Echeverria, C., Cantarello, E. and Bolados, G. (2011) 'Impacts of human disturbances on the dynamics of a dryland forest landscape', *Biological Conservation*, 144: 1949–60.

Newton, A.C., Cantarello, E., Tejedor, N. and Myers, G. (2012) 'Dynamics of a woodland landscape under high herbivore pressure: a partial test of Vera's theory', *Biological Conservation*, (in review).

Olff, H., Vera, F.W.M., Bokdam, J., Bakker, E.S., Gleichman, J.M., De Maeyer, K. and Smit, R. (1999) 'Shifting mosaics in grazed woodlands driven by the alternation of plant facilitation and competition', *Plant Biology*, 1: 127–137.

Perrella, L. (2009) 'A study of the determinants of tree invasion on lowland heath in the presence of large herbivores', unpublished BSc dissertation, Bournemouth University, Bournemouth.

Porte, A. and Bartelink, H.H. (2002) 'Modelling mixed forest growth: A review of models for forest management', *Ecological Modelling*, 150: 141–88.

Rackham, O. (2003) *Ancient Woodland: Its History, Vegetation and Uses in England*, Castlepoint Press, Dalbeattie.

Reimoser, F., Armstrong, H. and Suchant, R. (1999) 'Measuring forest damage of ungulates: What should be considered', *Forest Ecology and Management*, 120: 47–58.

Rubenstein, D.R., Rubenstein, D.I., Sherman, P.W. and Gavin T.A. (2006) 'Pleistocene park: Does re-wilding North America represent sound conservation in the 21st century?', *Biological Conservation*, 132: 232–8.

Scheller, R.M., Domingo, J.B., Sturtevant, B.R., Williams, J.S., Rudy, A., Gustafson, E.J. and Mladenoff, D.J. (2007) 'Design, development, and application of LANDIS-II, a spatial landscape simulation model with flexible temporal and spatial resolution', *Ecological Modelling*, 201: 409–19.

Small, D. and Haggett, G.M. (1972) 'A study of broadleaved woodland changes and natural regeneration of broadleaves in the Ancient and Ornamental woodlands from 1867–1963', *New Forest, Forestry Commission Management Plan 1972–1981*, Appendix D, Forestry Commission, Lyndhurst.

Smith, J. and Burke, L. (2010) 'Managing the New Forest's Crown lands', in A. C. Newton (ed.) *Biodiversity in the New Forest*, Pisces Publications, Newbury, Hampshire, pp212–17.

Sutherland, W.J., Armstrong-Brown, S., Armsworth, P.R., Tom, B., Brickland, J., Campbell, C.D., Chamberlain, D.E., Cooke, A.I., Dulvy, N.K., Dusic, N.R., Fitton, M., Freckleton, R.P., Godfray, H.C.J., Grout, N., Harvey, H.J., Hedley, C., Hopkins, J.J., Kift, N.B., Kirby, J., Kunin, W.E., Macdonald, D.W., Marker, B., Naura, M., Neale, A., Oliver, T., Osborn, D., Pullin, A.S., Shardlow, M.E.A., Showler, D.A., Smith, P.L., Smithers, R.J., Solandt, J.-L., Spencer, J., Spray, C.J., Thomas, C.D. and Thompson, J. (2006) 'The identification of 100 ecological questions of high policy relevance in the UK', *Journal of Applied Ecology*, 43: 617–27.

Svenning, J.-C. (2002) 'A review of natural vegetation openness in north-western Europe', *Biological Conservation*, 7: 290–96.

Tansley, A.G. (1935) 'The use and abuse of vegetational concepts and terms', *Ecology*, 16: 284–307.

Taylor, P. (2009) 'Re-wilding the grazers: Obstacles to the 'wild' in wildlife management', *British Wildlife*, June: 50–5.

Tubbs, C.R. (1968) *The New Forest: An Ecological History*, David & Charles, Newton Abbott.

Tubbs, C.R. (2001) *The New Forest. History, Ecology and Conservation*, New Forest Ninth Centenary Trust, Lyndhurst.

Vera, F.W.M. (2000) *Grazing Ecology and Forest History*, CABI Publishing, Wallingford, Oxon.

Vera, F.W.M. (2009) 'Large-scale nature development – the Oostvaardersplassen', *British Wildlife*, June: 28–36.

Weisberg, P.J, Coughenour, M.B. and Bugmann, H. (2006) 'Modelling of large herbivore-vegetation dynamics in a landscape context', in K. Danell, P. Duncan, R. Bergström and J. Pastor (eds) *Large Herbivore Ecology, Ecosystem Dynamics and Conservation*, Cambridge University Press, Cambridge, pp348–82.

Whitehouse, N.J. and Smith, D. (2010) 'How fragmented was the British Holocene wildwood? Perspectives on the 'Vera' grazing debate from the fossil beetle record', *Quaternary Science Reviews*, 29: 539–53.

13 Forest and land management options to prevent unwanted forest fires

Caroline Boström, Ana Sebastián, Carmen Hernando Lara, Rosa Planelles, Armando Buffoni, Rosario Alves, Marielle Jappiot and Jesús San Miguel Ayanz

Introduction

Fire is one of the environmental risks expected to increase in connection with climate change. Wildfires will be more common in regions where today they are considered less likely, and the implications for formerly grazed landscapes may be significant. The European Union FireSmart project is a project whose objectives are to identify obstacles that hinder the effectiveness of forest fire preventive measures and to derive recommendations to integrate prevention practices in regular forest management plans. The project tackles both European and local levels of addressing forest fire prevention. The local level has been covered mainly through the implementation of test areas in France, Italy, Spain and Portugal. Documents containing information about different methods and practices of forest fire prevention in Europe have been gathered in a database, which now contains more than 1,400 entries. The material available in this database has been analysed according to the strengths, weaknesses, opportunities and threats (SWOT) of different land-management practices and for each of the five different aspects studied; agroforestry, fire causes, preventive silviculture, wildland–urban interfaces and awareness raising and training. The analysis was then used to derive practical recommendations on how to turn current negative fire prevention factors into viable and proactive factors able to strengthen prevention methods. Agroforestry was found to be a strong preventive method and the strengths and weaknesses of the method as well as the practical recommendations for increasing the preventive values of agroforestry are presented. The implications and possibilities for future wood pasture are considered in this chapter.

Forests are an important resource in Europe and as such of importance to environmental, economic and social values. Covering about 40 per cent of European land area, forests impact a great number of people. Fire is an integral part of these ecosystems, but can also become a threat when it affects biodiversity, economy and social values. It is expected that with climate change the number of fires will increase from an already high number of about 40,000 fires per year. Today wildfires in Europe can be linked to socioeconomic developments and the consequent change of life-habits, with mismanagement and abandonment of rural areas identified as two of the main drivers.

The EU-FireSmart is a European Framework 7 support action research project, with the main objective of identifying obstacles that hinder forest fire prevention and preventive measures. The project worked mainly with different land-management options that can increase forest fire prevention and on identifying the weak points to see where more efforts could be put to improve fire prevention as a whole (Boström *et al.*, 2011).

The FireSmart project started its work in February 2010 and had funding for a period of two years. The project was coordinated by GMV in Spain and had participants from the countries of France (CEMAGREF), Italy (Ambiente Italia), Spain (EIMFOR, INIA) and Portugal (Forestis), as well as from the European level (CEPF, JRC). The geographical spread as well as the different types of organisations involved in the project have made it possible to include different aspects of forest fire prevention in the project. The geographical spread has also made it possible to closely study fire prevention in four areas chosen by the project to serve as 'test areas' in France, Italy, Spain and Portugal. In these test areas the local and regional aspects of forest fire prevention have been studied more closely. The European perspective has then been added to give the analyses a broader scope.

To begin with, the project gathered scientific and technical documents in a database that now contains more than 1,400 entries related to different topics. The documents were gathered from all around the world, but with focus on Europe and especially within the four test areas chosen by the project. The topics especially studied by the project were agroforestry, fire causes, preventive silviculture, awareness raising and training as well as wildland–urban interfaces. The database is available for the general public on the project's website even after the end of the project.

The FireSmart project involved a questionnaire sent out to land managers and experts in the test areas and in the rest of Europe. These put questions to the respondents about what they thought were the most efficient or inefficient methods for forest fire prevention, the analyses of the questionnaires showed the weak and strong points of the different subjects studied within the project, in regard to forest fire prevention. The results from the questionnaires confirmed the results of the SWOT analyses and also achieved a wider input and expert thoughts on the subject of forest fire prevention. This text further develops the results derived from the analyses made of the strengths, weaknesses, opportunities and strengths found in the area of agroforestry as a method to prevent forest fires.

Agroforestry was one of the topics studied in particular by the FireSmart project as an option to prevent forest fires. Agroforestry can be defined as a sustainable land management approach that integrates both agricultural and forestry management practices in the same land area. Agroforestry and its practices have been defined by different authors as practices that involve the deliberate integration of trees with agricultural crops and/or livestock, either simultaneously or sequentially on the same plot of land. All agroforestry ecosystems integrate people as part of the systems, as they are artificial systems to a higher or lower extent. Agroforestry is a land-management option where one of the components, either forestry or agriculture, can be promoted over the other, or both simultaneously trying to reach an equilibrium (Mosquera *et al.*, 2009)

Agroforestry ecosystems are recognised as beneficial to the prevention of forest fires (Moreira *et al.*, 2009) since these ecosystems often have a lower fuel load due to the fact that the area below the trees is cleared of undergrowth. The lack of a complete canopy cover reduces the risk of intense crown fires. Agroforestry ecosystems also have the capacity to reduce fire spread and energy release in relation to fires, which allows forest fire-fighting personnel to close in on the fire and work in a safer environment. Lower risk exposure on fire-fighting occasions is something that helps to reduce damages caused by fires and can thus be considered a part of forest fire prevention measures.

Materials and methods

The documents collected by the Project Consortium in the project database were analysed by means of SWOT analysis. Strengths, weaknesses, opportunities and threats were identified and listed for the five topics. The subjects were first analysed from a general point of view and then more closely analysed from the legislative, institutional and socioeconomic aspects. The statements that were listed in the primary SWOT analyses were then quantitatively evaluated so as to get a clearer picture of which statements were of higher importance and which were not.

Through using the quantitative SWOT analysis it was possible to see the overall importance of the strengths, weaknesses, opportunities and threats for each of the subjects and themes analysed. By going through them the project could identify areas where more efforts should be put, as well as areas where fire prevention is already quite strong.

In addition to the SWOT analyses, a 96-item questionnaire had previously been designed to obtain information from the respondents about the efficiency and consequences of current management practices. The questions covered subjects such as restrictions, legal issues, social and communication-related activities in terms of wildfire prevention. The questionnaire was addressed to forest managers and scientists. From the survey 460 completed questionnaires were gathered, mainly from Spain (62 per cent), followed by Italy (13 per cent) and Portugal (10 per cent). Results concerning agroforestry and grazing as forest fire prevention measures are the focus of this chapter.

The results from the SWOT analyses were then further transformed into a comprehensive list of identified obstacles. This list was then used as one of the inputs to a list of derived recommendations, together with the SWOT analyses. The final list of recommendations also took the results from the analysis of the questionnaires into account.

Results

As noted, a lower density of undergrowth in agroforestry ecosystems leads to a lower fuel level and thus reduces the risk for intense fire events. Agroforestry ecosystems also have a low level of intense crown fires, and they can through their structure impact forest fire behaviour and reduce the risks of intense fires.

In addition, fires in agroforestry ecosystems can often be detected at an early stage since there are often shepherds or cattle-herders following their animals through these areas. Human presence is also higher in some regions with a high amount of agroforestry ecosystems because of the long tradition of agroforestry practices that has made the landscape and environment attractive to tourists and for touristic activities. The presence of tourists and tour guides can also contribute to an earlier detection of fires.

Human presence in an area is however double-edged, since over 95 per cent of the fires in the Mediterranean region are caused by humans. Many times this firing is in connection to the renewal of pastures and tourism. Legislation that governs the use of fire is in place throughout the Mediterranean region, but these laws are not always followed.

Multi-functional and sustainable use of agroforestry ecosystems can contribute to forest fire prevention. This is made not only by creating employment opportunities in rural areas, but also by contributing to landowners' incomes through the extraction of, for example, wood, firewood, cork or acorns. At the same time, a continuous income stream is maintained through the production of meat or agricultural crops.

In the FireSmart database most of the material gathered connected to agroforestry is focused on fuel management through grazing in fuel breaks. This can be an effective way of reducing the need for regular mechanical cleaning of these areas.

The FireSmart SWOT analysis found that one of the crucial weaknesses of agroforestry as a method to prevent forest fires was low profitability, which has led to decreased use of the method land management. Reduced use of agroforestry and land abandonment following this decrease leaves many areas with increased fuel loads at the same time as human presence in the areas decreases.

Maintenance of grazing lands

Renewal of pasture lands was also found to be a weakness in the SWOT analysis since this is often done without legal permission, and when pasture renewals get out of control due to lack of knowledge or experience they easily start forest fires. The involvement of land managers in forest fire prevention is important for achieving a favourable impact from different measures but it is often low. Increasing the participation of land managers in actions that aim to prevent forest fires could be greatly improved, and could in turn improve the efficiency of the preventive actions. Information that is aimed towards land managers, and that describe fire preventive measures in an understandable way, is needed. Land managers could also be involved more in the different schemes that aim to prevent forest fires. Since it is difficult to make this like other traditional agricultural methods of land management practice profitable, it is not easy to get younger people interested in starting up businesses. In order to address this, information should be provided about how land managers practicing agroforestry contribute to society, ecology and conservation simply by managing to stay in business.

Marketing, branding and opportunities

Branding schemes was something that emerged as a highly rated opportunity in the SWOT analysis. The use of different branding schemes so as to, for example, indicate the geographical origin of a product could be used in areas where agroforestry is a traditional land use with favourable impacts. Through increasing the economic benefit of keeping livestock or crops under trees, the abandonment of these lands could decrease and thereby help with the prevention of fires in these areas. Human and ecological values can thus be preserved without extra subsidy input, benefiting all parties involved.

In the agroforestry SWOT analysis of the FireSmart project, the lack of studies on agroforestry and how to adapt grazing practices at regional and local levels in the test area countries was identified as a weakness. In the areas studied by the project, another weakness was the lack of information about how agroforestry fits within the framework of sustainable land management.

The biggest opportunity found in the FireSmart SWOT analysis in relation to agroforestry areas was their capacity to decrease the speed of forest fires already ablaze. A greater areas used for agroforestry could therefore be very good for the overall prevention of forest fires in a region, especially if they were dispersed in a strategic way on the landscape level.

Land management in fire prone regions is important, and to increase the use of agroforestry as a land management method a higher value for the products from these ecosystems is vital. Another way to increase the value of agroforestry areas could be to create extra income for land managers through developing tourism activities. This could widen his or her income base, and increase the overall profitability of the land use.

Ecosystem services, biodiversity and land use as fire prevention

Biodiversity and environmental services are 'side-effects' of good agroforestry management. To give subsidies to aid landowners to develop these societal services could be an option to keep land management on these lands. Subsidies already exist for agroforestry as such in the Rural Development Programme of the European Union, and could be used more and in a more efficient way at member-state level.

When it comes to identified threats in the FireSmart SWOT analysis on agroforestry, the abandonment of rural areas and specifically agroforestry as a land management practice were considered most important.

The weaknesses of agroforestry and grazing as forest fire prevention measures were dominant with 47 per cent of the overall importance in the analysis. When analysed quantitatively, some of the strengths found in the general agroforestry SWOT analysis were in fact only quite weak, whereas the weaknesses found by the analysis were more significant. Strengths and opportunities do however make up almost 40 per cent of the overall importance. If these strengths and weaknesses are considered and the findings applied to improve management, they could make a significant difference. This in turn might give agroforestry a better chance to be a successful forest fire prevention method.

The institutional SWOT analysis on agroforestry and grazing as forest fire prevention methods showed the strengths and opportunities given a higher importance than the weaknesses and threats. The general importance of each of the strengths was also higher than for the other statements in the analysis. The most significant strength was found to be the encouragement given by the Spanish government to increase rural activities. This encouragement is based on trying to increase the participation of shepherds in fire prevention. In the Spanish test area this has led to agreements between shepherds and the land managers. The most significant weakness in the general analysis was also found to be in Spain, and concerned the lack of communication on agroforestry practices between different regions and regional institutions within the country. Conflicts and disagreements between shepherds and land managers in Spain were identified as one of the most important threats to forest fire prevention in this region. Only two statements in the SWOT analysis were related to opportunities, but they were both strong. The strongest opportunity of the two was considered to be the possibility of getting support from the EU Common Agricultural Policy for different agroforestry-related land-management practices.

In the legislative SWOT analysis the strengths and opportunities both got about 35 per cent of the weighted importance. The statements in both of these categories were of an overall high strength. The strongest statement from the legislative point of view was considered to be the existence of legislation and regulation connected to the practices of using fire for pasture renewal. Also in the legislative analysis it was clear that the conflicts between shepherds, farmers and local authorities were important weaknesses. These conflicts were observed on a local level in the Spanish test area and most often they relate to agricultural practices and methods. Integrated grazing in forested areas was the strongest opportunity found in the legislative SWOT analysis, mainly because it can reduce the impact of forest fires if the grazing is used correctly. Overgrazing and other livestock-keeping practices that can harm the ecosystems were at the same time considered to be the most important threat.

The socioeconomic aspects of the SWOT analysis showed that this is an area where weaknesses dominate. The weighted importance of the weaknesses in this area add up to no less than 68 per cent of the total score. However, the strengths and opportunities were still considered important. The social and economic values generated by agroforestry as a land-management practice were considered the most important strength. Agroforestry is a sustainable land was management and should as such be economically viable. The fact that agroforestry is not a financially and/or socially attractive way of managing land the most significant weakness identified by the project. The low profitability of livestock keeping was another strong weakness, whereas the possible income from tourists and touristic activities was considered to be the most important opportunity. The greatest threat identified in the socioeconomic SWOT analysis again related to the conflicts between different groups with an interest in land management, for example shepherds, land managers and local authorities.

In general, most of the strengths were found within the institutional and the legislative SWOT analyses, whereas the weaknesses were clearly dominating in the socioeconomic SWOT analysis. To improve the overall picture of agroforestry in regard to forest fire prevention, it was suggested that the weaknesses of the socioeconomic SWOT analysis should be addressed to see if they can be improved so as to reduce their impact. By targeting improvements on the weaknesses identified in the socioeconomic SWOT analysis, the overall general picture could be greatly enhanced.

Controlled grazing was something that scored higher than other fire prevention techniques in the analysis of the questionnaire. Controlled grazing scored higher in Greece and Spain than in the other countries answering the questionnaire. Associations of livestock owners were considered to be important, mostly in France and Spain, but the overall questionnaire respondents indicated that this was a good way to enforce fire prevention. Conflicting interests were considered to make the management of forests in a preventive way more difficult, as was seen in the SWOT analyses.

The SWOT analysis aimed to identify obstacles that hinder fire prevention. For agroforestry these obstacles were found to be the low profitability of the land-management practice. The reason that this was selected as the main obstacle was that it contributes to the other threats to agroforestry, namely land abandonment and the consequent build-up of fuel loads in previously managed areas. These obstacles were something that all the studied subjects had in common and the main obstacles for fire prevention in general could thus be said to be the non-profitability of land management and the consequent abandonment of rural areas. These impacted through factors such as the subsequent build-up of fuel loads in previously managed areas. The main obstacles are thus intermingled and it is impossible to make a clear distinction between which is the starting point for the current problems.

From the legislative point of view, the obstacles to agroforestry were found to be the low priority of agroforestry and forestry in politics and a short budgetary period that does not adequately reflect management priorities. The conflicts between different institutions involved in forest fire prevention, as well as between different groups of land managers, and the reluctance from different institutions and groups to cooperate, were found to be major barriers. This was reflected also in the socioeconomic obstacle of conflicts between shepherds and forest managers in Spain. This was found to be the main obstacle to forest fire prevention in the Spanish test-area.

Conclusions

The results from the SWOT analyses and the FireSmart questionnaire show that the most efficient way to improve agroforestry in regard to forest fire prevention would be to try to tackle the socioeconomic weaknesses found. The SWOT analysis gives clear indications of where efforts are needed.

It is clear that one of the most important things to tackle is the abandonment of agroforestry practices. To increase the profitability of agroforestry as a land management method would be one of the most efficient methods to prevent further abandonment of the practice as this would increase the attractiveness of using this land management method. Many different examples on how this can be done exist in the Mediterranean region. The diversification of income has been identified by the FireSmart Consortium as one of the main possibilities to explore. Through increasing the possibilities of multi-functional land use, the impact of different changes would be lessened. Tourism activities boosting local income is one of the possibilities for diversification, but there are other options. Such as keeping different kinds of animals, or growing trees that can give continuous or regular income, such as cork-oaks or olive trees.

However, one of the most important effects of keeping agroforestry as a land management practice would be the continuous employment opportunities for the population of rural areas. With a populated landscape fires could be detected at an earlier stage and thereby easier to control and extinguish. To achieve more effective land management and the continuation of agroforestry practices, more information about agroforestry and its benefits is needed for both land managers and the general public. In this respect, branding schemes for local products could be improved and marketed. The general public should be informed about all the services provided by agroforestry ecosystems. This would greatly help to keep profitability in these *ecosystems*.

The results from the analysis of the opinions gathered through the FireSmart questionnaire confirmed the SWOT analysis in terms of benefits generated by agroforestry and grazing as forest fire preventive measures. The questionnaire analysis also confirmed that conflicts of interest are important obstacles to implementing agroforestry and grazing as forest fire preventive measures.

To summarise there are two main obstacles to forest fire prevention. The first is the low profitability of land management and the second is the low awareness of the importance of forests and forestry and their economic, social and environmental benefits. The abandonment of rural areas and rural land-management methods are the key driving factors in increasing fire risk. The low profitability and the subsequent abandonment of land gives rise to increases in fuel loads in previously managed areas and thus increases fire risk. The main obstacles for action for a better overall fire prevention include conflicts between stakeholders.

References

Boström, C., Yague, M.J., Hernando, C., Planelles, R., Buffoni, R., Alves, R., Jappiot, M. and Miguel, J.S. (2011) *EU-FIRESMART, Forest and Land Management Options to Prevent Unwanted Forest Fires: SWOT Analyses in Agroforestry Systems*, Wildtrack Publishing, Sheffield, 84–90.

Moreira, F., Vaz, P., Catry, F. and Silva, J.S. (2009) 'Regional variations in wildfire susceptibility of land-cover types in Portugal: Implications for landscape management to minimize fire hazard', *International Journal of Wildland Fire*, 18(5): 563–74.

Mosquera, M.R., McAdam, J.H., Romero, R., Santiago, J.J. and Rigueiro, A. (2009) 'Definitions and components of agroforestry practices in Europe', in Rigueiro, A., McAdam, J. and Mosquera, M.R. (eds) *Agroforestry in Europe: Current Status and Future Prospects*, Advances in Agroforestry, Vol. 6, Springer, The Netherlands, pp3–20.

Part IV
Case studies

14 Grazing Refuge Habitats and their importance for woody plants in the west of Scotland

Richard Gulliver

Introduction

The activities of mammalian grazers, especially domestic stock, are partly or completely restricted at certain landforms, for example crags. These 'Grazing Refuge Habitats' and their woody plant species are significant for two main reasons. First, they allow the continued maintenance of the small populations that are infrequent locally, regionally or nationally, collectively maintaining the genetic diversity of species. Second, they provide a potential source of seed for the establishment of woody plants into elements of the surrounding landscape. Additionally, trees and shrubs planted or seeded into such habitats are unlikely to require artificial protection from grazing.

Set against the advantages of freedom from grazing are increased exposure to wind, proneness to drought and limited resources, for example soil water and nutrients. Critical issues for woody species growing in refuge habitats include the level of production of propagules; the genetic diversity of the seed; the effectiveness of the distribution system; and the nature and availability of suitable receptor habitats. There is often a dearth of information on such topics. Species are the basic units of conservation activity. Knowledge of their biology and its regional variability, their role in communities and their relationship to landscape studies is of vital importance.

As well as enriching the landscape, Grazing Refuge Habitats can stimulate an appreciation and discussion of their origins, their naturalness and their future relationship with the surrounding landscape.

Woody plants may be affected at both seedling and young plant stage by a range of mammalian grazers. Consumption may result in immediate plant death. Alternatively, initial grazing may be followed by re-growth, although successive grazing over a period of years can result in death. Occasionally repeatedly grazed plants live for a large number of years, confined to a dwarf form. Woody plants may be targeted, for example when protruding above snow; or they may eaten as part of a general consumption of herbage. Grazing may occur in woodland.

A variety of landforms preclude access to some or all mammalian grazers. Near-vertical sea cliffs are particularly inaccessible. Among the grazers, small mammals are the most pervasive. They consume the woody tissue at the base of young trees and shrubs. Grazers include native mammals, feral mammals and domestic stock. In many situations, for example hill land in Scotland, all three categories can be present.

The main theme in this account is the consideration of the landscape value of Grazing Refuge Habitats and their power to stimulate interest in their origin and future. This simultaneously draws on research and perspectives from 1) field botany, for example Evans *et al.* (2002); 2) plant ecology, for example the *Biological Flora of the British Isles* accounts (asterisked in Table 14.3); 3) forestry (including tree diseases), for example Mutch (1988), Forestry Commission publications and the journal *Scottish Forestry*; 4) environmental history, for example Smout (1997a and 2003), Smout *et al.* (2005); and 5) palynological studies for example Davies (2011), Sansum (2004 and 2005).

Categories of landform acting as Grazing Refuge Habitats

The main categories of landform that partly or completely restrict the activities of mammalian grazers, i.e. Grazing Refuge Habitats, are shown in Table 14.1, using the west of Scotland as the study region, with examples presented from the woody plant flora. The main types of mammalian grazer are also shown in six categories. The presentation is a simplification for the purposes of providing an overview and should be treated as such. It is based on the general assumption that all categories are 'unavailable' to domestic stock i.e. sheep (*Ovis aries*) and cattle (*Bos taurus*), although some of the listed habitats will be accessible to either wild or feral grazers. However, in each habitat category there will be some examples on the ground where grazing by domestic stock never occurs, some examples where it occurs extremely rarely and some where it occurs at infrequent intervals. For example, small changes in verticality in cliffs and crags may affect animal accessibility greatly. Very hungry animals may graze in areas that would normally be avoided. The deer category includes native and feral species, as does the lagomorph category.

In this account reference to one, some or all of the landforms in Table 14.1 has the first letter in each word capitalised, i.e. Grazing Refuge Habitats; references to grazing refuges in the loose or general sense are not capitalised. The word 'gorge' is here used to mean a very steep-sided valley (often with vertical sides) with a watercourse in the bottom; and 'ravine' to indicate a very steep-sided valley with no watercourse in the bottom. Other authors may use the words in a somewhat different sense.

Boulder fields are a Grazing Refuge Habitat that is not included in Table 14.1. They vary in extent, altitude, height of crag or cliff that generates the boulders, width of spaces between the boulders, soil conditions and presence/absence of vegetational cover between the boulders. Hence their potential as a refugium varies from site to site.

Table 14.1 Overview of broad trends of susceptibility of Grazing Refuge Habitats to grazing

	Cattle	Sheep	Goats	Deer	Lago-morphs	Small mammals
River bank – mineral soil	WWLL	(WWLL)	(WWLL)	(WWLL)	GP	GP
Gorge with watercourse					GP	GP
Ravine, no watercourse			(GP)	(GP)	GP	GP
Isolated large boulders					GP	GP
Crags and inland cliffs moderate slope			GP	GP	GP	GP
Crags and inland cliffs near vertical					GP	GP
Sea cliffs moderate slope			GP	GP	GP	GP
Sea cliffs near vertical					GP	GP
Islands in lochs	ITBM	ITBM	(GPSIC)	(GPSIC)	(GPSIC)	(GP)

Notes: GP = grazing possible, GPSIC = grazing possible by swimming or in icy conditions, ITBM = if transported by man, WWLL = when water level is low.
Bracketed entries signify may occur on rare occasions.

Grazing Refuge Habitats in woods

If grazing pressure becomes less intense, Grazing Refuge Habitats may act as locations from which colonisation of the wood by less-common species can occur. If grazing pressure in the wood becomes more intense, their status as a refugium assumes an even greater importance.

Several authors have concluded that in many semi-natural woodlands in upland Scotland the woody plant complement had been reduced by long periods of management for economically useful tree species, for example Cosgrove *et al.* (2005); Smout (1997b); Smout and Watson (1997); Stewart (2003a and b). Grazing Refuge Habitats have a potential role to play in reversing this trend.

Relationship to surrounding land

In largely unwooded land, Grazing Refuge Habitats may represent the last remnants of the former woody cover, now lost over many years due to grazing, sometimes coupled with burning. The component woody species of grazing refuges may be periodically enriched by new species establishing from bird-deposited or wind-carried seed. However, local extinction of some of the woody plant species may occur from time to time. In hill land the refuges may exist in a matrix from which all typically woodland plants have been absent for many centuries.

A postulated natural origin for open (non-wooded) Scottish upland landscapes (below the tree line) has been proposed by Fenton (2008).

Bennett (2009) considers Fenton's exclusion of anthropogenic pressures over the millennia led to an 'incomplete and misleading' account of landscape changes. Peterken (2009) proposed that a forest habitat network was compatible with moorland survival, cf. Fenton (2008).

Susceptibility of Grazing Refuge Habitats to environmental factors

Table 14.2 provides a subjective assessment of the susceptibility of Grazing Refuge Habitats in the west of Scotland to key environmental factors. Strong winds can cause tissue damage; loss of plant parts, for example branches; uprooting with some roots retaining a connection with the substrate; and 100 per cent uprooting. Similarly drought can cause tissue damage, damage to branches or total death. The two factors can act together during dry, windy periods. Each factor in Table 14.2 can vary greatly in degree from site to site and within sites, for example vertically in different parts of cliff faces and laterally if the orientation of the cliff changes.

Grazing refuges can occur in either the shaded or unshaded state (column 4 in Table 14.2); the presence of an established tree canopy strongly affects the environmental conditions. For example, a narrow gorge may be subject to intense shade, which may be a negative factor for seed output by shrubs and small trees on the sides of the gorge. However, the overall canopy may provide protection for shrubs and small trees from the desiccating and potentially uprooting effect of strong winds. An equivalent section in the open (i.e. without tree cover) will have high light levels, permitting vigorous growth and reproduction, and there may be more likelihood of wind blown propagules arriving; but the negative impact of wind movement may be considerable.

Heavy snowfall can result in the breaking of branches or the toppling of snow-laden trees. Depending on its scale, rock fracture can cause damage or death to individuals, or groups of trees and shrubs.

Column 7 in Table 14.2 gives an indication of general soil conditions. Compared with the surrounding area, growth conditions may be less good in small pockets of 'mor' humus on crags; however, they may be better on some non-vertical faces with a little water flow and frequent fracture of the rock into small pieces. The weathering and fragmentation processes liberate minerals and nutrients into the substrate. Mineral-rich substrates may provide good germination conditions for some woody plants. Boulder fields are not included in Table 14.2 (and 14.1) because their form and soil conditions may differ between, and sometimes within, sites.

In some instances Grazing Refuge Habitats may provide a refuge from fire that can negatively affect both plant life and substrate. However, the woody nature of the ericaceous vegetation (Figure 14.1), which is sometimes present, could make some grazing refuges as susceptible to fire as the surrounding land.

Table 14.2 Overview of broad trends of susceptibility of Grazing Refuge Habitats to environmental factors plus nature of soil

	Wind	Drought	Shade from established canopy	Heavy snowfall	Catastrophic event	Soil: general trends
River bank – mineral soil	P-OS		P		BE	MS
Gorge with watercourse	P-OS	(P-OS)	P		BE, RF	NSD*
Ravine, no watercourse	P-OS	P-OS	P		BS, RF	NSD*
Isolated large boulders	P-OS	P-OS	P	P	RF	CR,CV,ML
Crags and inland cliffs moderate slope	P-OS	P-OS	P	P	RF	CR,CV,ML
Crags and inland cliffs near vertical	P-OS	P-OS	(P)	P	RF	CR,CV,ML
Sea cliffs moderate slope	P-OS	P-OS	P		RF	CR,CV,ML
Sea cliffs near vertical	P-OS	P-OS	(P)		RF	CR,CV,ML
Islands in lochs	P-OS	(P-OS)	P	P		NSD

Note: In general the larger the tree or shrub, the greater the degree of susceptibility. Shaded examples of the various Grazing Refuge Habitats may be uncommon. BE = bank erosion, BS = boulder shower, CR = cracks, CV = crevices, ML = mini-ledges, MS = mineral soil, NSD = normal soil development, * = for steep slopes, P = possible, P-OS = possible open sites, RF = rock fracture. Bracketed entries signify may occur on rare occasions.

Case study: The effects of grazing on an aspen on a crag on Colonsay (Inner Hebrides)

The study site (GR NR 40739 98316) contained a single fallen aspen (*Populus tremula*) tree; the distal ends of the roots having maintained their purchase in a fissure in the rock. The tree may have been blown over, or may have fallen as a result of one section of the crag breaking away from the rest. The more-or-less horizontal trunk is supported a little above ground level by the major branches (see Figure 2 in Gulliver, 2011). Two groups of short shoots that were growing from an exposed major root were examined on 12 April 2010. These were accessible to sheep and cattle (deer are absent on Colonsay but feral goats are present). The end of every example of new growth in 2010 had been truncated by mammals. All that remained of the shoots produced pre-2010 were woody protuberances i.e. the shoot bases (n = 67 and 81). Death and loss of all the pre-2010 shoots was attributed to consumption by sheep and cattle and not to wind exposure. However, the possibility that pre-2010 shoots died from desiccation and subsequent breakage cannot be entirely ruled out. Observations of shoots that have been consumed when leaf and stem material are accessible to stock are widespread.

Study species

The species considered in this account are listed in Table 14.3; common and widespread woody species, for example rowan (*Sorbus aucuparia*), downy

Figure 14.1 Four small aspen trees (*Populus tremula*) on a linear crag near the east coast of Colonsay
Note: GR NR 40737 95429. The oblique angle of the left-hand (seaward) tree may have been caused by strong winds from the west. The more prostrate tree nearer the east coast, with a thinner canopy, is a rowan (*Sorbus aucuparia*).
Source: Richard Gulliver

birch (*Betula pubescens*) and hybrid birch (*Betula pendula* x *pubescens*) have been excluded. Habitats of the study species in Main Argyll (vice-county [biological recording district] 98), many of which are partly or wholly inaccessible to grazers are included in Table 14.3. Juniper (*Juniperus communis*) includes dwarf juniper (*Juniperus communis* subspecies *nana*) and upright juniper (*Juniperus communis* subspecies *communis*) see chapter Appendix 1. Pedunculate oak (*Quercus robur*), sessile oak (*Quercus petraea*) and their hybrid (*Quercus* x *rosacea*) all occur in the west of Scotland. Many accounts group these three taxa together as oaks (*Quercus* species), and this approach has been adopted here when appropriate; see also chapter Appendix 2.

Scientific names and English names (though not capitalisation) follow Stace (2010). Allocation of the species to native or alien status is from Preston *et al.* (2002). This publication shows their distribution at the hectad (10 x 10km^2) level. Regional variation in native status is considered in the discussion. Evans *et al.* (2002) recorded species at the tetrad (2 x 2km^2) level in the parish of Assynt in Sutherland, which has 164 tetrads contained

Table 14.3 Study species

English name	Scientific name	Habitats in one sample area – Main Argyll (vice-county [biological recording district] 98) (Rothero and Thompson, 1994)	Publications on species biology	Ellenberg light values (Hill et al., 1999)
Aspen	Populus tremula	coastal cliffs, crags and ravines	MacKenzie (2010) and Worrell (1995)	6
Bird cherry	Prunus padus	wooded ravines	Leather (1996)*	5
Downy currant	Ribes spicatum	–	Richards (2011)	4
Guelder rose	Viburnum opulus	ravines and rocky woodlands on base-rich soils	Kollmann and Grubb (2002)*	6
Hazel	Corylus avellana	woodland, scrub coastal crags and in ravines	Sanderson (1958)a	4
Holly	Ilex aquifolium	rocky burn sides and crags	Peterken and Lloyd (1967)*	5
Juniper both ssp.b	Juniperus communis	see below	Thomas et al. (2007)* and Ward (2007)	8
Juniper upright ssp.b	~ ssp. communis	most frequent on dry soil over calcareous coastal rocks	as above	8
Juniper dwarf ssp.b	~ ssp. nana	a plant of exposed situations in the mountains	as above	8
Oak, pedunculate	Quercus robur	all areas	Jones (1959)* and Morris and Perring (1974)	7
Oak, sessile	Quercus petraea	present in all areas	(both accounts relate to pedunculate, sessile and hybrid oak)	5
Oak, hybrid	Quercus x rosacea	recorded from most areas		[6]c
Rock whitebeam	Sorbus rupicola	crags and coastal cliffs	Rich et al. (2010) and Stirling (1994)	8d

Note: ssp. = subspecies. * *Biological Flora of the British Isles* account published in the *Journal of Ecology*. Juniper and holly are evergreen, the rest are deciduous; Data on species biology specific to the west of Scotland is desirable in many instances. For British Ellenberg light values - a value of 4 indicates the plant is associated with shaded habitats; a value of 8 indicates the plant is associated with open habitats. Intermediate values indicate intermediate conditions. a = a summary of several important aspects of Sanderson's research appears in Vera (2000: 333–9); b = see chapter Appendix 1; c = not in Hill *et al.* (1999), intermediate value between 5 and 7; d = on Skye 'in woodland at sea level', as well as on 'basic or limestone cliffs' (Murray and Birks, 2005).

within 11 hectads. The apparent abundance of a species varies with scale of recording. Hence bird cherry (*Prunus padus*) occurs in 6 out of 11 (54.54 per cent) hectads in Assynt (Sutherland) giving an impression of moderate abundance, but at the tetrad level its frequency is 13 out of 164 (7.93 per cent), indicating local rarity.

National rarity criteria follow Stace (2010) and therefore relate to the whole of the British Isles, i.e. Britain, Ireland and the Channel Islands. An earlier scheme related to Britain, for example see Rose (2006). Species classed as rare occur in 1–15 10 x 10km squares; those classed as scarce in 16–100 10 x 10km squares; and those classed as uncommon occur in 101–250 10 x 10km squares. Rock whitebeam and downy currant (*Ribes spicatum*) are both scarce. The dwarf subspecies of juniper (subspecies *nana*) has the rarity designation, uncommon. Juniper (both subspecies pooled) is a UK Biodiversity Action Plan Species. Grazing Refuge Habitats can contain non-native as well as native species.

Based on personal observation, rock whitebeam and downy currant were most confined to Grazing Refuge Habitats and oak was least strongly associated with grazing refuges, for example also occurring in some quantity in woodlands, both enclosed and unenclosed.

The canopy of juniper (see chapter Appendix 1) is potentially the most prostrate of the study species (see Figure 5 in Gulliver, 2011); the low growth of any one individual plant almost certainly resulting from a mixture of genetic constitution and environmental factors. Cuttings of prostrate plants grown by the author and his wife in sheltered conditions remained prostrate (Gulliver and Gulliver, 2011). The canopies of oak are often markedly modified by exposure (see Figure 3 in Gulliver, 2011) and those of holly somewhat affected. For the remaining tree species, the canopies normally resemble stunted and wind-damaged versions of forms that would develop in unexposed locations. Hazel produces several woody shoots from its base. In response to exposure these shoots usually remain upright, but tend to be short due to repeated death by desiccation at their tips. This may be followed by dieback induced by pathogens. Grazing Refuge Habitats vary in the amount of exposure they experience. Subjective scales of the degree of canopy modification in oaks and the abundance of dead stems of hazel might be used to provide an indication of exposure levels. British Ellenberg light values (Hill *et al.*, 1999) show the species' preferences for open or shaded conditions (see notes to Table 14.3).

There are parallels between grazing refuges for woody plants in the west of Scotland and the English Lake District. As a consequence, some information is presented from this region. The habitats listed in Table 14.1 provide refuges for non-woody species as well as woody ones. Ghylls (steep gullies in the hillside) in the Coniston basin in the Lake District provide a refuge for the fern *Hymenophyllum wilsonii* and the grass *Festuca altissima* as well as the tree *Tilia cordata* (Barker, 1998).

Reproduction and dispersal: Study species

For long-term survival, woody plants need to both maintain populations in existing Grazing Refuge Habitats and to colonise new refuge and general habitats. However, for many species one or both of these mechanisms falls short of the ideal in terms of efficiency. A selection of examples is presented on a species-by-species basis below. The accounts of species biology listed in column 4 of Table 14.3 frequently provide extra data on these topics.

Aspen

Aspen is dioecious. Seed production has only been observed infrequently. Evans *et al.* (2002), describing the situation in Assynt, state 'although it does not flower or fruit freely, it must sometimes set seed successfully as the majority of its more isolated sites could not have been colonised in any other way'. 1996 was a good seed year following the hot, dry summer of 1995 (Parrott, 2010); no subsequent good seed years are described in Parrott (2010). As aspen seed is wind dispersed, the propagule number for an individual of any given size is greater than that for trees and shrubs with berries; but the output will be dispersed generally, i.e. there is no targeted seed deposition as can occur with species whose seed is defecated by perching birds. Once established, the number of individuals is often increased by means of suckering, which in Grazing Refuge Habitats may produce a line or small group of trees. A checklist of characters for recognising clones, drawn from American experience, is given in MacKenzie's literature review (2010). There are two possibilities with regard to clonal attributes and habitats: 1) the characters present match those that are required to perform extremely well in the habitat in question; and 2) there is no matching – characters vary randomly from one clone to the next. It would be useful to know which is the case. Work on aspen genetics by Easton (1997), summarised by Ennos (2003), reveals a complex picture. It seems likely that for Scotland as a whole differences have persisted by vegetative reproduction from a period of active sexual reproduction that occurred during better climatic conditions pertaining up to c.4,000 years ago; but that some sexual reproduction is currently taking place within local populations. Eadha Enterprises holds a large collection of Scottish aspen clones (Eadha, 2012).

Bird cherry

Bird cherry produces small black fruits that are bird dispersed. Seedlings are not often found (Leather, 1996). The plant produces root-suckers that can develop into dense thickets (Leather, 1996).

Downy currant

For downy currant 'fruit-set is usually very poor' (Richards, 2011). Opportunities for bird dispersal may therefore be very limited. Fruit is shown at Storr, Skye on 31 July 2004 in Farmer (2004). It was observed in the same year on 25 June on Islay – site visits by the author were not made every year. Downy currant produces clonal stands (Richards, 2011). On Islay, natural layering has been noted by the author in both riverbank habitats and the boulder field site. This site is shown as Figure 4 in Gulliver (2011). It is immediately above the high-tide mark. Downy currant was formerly grown in gardens. It produces a somewhat acid fruit. Its habitat in north-east England and the Pennines is the understorey of woods; in the west of Scotland it can be an open-land plant. However, its shade tolerance (Table 14.3) suggests an ability to add to the species complement of Scottish woodlands. A thin tree canopy is present at its riverbank sites on Islay.

Guelder rose

Guelder rose was believed by Kollmann and Grubb (2002) to have a relatively modest variation in fruit production from year to year, except in drought years, for example 1995. These observations on the influence of climate relate to southern England (Incidentally the hot dry year of 1995 in Scotland resulted in seed production by aspen in 1996). Kollmann and Grubb's (2002) extensive review of the biology of the species includes many personal observations and the results of a considerable body of research. Field data, especially from Scotland, are needed on levels of fruit and seed production.

Hazel

Young hazel plants are sometimes observed in woodland in the study area. Some former sea cliffs carry dense stands of hazel, suggesting successful establishment is possible, though some stands may be very old and expanding slowly over time. For southern England, Gurnell (1993) found that good seed production followed years with warm months in July and/or August.

Holly

McNeill (1910) describes the situation on Colonsay thus: 'trees in exposed situations rarely produce berries'. In shaded situations growth of holly may be reduced, especially in shaded gorges; and reproductive output could also be affected. Whereas the upright growth form of holly renders the species susceptible to wind damage, it also provides perching positions. Local and/or migrating birds, having fed on the berries of other individual plants, may defecate, ultimately resulting in new young trees. Assuming some regeneration from the berries that fall from the established tree, a genetically mixed small stand may result. Holly trees can be male or female, or sometimes hermaphrodite (Beckett and Beckett, 1979; Peterken and Lloyd, 1967). Trees in the wild but near to gardens and policy (estate) woods may have received pollen from ornamental cultivars (some of which have unusual forms of prickles and/or atypically coloured leaves). Any trees establishing from such seed will have a mixed genome. Holly can produce suckers from shallow lateral roots (cf. Gulliver, 2011; Peterken and Lloyd, 1967). To date this process has not been observed by the author on Islay or Colonsay.

Juniper

On Islay a considerable number of juniper plants visited in May in any one year show no signs of either male or female cones. The dwarf forms (chapter Appendix 1) are likely to be long lived. Dead individuals are not uncommonly observed in small populations on Islay.

Oaks

Oaks in Assynt (Sutherland) are stated to 'fruit only rarely this far north' (Evans *et al.*, 2002). The occurrence of hybrids at some locations suggests a greater degree of genetic variation in oaks than in the other study species. This may assist the evolution of forms that are adapted to exposed Grazing Refuge Habitats (see also chapter Appendix 2). Woody plants with berried seeds may be distributed from one grazing refuge to another by birds; wind-born seed may be dispersed by air currents. Acorns can be bird dispersed, but it seems possible that dispersal of light wind-born seed or of berries by birds may be more efficient than dispersal of acorns.

Rock whitebeam

Rock whitebeam is an apomictic tetraploid, unlike the other study species that all use sexual reproduction. In apomictic reproduction the generation of seed does not depend on the presence of pollinators and the offspring are genetically uniform. Stirling (1994) states, 'mature individuals readily produce flowers and fruit'. In this regard it differs from many of the other species discussed. Rich *et al.* (2010) have reviewed all the whitebeams in Britain and Ireland.

Reproduction and dispersal: Non-native examples

Rhododendron

Rhododendron ponticum may be considered to be an approximate analogue of holly, though with a more shrub-like growth form and with the added advantage of foliage that is unpalatable to mammals. Copious wind-dispersed seed is produced.

Cotoneasters

Horizontal-growing, small-leaved cotoneasters share some similarities with dwarf juniper, though their leaves are devoid of sharply pointed ends (Gulliver, 2011). Examples include thyme-leaved cotoneaster (*Cotoneaster thymifolius*), small-leaved cotoneaster (*Cotoneaster microphyllus*) and entire-leaved cotoneaster (*Cotoneaster integrifolius*). Typically they produce large quantities of berries. Farmer (2003) describes entire-leaved cotoneaster as 'an invasive pest species in some places, obliterating the native flora of a rock outcrop'.

Deducing nativeness

The issue of whether a species is 'native or introduced' needs to be considered on a species-by-species basis, and sometimes also on a sub-regional and/or site basis. For the west of Scotland, downy currant is presumed native at sites on Islay and Skye, but introduced on Mull (Jermy and Crabbe, 1978). Bird cherry is introduced on Islay but native in parts of Main Argyll. Guelder rose is native at one Islay site but introduced at others. On Mull it is introduced in one hectad and native at all others.

Rock whitebeam is likely to be native at all its sites, but the species has quite narrow habitat specificity. At some locations there may have been crossing between native forms of holly and introduced ornamental forms. Pedunculate oak may have been planted in sessile oak areas (see chapter Appendix 2) and vice versa. Aspen and juniper are probably native at most, possibly all, sites. Hazel may be somewhat linked to man as its nuts were exploited for food in the Mesolithic period.

The foregoing summary masks a great deal of the interest in working-out or speculating-on the processes that occur at each new Grazing Refuge Habitat as it is encountered. To take a hypothetical example, a guelder rose growing on the banks of a small river in policy (estate) woods, might superficially appear to be introduced. However, detailed investigation may show that a series of plants occur at increasing abundance up the watercourse into hill country, suggesting a natural origin. However, if population sizes diminish progressively from the estate, colonisation into the country from an originally planted individual or group of plants may be inferred.

Oaks are particularly intriguing. For example, a large number of hybrid individuals may indicate an attempted 'improvement' of the oaks in the eighteenth or nineteenth century by introduction of pedunculate oak, now totally replaced by hybrids. The original native sessile oaks may also now be uncommon, with hybrids in the majority (see also chapter Appendix 2). In studies of woodland history using pollen analysis, for example Davies (2011) and Sansum (2004 and 2005), it is not possible to distinguish between the species of oak generating the pollen.

The importance of Grazing Refuge Habitats

Grazing Refuge Habitats can be shown to have great importance in the following regards: 1) the continued maintenance of species that are infrequent locally, regionally or nationally in a zone with minimal or zero grazing by domestic stock; 2) as potential sources of seed for the establishment of woody species into currently unoccupied Grazing Refuge Habitats, species-poor woodlands and moorland; 3) contribution to the visual quality of the landscape; and 4) providing landscape-interpretation interest in deducing their origins.

Given sufficient resources, population genetics at the molecular level should be examined. Ennos (2003) reviews the subject and gives examples of genetic marker (isoenzyme) variation in aspen in Scotland and of genetic marker diversity in two isolated populations of rowan in the Scottish borders.

Colonisation of land adjacent to the refuge

If grazing by stock ceases on land adjacent to the refuge, colonisation could take place for species that are readily producing seed, subject to the availability of regeneration niches (Grubb, 1977). Observation suggests this occurs fairly readily in the case of holly. Scarification and other forestry processes (see for example Mutch, 1988; Stiven, 1997) will facilitate the process. In some of the study species, for example aspen, seed is rarely produced. In the case of aspen, individuals may spread from the refuge by root-suckers. Alternatively, they can be multiplied up from root-suckers in the nursery (Eadha, 2012; Livingstone, 2009 and 2011), and local forms planted into adjacent landscape. Such planting would be periodically reinforced by natural seeding.

Planting at the refuge itself

Extensive landscape enrichment with trees and shrubs could include Grazing Refuge Habitats as planting and/or seeding sites. Benefits from a lack of large mammals need to be balanced against exposure: limited planting pockets and accessibility difficulties for the planters at some sites, but not others, for example see Figure 14.1. Further factors relevant to site selection include water movement down the face after rain; apparent freedom from fire risk; and absence of invasive, non-native woody plants.

Habitats that are refuges from grazing sheep and cattle may still be utilised by several wild or feral grazers, and many will still be subject to some small mammal grazing of woody stems of young plants. The potential for community planting exercises forms part of an earlier account of Grazing Refuge Habitats (Gulliver, 2011).

Conclusion

In conclusion, one may state that the persistence up to the present day of Grazing Refuge Habitats, including examples rich in woody plant species, can be considered to be a pointer to an important potential for such refuges for conservation, site interpretation and landscape appreciation.

Appendix 1: Background information on dwarf and upright forms of juniper

Dwarf forms, i.e. subspecies *nana*, perform well in environments with high wind exposure. In Scotland they predominately occur in the north-west, including near the sea where high levels of exposure can occur at all altitudes. Upright forms i.e. subspecies *communis* are absent in most of north-west Scotland (Preston *et al.*, 2002). They can perform well in woodland situations – especially in eastern Scotland. In Britain as a whole, though, they are often to be found in open (unwooded) land. At some locations a complete spectrum of forms occurs. Accordingly a number of plant scientists identify juniper to the species level only (Preston *et al.*, 2002). Botanists taking this view in the west of Scotland include Evans *et al.* (2002) and this author. Rothero and Thompson (1994) state that 'intermediates between the two subspecies are not uncommon'.

Appendix 2: The relationship between pedunculate and sessile oak

Both pedunculate and sessile oak can occur as native species in parts of the region (further details in Gulliver, 2011). If at any one site, one species of oak is thought to be native and the other introduced, the hybrid might be considered to have an intermediate degree of nativeness. Heterogeneity at the local level was shown in a sample population of 12 trees on the east side of Loch Nedd, examined by Evans *et al.* (2002). One was assigned to pedunculate oak, one to sessile oak and the rest were intermediate. Difficulty in the identification of oaks becomes particularly apparent when producing species maps. The tetrad maps for Assynt in Evans *et al.* (2002) and the English Lake District in Halliday (1997) use a single category – oaks.

Sessile oak tends to be associated with the poorer soils and pedunculate oak with the richer ones (see for example Beckett and Beckett, 1979; Hadfield, 1957; Harmer *et al.*, 2010; Stace, 2010). However, one cannot universally assume that in upland Britain sessile oak will be the native species. For the English Lake District, 'the isolated high-level woods of *Q. petraea* [sessile oak] above Keskadale and Birkrigg in the Newland Valley (20.18, 20.20) although usually considered to be such fragments [of native woodland] are now thought more likely to represent plantings' (Halliday, 1997).

In discussing Scottish Atlantic oakwood s, Smout (2005) states 'gaps... were often planted up with acorns...There is no reason to think that the foresters would favour acorns of local provenance, and some reason to think they preferred, if they could get them, English acorns'. This represents a possible means of introduction of pedunculate oak into sessile oak areas.

Pedunculate oak is the main species in Vice County [Biological Recording District] 101 – Kintyre and is the species found in extremely exposed coastal locations where planting is unlikely to have ever occurred. Cunningham and Kenneth (1979) report 'The certainly native scrub oak populations of extreme exposure appear to belong here. These – as near Ormsary – grade from c. 3 ft. [c.0.9m] on the seaward to c. 15 ft [c.4.5m] on the landward side'.

Sessile oak tends to have a narrower crown and straighter branches than pedunculate oak (Hadfield, 1957; Mutch, 1988). At some locations these features might have caused sessile oak to be favoured for planting in pedunculate oak areas.

Acknowledgements

I would like to thank my wife, Mavis Gulliver, for help and support throughout this study, and Ms Christine Handley for care and diligence in handling the manuscript.

The administrators of The Paddy Coker Research Fund and the Stanley Smith Horticultural Trust are each thanked for a grant that helped to fund parts of this article; as are Mr Ian Evans, Dr Peter Grubb, Mr David Jardine, Mr Peter Livingstone, Dr Heather McHaffie, Mr John Parrott, Dr Malcolm Ogilvie, Dr Paul O'Hara, Mr Peter Quelch, Dr Tim Rich, Mr Gordon Rothero, Dr Peter Thomas and Dr Scott Wilson.

References

Barker, S. (1998) 'The history of the Coniston woodlands, Cumbria, UK', in K.J. Kirby and C. Watkins (eds) *The Ecological History of European Woods and Forests*, CABI, Wallingford, pp167–183.

Beckett, K. and Beckett, G. (1979) *Planting Native Trees and Shrubs*, Jarrold Colour Publications, Norwich.

Bennett, K.D. (2009) 'Woodland decline in upland Scotland', *Plant Ecology and Diversity*, 2(10): 91–3.

Cosgrove, P., Amphlett, A., Elliott, A., Ellis, C., Emmett, E., Prescott, T. and Watson Featherstone, A. (2005) 'Aspen: Britain's missing link with the boreal forest', *British Wildlife*, 17(2): 107–15.

Cunningham, M.H. and Kenneth, A.G. (1979) *The Flora of Kintyre*, EP Publishing, East Ardsley, Wakefield.

Davies, A.L. (2011) 'Long-term approaches to native woodland restoration: Palaeoecological and stakeholder perspectives on Atlantic forests of Northern Europe', *Forest Ecology and Management*, 261: 751–63.

Eadha (2012) 'Eadha Profile', www.energyshare.com/eadha, accessed 6 February 2012.

Easton, E.R. (1997) 'Genetic variation and conservation of the native aspen (*Populus tremula* L.) resource in Scotland', Unpublished PhD thesis, University of Edinburgh, Edinburgh.
Ennos, R. (2003) 'The contribution of population genetic studies to plant conservation', *Botanical Journal of Scotland*, 55(1): 89–100.
Evans, P.A., Evans, I.M. and Rothero, G.P. (2002) *Flora of Assynt*, P.A. Evans and I.M. Evans, Assynt, Scotland.
Farmer, C. (2003) 'West Highland flora, entire-leaved cotoneaster, *Cotoneaster integrifolius*', www.plant-identification.co.uk/skye/rosaceae/cotoneaster-integrifolius.htm, accessed 10 January 2012.
Farmer, C. (2004) 'West Highland Flora, Downy Currant, *Ribes spicatum*', www.plant-identification.co.uk/skye/grossulariaceae/ribes-spicatum.htm, accessed 10 January 2012.
Fenton, J.H.C. (2008) 'A postulated natural origin for the open landscape of upland Scotland', *Plant Ecology and Diversity*, 1(1): 115–27.
Grubb, P.J. (1977) 'The maintenance of species richness in plant communities: The importance of regeneration niche', *Biological Reviews*, 52: 107–45.
Gulliver, R. (2011) 'Observations on trees and grazing refuges in the west of Scotland', in I.D. Rotherham and C. Handley (eds) *Animal, Man and Treescapes*, Wildtrack Publishing, Sheffield, [NB please read scarce for rare in line 24 of the Introduction – General subsection], pp169–91.
Gulliver, R. and Gulliver, M. (2011) 'Juniper Re-establishment Project – Isle of Islay, Argyll, Scotland (poster)', in I.D. Rotherham and C. Handley (eds) *Animal, Man and Treescapes*, Wildtrack Publishing, Sheffield, pp192–5.
Gurnell, J. (1993) 'Tree seed production and food conditions for rodents in an oakwood in Southern England', *Forestry*, 66(3): 291–315.
Hadfield, M. (1957) *British Trees: A Guide for Everyman*, Dent, London.
Halliday, G. (1997) *A Flora of Cumbria*, Centre for North-West Regional Studies, University of Lancaster, Lancaster.
Harmer, R., Kerr, G. and Thompson, R. (2010) *Managing Native Broadleaved Woodland*, The Stationery Office, Edinburgh.
Hill, M.O., Mountford, J.O., Roy, D.B. and Bunce, R.G.H. (1999) *Ellenberg's Indicator Values for British Plants*, Ecofact Volume 2, Technical Annex, Institute of Terrestrial Ecology, Abbots Ripton, Huntingdon.
Jermy, A.C. and Crabbe, J.A. (1978) *The Island of Mull: A Survey of its Flora and Environment*, British Museum Natural History, London.
Jones, E.W. (1959) '*Quercus* L. Biological Flora of the British Isles', *Journal of Ecology*, 47(1): 169–222.
Kollmann, J. and Grubb, P.J. (2002) '*Viburnum lantana* L. and *Viburnum opulus* L. (*V. lobatum* Lam., *Opulus vulgaris* Borkh.), Biological Flora of the British Isles No.226', *Journal of Ecology*, 90: 1044–70.
Leather, S.R. (1996) '*Prunus padus* L. Biological Flora of the British Isles No.189', *Journal of Ecology*, 84(1): 125–32.
Livingstone, P. (2009) '*BULB Aspen project*', in Parrott, J. and MacKenzie, N. (eds) *Aspen in Scotland: Biodiversity and Management*, Highland Aspen Group, Kincraig, Scotland, pp67–8, also at *www.scottishaspen.org.uk*.
Livingstone, P. (2011) 'Aspen apothecary', *Reforesting Scotland*, 44: 12–15.
MacKenzie, N.A. (2010) *Ecology, conservation and management of Aspen – A Literature Review*, Scottish Native Woods, Aberfeldy, Scotland.

McNeill, M. (1910) *Colonsay: One of the Hebrides*, David Douglas, Edinburgh.

Morris, M.G. and Perring, F.H. (1974) *The British Oak – Its History and Natural History*, published for the Botanical Society of the British Isles by E.W. Classey, Farringdon, Berkshire.

Murray, C.W. and Birks, H.J.B. (2005) *The Botanist in Skye and Adjacent Islands*, C.W. Murray and H.J.B. Birks, Prabost, Scotland and Bergen, Norway.

Mutch, W. (1988) *Tall Trees and Small Woods: How to Grow and Tend Them*, Mainstream, Edinburgh.

Parrott, J. (2010) 'The production of Scottish Aspen: Strategies for meeting demand. Web base document', www.scottishnativewoods.org.uk/documents/2010 Aspen production strategy.pdf, accessed 1 December 2011.

Peterken, G. (2009) 'Response to "A postulated natural origin for the open landscape of upland Scotland"', *Plant Ecology and Diversity*, 2(1): 89–90.

Peterken, G.F. and Lloyd, P.S. (1967) '*Ilex aquifolium* L. Biological Flora of the British Isles', *Journal of Ecology*, 55(3): 841–58.

Preston, C.D., Pearman, D.A. and Dines, T.D. (2002) *New Atlas of the British and Irish Flora*, Oxford University Press, Oxford.

Rich, T.C.G., Houston, L., Robertson, A. and Proctor, M.C.F. (2010) *Whitebeams, Rowans and Service Trees of Britain and Ireland*, monograph of British and Irish *Sorbus* L. B.S.B.I. Handbook No. 14, Botanical Society of the British Isles, London.

Richards, A.J. (2011) 'Species account: *Ribes spicatum*', Botanical Society of the British Isles, http://sppaccounts.bsbi.org.uk/content/ribes-spicatum-and-ribes-rubrum-0, accessed 6 February 2012.

Rose, F. (2006) *The Wild Flower Key*, Warne, London.

Rothero, G. and Thompson, B. (1994) *An Annotated Checklist of the Flowering Plants and Ferns of Main Argyll*, Argyll Flora Project, Argyll, Scotland.

Sanderson, J.L. (1958) 'The autecology of *Corylus avellana* (L.) in the neighbourhood of Sheffield, with special reference to its regeneration', unpublished PhD thesis, University of Sheffield, Sheffield.

Sansum, P. (2004) 'Historical resource use and ecological change in semi-natural woodland: Western oakwoods in Argyll, Scotland', PhD thesis, University of Stirling

Sansum, P. (2005) 'Argyll oakwoods: Use and ecological change, 1000 to 2000 AD – a palynological-historical investigation', *Botanical Journal of Scotland*, 57(1–2): 83–97.

Smout, T.C. (ed.) (1997a) *Scottish Woodland History*, Scottish Cultural Press, Dalkeith

Smout, T.C. (1997b) 'Highland land use before 1800: Misconceptions, evidence and realities', in T.C. Smout (ed.) *Scottish Woodland History*, Scottish Cultural Press, Dalkeith, pp5–23.

Smout, T.C. (ed.) (2003) *People and Woods in Scotland: A History*, Edinburgh University Press, Edinburgh.

Smout, T.C. (2005) 'Oak as a commercial crop in the eighteenth and nineteenth Centuries', *Botanical Journal of Scotland*, 57: 107–14.

Smout, T.C., and Watson, F. (1997) 'Exploiting semi-natural woods 1600–1800', in T.C. Smout (ed.) *Scottish Woodland History*, Scottish Cultural Press, Dalkeith, pp86–100.

Smout, T.C., MacDonald, A.R. and Watson, F. (2005) *A History of the Native Woodlands of Scotland, 1500–1920*, Edinburgh University Press, Edinburgh.

Stace, C.A. (2010) *New Flora of the British Isles: Third Edition*, Cambridge University Press, Cambridge.

Stewart, M. (2003a) 'Using the woods, 1600–1850 (1) The community resource', in T.C. Smout (ed.) *People and Woods in Scotland – A History*, Edinburgh University Press, Edinburgh, pp82–104.

Stewart, M. (2003b) 'Using the woods, 1600-1850 (2) Managing for profit', in T. C. Smout (ed.) *People and Woods in Scotland – A History*, Edinburgh University Press, Edinburgh, pp105–27.

Stirling, A., McG. (1994) '*Sorbus rupicola* (Syme) Hedlund', in A. Stewart, D. Pearman and C.D. Preston (eds) *Scarce Plants in Britain*, Joint Nature Conservation Committee, Peterborough, p395.

Stiven, R. (1997) 'The environmental impacts and effectiveness of different forestry ground preparation practices', Scottish Natural Heritage Information and Advisory Note Number 76, www.snh.org.uk/publications/on-line/advisorynotes/76/76.html, accessed 7 February 2012.

Thomas, P.A., El-Barghathi, M. and Polwart, A. (2007) '*Juniperus communis* L. Biological Flora of the British Isles No. 248', *Journal of Ecology*, 95(6): 1404–40.

Vera, F.W.M. (2000) *Grazing Ecology and Forest History*, CABI, Wallingford, Oxford.

Ward, L.K. (2007) '*Juniperus communis* L. Plantlife Species Dossier, Vol. 1, 1–21, www.plantlife.org.uk/uploads/documents/Juniperus_communis__Dossier__part1.pdf, Vol. 2, 22–46, www.plantlife.org.uk/uploads/documents/Juniperus_communis_dossier_part2.pdf, accessed 4 July 2011.

Worrell, R. (1995) 'European aspen (*Populus tremula* L): A review with particular reference to Scotland, I, Distribution, ecology and genetic variation', *Forestry*, 68: 94–105.

15 Legacies of livestock grazing in the forest structure of Valonia oak landscapes in the Eastern Mediterranean

Tobias Plieninger, Harald Schaich and Thanasis Kizos

Introduction

Since the 1960s, most dryland ecosystems of the Mediterranean Basin have experienced a comprehensive land-use transition from complex and multi-functional agrosilvopastoral land-use systems to simplified and intensified forms of livestock husbandry and agriculture (Pinto-Correia and Vos, 2001). Intensified livestock husbandry is believed to shift rangeland ecosystems from equilibrium states and to initiate degradation processes (Iosifides and Politidis, 2005). In the Eastern Mediterranean, overgrazing has resulted in the removal of soil cover and the domination of undesirable plants, mostly *Sarcopoterium spinosum* (Bakker *et al.*, 2005). These processes have been exacerbated by changes in the spatial configuration of grazing, which is now largely uncontrolled, continuous and all-season (Giourga *et al.*, 1998). A remote-sensing survey showed that between 1977 and 1996 40 per cent of rangelands on Crete (Greece) suffered declining vegetation cover in consequence of increased grazing pressure (Hostert *et al.*, 2003). However, grazing pressure is highly heterogeneous, so that over- and under-grazing can be observed even in immediate proximity of each other (Röder *et al.*, 2007). Both over- and under-use can modify ecosystem structure and functions, as Mediterranean drylands are tightly coupled human–environment systems (Aranzabal *et al.*, 2008). For example, a comprehensive cessation of livestock may involve loss of biodiversity and devastating wildfires (Papanastasis, 2009).

(Agro)silvopastoral woodlands are common Mediterranean vegetation complexes. They have been shaped by human uses and correspond to different stages of regressive succession of the Mediterranean climax forests, which have virtually disappeared (Scarascia-Mugnozza *et al.*, 2000). Many of these complexes can be considered 'legacies' of past land uses. Land-use legacies persist and continue to influence ecosystem structures and functions, though former uses may have been abandoned decades or centuries ago. Land-use legacies express themselves in diverse ecological phenomena, ranging from biodiversity, vegetation structure and soil properties to biogeochemical cycles (Foster *et al.*, 2003). The legacy of traditional grazing and cultivation practices in Mediterranean woodlands is not well understood, but it is assumed to manifest itself in elevated morphological plasticity, stress tolerance and ecological resilience (Bergmeier, 2008). Traditionally, silvopastoral woodlands have delivered a large variety of ecosystem goods, including firewood, charcoal, food for humans and animals, gums, resins, dyes, pharmaceuticals, cork and aromatic plants. Moreover, they provide intangible ecosystem services, such as soil protection, stabilisation of soils, reduction of water runoff in mountainous and hilly watersheds, maintenance of landscape beauty and microclimate amelioration (Scarascia-Mugnozza *et al.*, 2000). Loss of resilience in the silvopastoral systems in the Western Mediterranean has been expressed in a geographically widespread and profound lack of tree regeneration, accompanied by a gradual ageing and dieback of existing forest stands (Moreno and Pulido, 2008).

Silvopastoral woodland dominated by *Q. macrolepis* Kotschy is a particularly under-studied type of oak woodland (Figure 15.1). In consequence of

Figure 15.1 Silvopastoral oak woodland on formerly arable terraces in Filia municipality, Greece
Source: Plieninger *et al.*, 2011

conversion of forests to agricultural land, illegal lumbering, overgrazing and forest fires, *Q. macrolepis* stands have become marginal and fragmented into small forested units or isolated individuals in various locations (estimated remaining area in Greece: 29,600ha (Pantera *et al.*, 2008)). *Q. macrolepis* is frequently mixed with *Q. pubescens* Willd. and *Q. cerris* L. Being one of the few deciduous oak species in the Eastern Mediterranean zone, *Q. macrolepis* is being increasingly appreciated as a means to control desertification, for its ability to survive after wildfires and to thrive under conditions unfavourable to other oak species, and for an acorn mast that can support both wild fauna and domestic pigs (Pantera *et al.*, 2008). Since the European Union has declared *Q. macrolepis* forests to be a natural habitat type of community interest (Habitat Directive 92/43EEC), efforts to protect, manage and expand their populations by afforestation are underway. The overall aim of the following landscape-level study is to contribute towards conservation efforts by investigating land-use history, monitoring forest structure and regeneration of *Quercus* woodlands, and modelling ecological and management factors that determine habitats for oak regeneration on Lesvos Island, in the Eastern Aegean (Greece). In particular, we aim to answer the following research questions:

- What land-use changes have taken place since the early twentieth century that may be influential on silvopastoral oak woodlands?
- Does forest structure (especially long-term regeneration and short-term recruitment of *Q. macrolepis* and accompanying oak species) reflect local land-use legacies?
- How are patterns of oak regeneration related to current grazing intensities, environmental site characteristics, and overstorey and understorey plant community structure?

Materials and methods

Study area

As study sites (Figure 15.2), we selected silvopastoral oak woodlands in the municipality of Filia (39° 15'–39° 20' N, 26° 05'–26° 10' E; 2,190ha area) in north-central Lesvos Island. The area covers the valley of a small seasonal stream, and the village is located on a small plateau upwards from the outlet of the stream. The three hills that form the valley are 466m, 682m and 785m high; with the exception of the plateau and the pocket plain in the outlet of the stream, the area is sloping and steep. Soils are stony, shallow and severely eroded. The area has mild, humid winters that alternate with hot and dry summers, with a mean annual air temperature of 17.7°C and an average rainfall of 670mm.

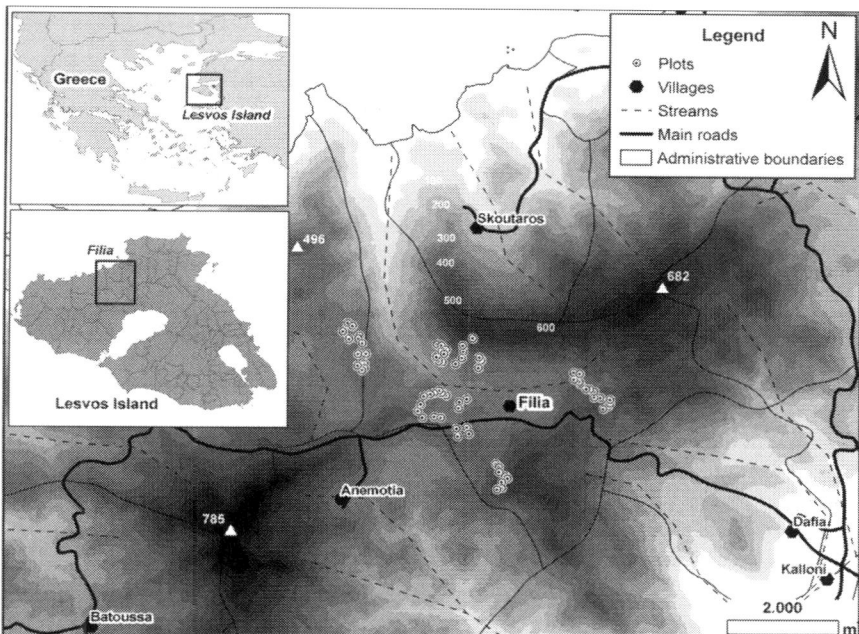

Figure 15.2 Map of Greece and location of the plots in the study area
Source: Plieninger *et al.*, 2011

Land-use statistics

Statistical information on land use and livestock for the Filia municipality was derived from the official data of the census for agriculture and animal husbandry, available at the settlement level from 1961 onwards. Earlier data, not available at settlement level, were taken at the level of the entire island. In Greek agricultural statistics, cultivated area includes arable land, vines, tree crops and fallow, while the utilised agricultural area (UAA) includes both cultivated and grazing lands.

Field measurements of forest structure and oak regeneration

As the sample for this study, we selected 70 parcels of silvopastoral woodland (see Figure 15.2). The parcels included in each cluster were identified by a random-walk procedure (Kent and Coker, 2000) in May and June 2009. We placed our sampling plot in the centre of each parcel. Each plot consisted of three concentric circles. Mature trees were recorded within a 15m radius (area: 706.9m^2), saplings and shrub species within a 5m radius (area: 78.5m^2), and seedlings within a 2.5m radius (area: 19.6m^2).

For the analysis of forest stand structure, the number, species and diameter at breast height (DBH) of all woody plants ≥10cm DBH were recorded within the outer circle. The frequencies of stems in 5cm diameter classes were determined for the three oak species in each plot. Size-class frequency diagrams were evaluated to indirectly assess long-term regeneration, as recruitment failure translates into 'gaps' in the current age structures of tree populations (Ramírez and Díaz, 2008). Short-term recruitment was assessed directly by measuring two juvenile life stages of oaks delimited by size: seedlings (base diameter <1cm) and saplings (base diameter: 1cm to <10cm). These were used as an indication of recent recruitment, while long-term regeneration was assessed by means of the size structure of adult oak populations. We counted the number of all saplings of the three oak species within the 5m radius. For each sapling, we measured base diameter of the main stem, height and the number of stems in cases of stem aggregations. Within the inner circle, the numbers of both individuals and aggregations of oak seedlings were counted.

Cover of canopy, shrubs (separately for each species), forbs, rocks, litter and bare soil were estimated by the step-point method (Evans and Love, 1957). Ground and canopy cover was noted down for 100 points, situated at 1-step intervals in a 50-step transect perpendicular to the slope and another 50-step transect parallel to contours. Short-term grazing intensity was estimated by counting the number of observations of faecal pellets along the steps of the two transects and is also expressed in percent of observations. Presence of shrub species within the 5m radius was also noted down.

We applied bivariate logistic regression to explore simple patterns of oak recruitment and thus potential drivers among site conditions, vegetation structure and grazing intensity parameters. The presence on a plot of at least one seedling or sapling (coded as 1) or no presence at all (coded as 0) was used as the dependent variable. An array of continuous parameters describing site conditions, vegetation structure and livestock grazing were used as independent variables.

Results

Land-use change

In the early twentieth century, most of the cultivated area on western Lesvos was covered by cereals and arable crops, followed by olives and vines. An increase of cultivated land followed the population increases of the 1930s and 1940s up until the 1950s; in the Mithimna district, where Filia is located, this increase led to arable land taking more than 70 per cent of the cultivated area, 26 per cent of which was in areas with 'scattered trees'. After 1961, the UAA in Filia increased slightly, due to the increase of grazing lands by 44 per cent and that of tree crops (practically speaking, olives) by 81 per cent until 2001 (Table 15.1). All arable uses – which were associated with extensive terracing to accommodate ploughing – have decreased, practically disappearing from the 1991 and 2001 statistics. The period from the 1950s to the 2000s was characterised by a more than twofold increase in the sheep population, while the number of sheep farmers has sharply decreased (Table 15.1). Even more dramatic livestock increases are to be found for the whole of Lesvos. In 1918, the total number of sheep was 70,000, a figure that remained more or less constant until the 1950s, when it went up to 93,000; but after that the sheep population effectively tripled on the island (Figure 15.3).

Table 15.1 Farm and land-use statistics for Filia village

	1961	*1971*	*1991*	*2001*
Farms (n)	404	322	196	192
UAA (ha)	1,968.1		1,990.6	2,129.4
Cultivated land (ha)	888.5	951.3	503.5	571.3
Arable land (ha)	256.5	56.1	4.3	3.9
Fallow (ha)	295.6	528.4		0
Tree crops (ha)	302.7	343.1	479.1	549.9
Grazing land (ha)	1,079.6		1,487.1	1,558.1
Farms with sheep (n)	177	166	111	101
Sheep (n)	3,971	6,846	6,886	8,064
UAA/farm (ha)	2.2	2.9	10.2	11.1
Average plot size (ha)	0.55	0.78	2.00	2.26
Plots per farm (n)	4.0	3.8	5.1	4.2

Source: Plieninger *et al.*, 2011

Forest structure

In total, 1,312 trees ≥10cm DBH belonging to 14 species were recorded. *Q. macrolepis* (44.4 per cent of all trees) and *Q. pubescens* (30.1 per cent) were by far the most frequent tree species. Total tree densities ranged from 28.3 to 820.5 trees ha^{-1}. Basal area extended from 2.2 to a maximum of 24.6m^2 ha^{-1}. Between one and seven tree species ≥10cm DBH were found within one plot, resulting in a mean of 3.4 species (±1.4) per plot. In total, 24 woody species (above and below 10cm DBH) were recorded (Table 15.2).

Long-term regeneration of oak populations

The three populations of oaks were composed of 583 *Q. macrolepis*, 395 *Q. pubescens* and 150 *Q. cerris* trees. Of these, 80.8 per cent (*Q. macrolepis*), 80.3 per cent (*Q. pubescens*) and 71.3 per cent (*Q. cerris*) were composed of one

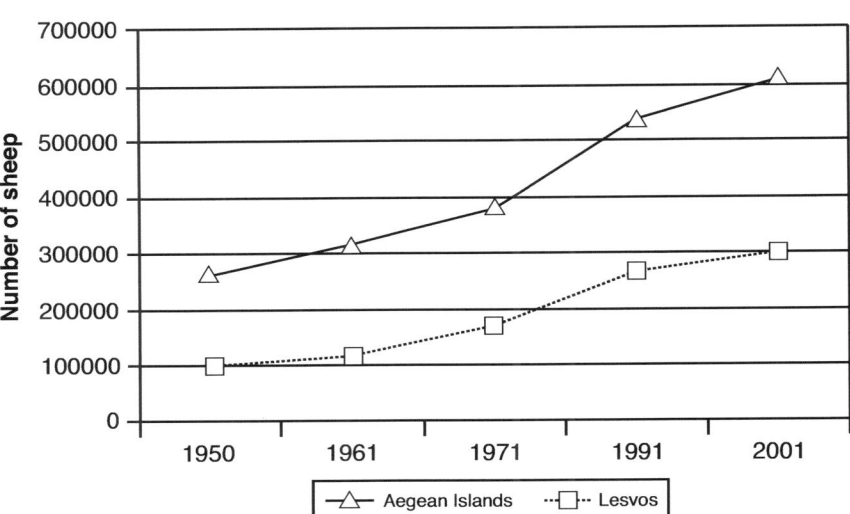

Figure 15.3 Number of sheep and number of sheep farms on the Aegean Islands and Lesvos, 1950–2001
Source: Plieninger *et al.*, 2011

major stem, whereas 19.2 per cent, 19.7 per cent and 28.7 per cent were multi-stemmed associations, indicating coppice-like stand structures with a high extent of vegetative regeneration. *Q. macrolepis* populations showed an inverse J-shaped distribution (Figure 15.4). Among all mature oaks recorded, the 10cm and 15cm size classes were most frequent, comprising on average 61.1 per cent (*Q. macrolepis*), 52.1 per cent (*Q. pubescens*) and 62.5 per cent (*Q. cerris*) of all trees. In *Q. pubescens* and *Q. cerris* stands, the 15cm class was more strongly represented (26.5 per cent and 34.9 per cent) than the 10cm class (25.6 per cent and 27.6 per cent), indicating a potential transition from an inverse J-shaped towards a bell-shaped size distribution.

Short-term oak recruitment

All in all, 353 oak seedlings were counted in the sampling plots, of which 276 could be categorised as isolated shoots and 77 as seedling aggregations. Seedlings were absent from 31 per cent of the plots. In those plots where seedlings occurred, densities ranged from 509.2 to 20,366.8 seedlings ha^{-1} (mean: 2,567.7, SD: 4,080.2). We counted 163 oak saplings in the 70 plots, with sapling densities ranging from 127.3 to 4,456.3 saplings ha^{-1} (mean: 296.5, SD: 605.7), where they occurred at all. Oak saplings were completely absent from 44 per cent of the plots. Sapling stem diameters ranged widely:

from 1cm to a maximum of 21cm (mean: 4.44cm, SD: 4.76cm). This distribution was skewed, with 47.2 per cent of oak saplings being in the 1cm size class.

Table 15.2 Overall occurrence of woody plants in the 70 plots, cover of woody species (where present) and density and basal area of the tree layer (>10cm DBH, where present) (mean values ± SD)

Species	Woody plant presence (%)	Shrub cover (%)	Tree density ($n\ ha^{-1}$)	Basal area ($m^2\ ha^{-1}$)
Asparagus acutifolius L.	57	1.62±0.77		
Ballota acetabulosa (L.) Benth.	36	2.44±1.60		
Cistus creticus L.	41	15.26±14.36		
Crataegus monogyna Jacq.	21	1.00±0.00	14.10±0.00	0.29±0.07
Fraxinus ornus L.	1		42.40	2.77
Juniperus oxycedrus L.	1		14.10	0.20
Lonicera etrusca Santi	4			
Olea europaea L.	24	2.50±0.71	22.20±13.28	1.53±1.36
Origanum vulgare L.	31	1.40±0.89		
Phillyrea latifolia L.	51	2.00±1.73	32.27±34.62	0.70±0.74
Pinus brutia Ten.	3		14.10	4.77
Pistacia terebinthus L.	27		14.10±0.00	0.20±0.09
Prunus domestica L.	51	2.53±2.62	29.28±21.12	0.46±0.38
Prunus dulcis (Mill.) D.A.Webb	1		14.10	0.35
Pyrus amygdaliformis Vill.	49	1.80±0.84	30.53±20.71	0.83±0.59
Pyrus communis L.	4		33.00±21.64	1.23±0.49
Quercus cerris L.	36	1.00±0.00	85.84±77.63	3.83±2.93
Quercus coccifera L.	7	1.33±0.58		
Quercus macrolepis Kotschy	87	1.39±0.51	135.21±126.07	5.78±4.56
Quercus pubescens L.	91	1.00±0.00	87.30±89.15	4.23±3.61
Rosa canina L.	9	1.00		
Rubus sp.	4	2.00±1.73		
Ruscus aculeatus L.	3			
Sarcopoterium spinosum L.	46	15.31±11.67		
Woody plants total	100	12.69±14.64	265.16±166.24	11.34±4.85

Source: Plieninger et al., 2011

Influences of grazing and other ecological factors

The occurrence of oak seedlings and saplings was significantly associated with several site, vegetation and management parameters (Table 15.3).

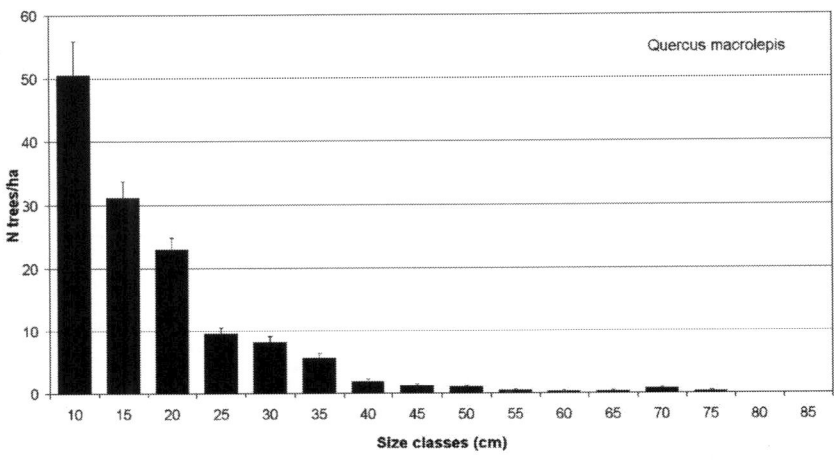

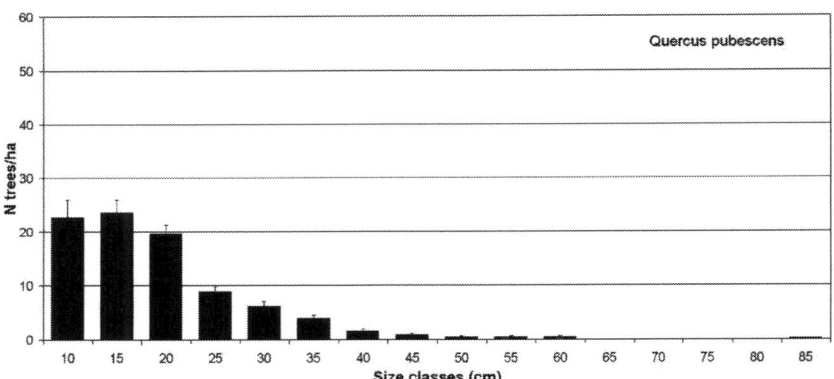

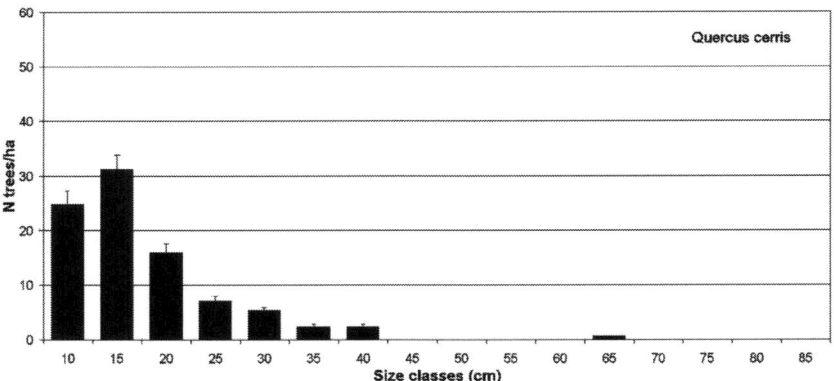

Figure 15.4 Size structure of mature a) *Q. macrolepis* (n = 61), b) *Q. pubescens* (n = 64) and c) *Q. cerris* (n = 25) stands (mean ± SD)
Source: Plieninger *et al.*, 2011

Dung frequency was negatively associated with both size classes. Forest cover, density and basal area of oak stands, woody plant richness and litter cover showed significant positive relationships for both seedlings and saplings. We did not detect any significant associations (threshold: p<0.05) for slope, northness, rockiness or shrub cover. Seedlings were negatively related to bare soil; saplings were negatively related to herbaceous cover. Occurrence of *Cistus creticus* shrubs was significantly and positively associated with both seedlings and saplings. *Asparagus acutifolius* was positively associated only with saplings. Associations with other shrubs were insignificant.

Discussion

Land-use transition

Available statistical data on land-use and human population changes reveal the great rupture that most Mediterranean land uses experienced from the 1950s to the 1970s (Pinto-Correia and Vos, 2001). A complex and multi-functional agrosilvopastoral land-use system was simplified into a pure livestock-raising system. In landscape terms, the former mosaic of terraced arable land, tree crops and pastures was replaced by more homogeneous grazed woodland. The abandonment of cereal cultivation was especially widespread on the terraces in the hill and mountain zones of the study area (Papanastasis, 2007). This land-use transition is the result of several parallel social and economic developments. The opening of local markets to competition from more productive areas had made most extensive cultivation of the Aegean Islands unprofitable since the 1950s (Papanastasis, 2007). A rural exodus reduced the population of the study area by more than 60 per cent (1940s–2000s) and the number of farms by around 50 per cent (1960s–2000s). However, an increased nationwide demand for dairy and meat products provided a powerful incentive for intensified livestock-raising in the 1970s and 1980s. European agricultural policies that granted per capita subsidies for sheep and goats in the 1980s and 1990s additionally contributed towards increasing herd sizes. However, the current ongoing intensification of livestock husbandry mainly seems to be the consequence of the reduced profitability of animal production: those that choose to keep their herds face diminishing per-head profit (if any) and increasing dependence on imported feed. To maintain total revenues, these farmers have been tending to further increase their herds and, thus, have more and more decoupled livestock-raising from the land (Iosifides and Politides, 2005).

Land-use legacies in forest structure and regeneration

In this study we asked whether the land-use transition outlined above has left specific legacies in terms of forest vegetation patterns, as these have

Table 15.3 Results of bivariate logistic regression analysis testing for effects of site parameters, vegetation structure and livestock grazing on oak seedling and sapling occurrence

	Seedlings			Saplings		
	sign	R^2	p	sign	R^2	p
Site conditions						
Slope			n.s.			n.s.
Northness			n.s.			n.s.
Bare soil	–	0.105	0.028			n.s.
Rock cover			n.s.			n.s.
Vegetation structure						
Forest cover	+	0.249	0.001	+	0.259	0.001
Quercus sp. density	+	0.094	0.048	+	0.273	0.001
Quercus sp. basal area	+	0.156	0.011	+	0.242	0.001
Woody plants richness	+	0.105	0.026	+	0.319	<0.001
Shrub cover			n.s.			n.s.
Herbaceous cover			n.s.	–	0.119	0.015
Litter cover	+	0.139	0.017	+	0.211	0.003
Livestock						
Dung frequency	–	0.106	0.025	–	0.138	0.015

Note: $n=70$, n.s.: $p \geq 0.05$, r indicates Nagelkerke R^2 as measure of goodness of fit.
Source: Plieninger *et al.*, 2011

also been found in deforestation–reforestation sequences elsewhere (Foster *et al.*, 2003). In general, legacies may be imprinted on several components of forest structure. First, tree species composition may be modified compared to ancient forests (Hermy and Verheyen, 2007). Second, agricultural legacies may leave biotic and abiotic characteristics, such as multiple-stemmed trees, whereas elements of ancient forests, such as large trees, dead snags and uproot mounds and pits, are rare (Foster *et al.*, 2003). Third, and most importantly, agricultural abandonment may generate specific age and size structures of forests (Marks and Gardescu, 2001). We recorded many species and elements in our plots that bear witness to an agrosilvopastoral past, for example cultivated tree species such as *Olea europea*, *Prunus domestica*, *Pyrus communis* and *Prunus dulcis*. Indeed, even the wide distribution of *Quercus macrolepis* can be considered an enduring land-use legacy, as oaks populations were deliberately expanded by humans at the end of the nineteenth century to use oak cupula for extraction of tanning agents (Bergmeier, 2008). Frequent terraces (Figure 15.1), threshing floors, stone walls, farm infrastructural elements and a proportion of 19 per cent to 29 per cent multi-stemmed mature trees present further evidence of past land uses.

We have documented a major land-use legacy in the diameter structures of silvopastoral oak woodlands, which we interpret as an indicator for long-term regeneration. The size structure of the *Q. macrolepis* population is similar to the inverse J-shaped distribution typical for natural Mediterranean oak forests (Pulido *et al.*, 2001). Assuming a close relationship between tree rings and diameter – as has been demonstrated for *Q. ilex* stands in the Mediterranean for homogeneous environments (Pulido *et al.*, 2001) – the population structures of both oak species correspond to those of natural multi-age forest stands. A positively skewed age distribution indicates continuous recruitment with a constant mortality rate of mature individuals, resulting in long-term persistence of the tree populations in the absence of exogenous disturbance (Oliver and Larson, 1996). The size distribution of the studied oak populations suggests that the traditional land-use system has supported continuous regeneration. Our data further show that short-term recruitment, as determined by seedling and sapling counts, is highly variable but generally low in relation to adult oak densities. In conclusion, the legacy of the land-use transition that we studied becomes evident in a major discrepancy between the successful long-term regeneration (as indicated by size-structure analysis) and the less successful short-term recruitment of oaks. Oak trees are mere legacies of the former, rather than active components of the current, land-use system.

Grazing and other ecological factors related to recruitment

Both oak seedlings and saplings are in consistent negative association with dung frequency, confirming the barrier that grazing can impose on oak regeneration. The significant impact of grazing on regeneration of *Q. ithaburensis* – a near relative species to *Q. macrolepis* – has been demonstrated by a study in Israel (Dufour-Dror, 2007). In a comparison between a grazed and an ungrazed oak stand, the densities of seedlings were 61 per cent and of sapling 67 per cent lower in the grazed treatment. This finding shows that the intensification of livestock grazing has been influential on forest regeneration and supports the view that Mediterranean vegetation is generally resilient against light and moderate grazing, but that impacts increase sharply beyond a certain threshold of grazing intensity (Plieninger, 2007). Recruitment of oak seedlings and saplings is also related to determinants such as forest cover, adult oak density and basal area, woody plant richness and litter cover. A relationship of basal area with occurrence or density of recruitment has also been found for other oak species in the Western Mediterranean (Pausas *et al.*, 2006). On the one hand, this can be interpreted as an effect of the elevated acorn input from mature oaks. On the other hand, it could be a consequence of more favourable microclimatic conditions in terms of the radiation and soil moisture of dense stands. The association of higher seedling and sapling densities with elevated woody plant diversity is difficult to interpret. Both may be related to a common cause, such as livestock grazing, which may hinder oak recruitment and the establishment of grazing-sensitive woody species at the same time. Together with the finding that forest cover (and not only oak cover) is a significant variable, this indicates that mixed woodlands are appropriate or even better habitats for oak recruitment than mono-specific oak woodlands. Litter cover seems to support recruitment by increasing soil moisture and, thus, water availability for oak seedlings (Espelta *et al.*, 1995). Bare soil is negatively associated with seedling recruitment, while herbaceous cover negatively impacts sapling occurrence. Thus, it seems that seedlings are more sensitive to the adverse conditions of open micro-sites, such as livestock trampling and insulation, whereas saplings suffer more from the competition of herbs.

Conservation and management implications

Low recruitment rates and current land-use trends indicate that the long-term persistence of silvopastoral oak woodlands may be at risk. Socioeconomic developments, such as increasing production costs (mostly due to high forage prices) and decreasing sales prices for livestock, have been leading to further increases in grazing intensities. An additional stress that will be testing the resilience of Eastern Mediterranean grazing systems is increasing drought conditions (as predicted by the scenarios of the Intergovernmental Panel on Climate Change). The past decades have shown a gradual reduction in rainfall and a subsequent reduction of tree growth in the Aegean Islands (Körner *et al.*, 2005). The predicted climate changes are likely to interact with grazing impacts, which may trigger a negative feedback cycle that increases soil erosion, reduces carrying capacity and in the end undermines the capacity of rangelands to sustain ecosystem services (Körner *et al.*, 2005). The inclusion of *Q. macrolepis* forests into the European Union's Habitat Directive as a habitat type of communal interest entails the ensuring of favourable conservation status through the formulation of a sound management plan. Our results suggest that oak regeneration can be enhanced by adaptive management of livestock grazing. For *Q. ithaburensis*, Dufour-Dror (2007) proposes a stocking rate limit of 0.7 livestock units ha^{-1}. Livestock management could also be improved by a controlled rotational system of grazing over several parcels and a seasonal steering of herds according to forage availability and depletion (Gutman *et al.*, 1999). A highly important task, however, will be to restore lost traditional knowledge on oak regeneration in cultivated and grazed landscapes through employing participatory approaches among resource users. This knowledge may help to foster resilience by promoting appropriate resource usage and oak regeneration at the same time.

Acknowledgements

This chapter is an edited version of an article that was originally published in 2011 in *Regional Environmental Change*, 11: 603–15. It is reproduced here with kind permission from Springer Science+Business Media.

References

Aranzabal, I., Schnitz, M.F., Aquilera, P. and Pineda, F.D. (2008) 'Modelling of landscape changes derived from the dynamics of socio-ecological systems: A case of study in a semiarid Mediterranean landscape', *Ecogical Indicators*, 8(5): 672–85.

Bakker, M.M., Govers, G., Kosmas, C., Vanacker, V., van Oost, K. and Rounsevell, M. (2005) 'Soil erosion as a driver of land-use change', *Agriculture Ecosystem and Environment*, 105(3): 467–81.

Bergmeier, E. (2008) 'Xero-thermophilous broadleaved forests and wooded pastures in the EU Habitats Directive: What is a favourable conservation status?', *Berichte der Reinhold-Tüxen-Gesellschaft*, 20: 108–24.

Dufour-Dror, J.M. (2007) 'Influence of cattle grazing on the density of oak seedlings and saplings in a Tabor oak forest in Israel', *Acta Oecologica*, 31(2): 223–8.

Espelta, J.M., Riba, M. and Retana, J. (1995) 'Patterns of seedling recruitment in West-Mediterranean *Quercus ilex* forests influenced by canopy development', *Journal of Vegetation Science*, 6(4): 465–72.

Evans, R.A. and Love, R.M. (1957) 'The step-point method of sampling – a practical tool in range research', *Journal of Range Management*, 10(5): 208–12.

Foster, D., Swanson, F., Aber, J., Burke, I., Brokaw, N., Tilman, D. and Knapp, A. (2003) 'The importance of land-use legacies to ecology and conservation', *BioScience*, 53(1): 77–88.

Giourga, H., Margaris, N.S. and Vokou, D. (1998) 'Effects of grazing pressure on succession process and productivity of old fields on Mediterranean Islands', *Environmental Management*, 22(4): 589–96.

Gutman, M., Holzer, Z., Baram, H., Noy-Meir, I. and Seligman, N.G. (1999) 'Heavy stocking of beef cattle and early season deferment of grazing on Mediterranean-type grassland', *Journal of Range Management*, 52(6): 590–9.

Hermy, M. and Verheyen, K. (2007) 'Legacies of the past in the present-day forest biodiversity: A review of past land-use effects on forest plant species composition and diversity', *Ecological Research*, 22(3): 361–71.

Hostert, P., Röder, A., Hill, J., Udelhoven, T. and Tsiourlis, G. (2003) 'Retrospective studies of grazing-induced land degradation: A case study in Central Crete, Greece', *International Journal of Remote Sensing*, 24(20): 4019–34.

Iosifides, T. and Politidis, T. (2005) 'Socio-economic dynamics, local development and desertification in Western Lesvos, Greece', *Local Environment*, 10(5): 487–99.

Kent, M. and Coker, P. (2000) *Vegetation Description and Analysis: A Practical Approach*, Wiley, New York, NY

Körner, C., Sarris, D. and Christodoulaiks, D. (2005) 'Long-term increase in climatic dryness in the East-Mediterranean as evidenced for the island of Samos', *Regional Environmental Change*, 5(1): 27–36.

Marks, P.L. and Gardescu, S. (2001) 'Inferring forest stand history from observational field evidence', in Egan, D. and Howell, E.A. (eds) *The Historical Ecology Handbook*, Island Press, Washington DC, pp177–98.

Moreno, G. and Pulido, F.J. (2008) 'The function, management and persistence of dehesas', in Rigueiro, A., Mosquera, M.R. and McAdam, J. (eds) *Agroforestry Systems in Europe Current Status and Future Prospect*, Springer, Heidelberg, Berlin, New York, pp127–60.

Oliver, C.D. and Larson, B.C. (1996) *Forest Stand Dynamic*, Wiley, New York.

Pantera, A., Papadopoulos, A.M., Fotiadis, G. and Papanastasis, V.P. (2008) 'Distribution and phytogeographical analysis of *Quercus ithaburensis* ssp. macrolepis in Greece', *Ecologia Mediterranea*, 34: 73–82.

Papanastasis, V.P. (2007) 'Land abandonment and old field dynamics in Greece', in Cramer, V.A. and Hobbs, R.J. (eds) *Old Fields: Dynamics and Restoration of Abandoned Farmland*, Island Press, Washington DC, pp225–46.

Papanastasis, V.P. (2009) 'Restoration of degraded grazing lands through grazing management: Can it work?', *Restoration Ecology*, 17(4): 441–5.

Pausas, J.G., Ribeiro, E., Dias, S.G., Pons, J. and Beseler, C. (2006) 'Regeneration of a marginal *Quercus suber* forest in the Eastern Iberian Peninsula', *Journal of Vegetation Science*, 17(6): 729–38.

Pinto-Correia, T. and Vos, W. (2001) 'Multifunctionality in Mediterranean landscapes – past and future', in Jongman, R.H.G. (ed.) *The New Dimensions of the European Landscape*, Springer, Dordrecht, pp135–64.

Plieninger, T. (2007) 'Compatibility of livestock grazing with stand regeneration in Mediterranean holm oak parklands', *Journal for Nature Conservation*, 15(1): 1–9.

Plieninger, T., Schaich, H. and Kizos, T. (2011) 'Land-use legacies in the forest structure of silvopastoral oak woodlands in the Eastern Mediterranean', *Regional Environmental Change*, 11(3): 603–15.

Pulido, F.J., Díaz, M. and Hidalgo de Trucios, S. (2001) 'Size-structure and regeneration of Spanish holm oak *Quercus ilex* forests and dehesas: Effects of agroforestry use on their long-term sustainability', *Forest Ecology and Management*, 146(1–3): 1–13.

Ramírez, J.A. and Díaz, M. (2008) 'The role of temporal shrub encroachment for the maintenance of Spanish holm oak *Quercus ilex* dehesas', *Forest Ecology and Management*, 255(5–6): 1976–83.

Röder, A., Kuemmerle, T., Hill, J., Papanastasis, V.P. and Tsiourlis, G.M. (2007) 'Adaptation of a grazing gradient concept to heterogeneous Mediterranean rangelands using cost surface modelling', *Ecological Modelling*, 204(3–4): 387–98.

Scarascia-Mugnozza, G., Oswald, H., Piussi, P. and Radoglou, K. (2000) 'Forests of the Mediterranean region: Gaps in knowledge and research needs', *Forest Ecology and Management*, 132(1): 97–109.

16 Palaeoecological records of woodland history during recent centuries of grazing and management examples from Glen Affric, Scotland and Ribblesdale, North Yorkshire

Helen Shaw and Ian Whyte

Introduction

Future woodland management and regeneration rely on decisions that are often driven by current cultural choices and ecological knowledge. What is the contribution of historical data to such management? While some argue that palaeoecological and historical knowledge is irrelevant due to the no-analogue climate and human land-use scenarios of the future, we suggest that historical knowledge must inform ecological knowledge of the past and modern cultural debate. However, our historical picture often lacks detail, forming a mere backdrop to decision making, or worse, developing into 'myths' (Smout, 2000), through which rhetorical viewpoints can be argued. Using case studies, we conclude that arboreal conservation needs to take a broad approach with regeneration in open areas and patchy clear areas in wooded environments.

Throughout much of the twentieth century nature conservation has been based upon static, deterministic models of ecology. Woodland, heathland and pastoral conservation fitted neatly into these views with nature conservation 'engaged in controlling nature' (Adams, 1997: 278) to maintain selected examples of these systems in perpetuity. The role of this reserve-based management in preserving at least some ecologically interesting areas of semi-natural biodiversity from the rigours of the twentieth century should not be underestimated. However, the use of the Clementsian climax community framework as the ecological basis for controlling and maintaining nature is a construct with its own challenges, a convenience for nature conservation policy rather than a long-accepted, unchallenged ecological paradigm.

In reality ecosystems often display non-equilibrium dynamics driven by stochastic processes and comprise species assemblages that shift in composition. This alternative understanding of ecology as dynamic was always present alongside Clementsian models in the twentieth century (e.g. Botkin, 1992; Gleason, 1926) but has gathered pace in nature conservation thinking (Adams, 1997) with the development of the ecosystem approach advocated by the Convention on Biological Diversity (CBD) (CBD, 1992; COP2, 1995).

In woodland ecosystems consideration of dynamic processes has highlighted two major areas of uncertainty: first, challenges to our understanding of the post-glacial woodland structure and its relevance to future woodland restoration; and second, our appreciation of the role of grazing history and human land use in the maintenance and destruction of trees and woodland.

For woodlands, the end point in Clements' climax community theory, the challenge to the deterministic ecological model of closed woodland was galvanised by Vera (2000). His thesis is that lowland woodlands were relatively open at the peak of their post-glacial development and that the pervading idea of vast closed-canopy woodland prior to human influence is erroneous. Vera proposes that processes occurred between open and closed canopy woodland stands and open grazed landscapes that created a shifting mosaic structure through space and time. Vera argues that the over-representation of pollen from woodland taxa has contributed to a flawed view of a post-glacial succession to a climax community of closed forest in lowland Europe. He suggests that the forest into which human populations first migrated was open and patchy, created by grazing animals in natural populations, and that palaeoecologists have difficulties in understanding the pollen signal from patchy woodland glades due to the lack of pollen productivity of plants in the grazed areas.

Palaeoecology, particularly palynology, has consequently come under much scrutiny. Many critiques of Vera's model have been put forward, with evidence from within palaeoecology used to refute his assumptions (e.g. Mitchell, 2005; Svenning, 2002). On balance, reviews (Kirby, 2003; Kreuz, 2008) find that there may be some evidence for open areas within natural woodlands but that this is probably overstated by Vera. However, Hodder *et al.* (2009) find that there has been a tendency to polarise thinking around 'false dichotomies' of either open or closed landscapes rather than the more likely scenario of a mosaic of differing levels of openness.

Vera has generated a valuable and interesting debate on the post-glacial structure of natural woodland. Within this he has shown the need for long-term understanding of woodland history to ensure an appreciation of dynamics and its importance in modern management. However, he has also highlighted possible errors in the very tools upon which knowledge of that history relies: the interpretation of historical and palaeoecological datasets. For thinking polarised upon the models of openness postulated by Vera, the natural progression is to conclude that palaeohistorical ecological research does not work, that its results are erroneous and should therefore be rejected. Given the need for a process-based understanding in land management, both in validation of the scale of dynamics and regional variation in the Vera hypothesis (Kirby, 2003 and 2004) and in the understanding of dynamics in space and time in cultural landscapes (Bignal and McCracken, 2009; Smout, 2010), this is unfortunate.

Despite the challenges, problems in reconstructing plant abundance from the pollen record continue to be a major consideration within palaeoecology addressed via methodological developments (e.g. Gaillard *et al.*, 2008 and references therein). An increased understanding of pollen source areas (Sugita, 2007a and b) and pollen productivity estimates (Broström *et al.*, 2004 and 2008) is being used to adjust the interpretation of output from traditional pollen analyses (e.g. Hellman *et al.*, 2008). Small openings within woodlands may still be difficult to sense, but McLauchlan *et al.* (2007) demonstrate that openings in up to 75 per cent canopy cover can be sensed via pollen analysis. However, Vera proposes landscapes with extensive open glades that are habitually visible to pollen analysts using multiple local-scale sites (Kirby, 2003).

Pollen analysis has long been known to work at various scales dependant on careful selection of the pollen sources (Jacobson and Bradshaw, 1981). The synthesis of many larger-scale pollen analyses has contributed to the myth of closed forest with the spread of post-glacial woodland (e.g. Bennett, 1989; Birks, 1989; Tipping, 1994). Many of the early analyses may have been used out of context or misunderstood. The maps of tree spreading and tree cover relied on interpolations over sometimes quite wide areas between palaeoecological sampling points and therefore displayed the woodland potential (the major tree species likely to be present) in different regions. The extent to which open space was a part of this woodland system is always a matter of speculation until confirmed on a site-by-site basis at a more local scale.

Given new dynamic conservation objectives, there is increasing need to understand community patterns and processes in ecology. To achieve this more attention needs to be paid to multiple local-scale palaeoecological analyses in individual case-study areas. It is only by this approach that we can develop an understanding of the variability in the structural composition and character of natural woodland. However, the focus of ancient woodland management has been on recreating some pre-anthropogenic vegetation state (with or without dynamics), and many question if c.5,000 years later, with climate change, human land use and a different species mix (including archaeophytes and aliens), the restoration of post-glacial woodland is even relevant. Cultural landscapes may have more relevance and importance, as Smout (2010) argues. Indeed traditional management and indigenous communities are given special mention in the ecosystems approach (CBD, 1992; COP2, 1995).

The development of nature conservation from stasis to dynamics should also provide a lesson for the understanding of our cultural past and its relationship with biodiversity. In the development of his thesis, Vera borrows from fisheries science to demonstrate a problem of shifting baselines in ecology (Pauly, 1995; Sheppard, 1995 in Vera, 2009). Shifting baselines are evident in the perception of nature, in that every generation tends to visualise as desirable the nature state of their youth, leading to a lowering of the baseline of what is natural and a tolerance of a creeping loss of biodiversity (Vera, 2009: 29). This highlights that, as ideas of cultural ecology develop, we are again in danger of missing the bigger picture; searching for baselines rather than examining the processes of cultural shifts and stasis, and the impact of these on biodiversity.

Palaeoecology can, we argue, add to our understanding of process in the recent past (Shaw and Whyte, 2008) and we can also integrate environmental histories with palaeoecological data to understand in more detail the changes that have occurred and the human activities that may have contributed to them. From this we can better appreciate the role of elements of traditional land management in driving ecosystem function, sustainability and resilience.

This chapter presents the results from two areas where palaeoecological results have demonstrated the need for greater knowledge of ecological and management histories and argues that site-specific palaeohistories should, where possible, form a basis for understanding all woodland regeneration and conservation.

Two case studies of palaeoecological research from very different UK upland environments are presented: Glen Affric in northern Scotland (Shaw, 2011; Shaw and Tipping, 2006) and Ribblesdale in northern England (Shaw and Whyte, 2010). A key question is whether these areas display stability or cyclicity through time that makes a deterministic conservation policy relevant for woodlands? Furthermore, to what extent does traditional management play a part in the current arboreal component in the landscape?

Glen Affric: Resilient woodland in a sea of vegetation openness and change

The wider context to the pinewoods of Scotland

Native pine woodlands have an emblematic status in Scottish forestry (e.g. Miles and Jackman, 1991). Pine woodland in Scotland is thought to represent, in many areas, the climatic climax of woodland in the post-glacial forest. This view of Scottish pine woods has led, over recent decades, to an aim to expand the woodland range in Scotland from one that is much reduced (Scottish Executive, 2006). Increases in numbers of deer, driven since the nineteenth century by winter feeding as part of estate shooting management, have prevented regeneration (Staines *et al.*, 1995). Additionally forestry has taken its toll, with use of timber for shipbuilding, construction and even drainage leading to a reduction of tracts of native forest in the nineteenth century followed by twentieth-century replanting with non-native conifers.

While some reduction in forest has, no doubt, taken place, the extent of the reduction of the woodlands over recent centuries is debateable. Historical research discloses an open landscape, compelling Breeze (1992) and Smout (2000) to conclude that the post-Roman wooded state of Scotland was essentially a myth. In addition, although, to some extent palaeoecology has contributed to this myth of a forested Scotland – via for example composite pollen maps showing potential forest zones (Bennett, 1984; Tipping, 1994) – the timing of this reduction, and the former extent, species composition and structure of pine woodlands, is less clear than is widely assumed. In 1995, Bennett identified that, although we can be assured of the *presence* of pine woodland in the earlier Holocene, we have limited understanding of the detailed structure of this woodland. Similarly Tipping (1994 and 2003) did not make any claim to a full understanding of the structure and canopy openness of the woodland within the zones identified by pollen analysis; merely a potential for dominant woodland taxa within each zone at that particular period. Without further widespread detailed pollen analyses, the structure and dynamics could not be interpreted in any detail.

Around Loch Sionascaig intensive pollen analyses suggest mosaics of vegetation within the Inverpolly (north-west Highlands) area through much of the Holocene (Birks, 1997 and references therein). Further suggestions of naturally open landscapes exist from a re-analysis of the sub-fossil remains of *Pinus sylvestris* (Bridge *et al.*, 1990; Dubois and Ferguson, 1985) and a review of the effects of paludification on forest growth (Crawford, 2000). Fenton (1997) proposes that the 'wet desert', much maligned by Fraser Darling in the 1950s (Stewart, 2010), is actually the true climatic climax community for much of north-west Scotland, the shift to it having been triggered by the onset of the Atlantic climate in the mid-Holocene, shown as a pine decline in many pollen diagrams (Bennett, 1984; Gear and Huntley, 1991).

Glen Affric in myth and reality

Glen Affric, the 'jewel' of Scotland's woodlands, lies in the north-central pine zone (Forestry Commission, 1998) in the Scottish Highlands. The area has a long history of conservation management (e.g. Aldhous, 1995; Bunce and Jeffers, 1977; Humphrey, 2006; Steven and Carlisle, 1959). In 1957, Steven and Carlisle identified the Glen Affric woodlands as one of the relict pine woods in Scotland with a history going back to the post-glacial woodland. In line with the wider aims of Scottish forestry, the main management aim over recent decades has been to expand the pine woodland area. However, as with the rest of Scotland, the structure and longevity of the woodland has, until recently, been a matter of hypothesis from ecological survey. The Glen Affric woodland currently becomes patchy and grades into open heath to the west of Glen Affric (Figure 16.1). This patchiness of woodland has been assumed to have been part of recent degradation due to forestry and deer grazing. In reality this assumption lacks supporting evidence.

Figure 16.1 Currently open area with scattered woodland to the south side of Loch Affric in Glen Affric
Note: Observe the structure of some of the pine trees indicating development in the open.
Source: Helen Shaw

It is clear that in the recent past, planted non-native forestry has spread across areas of the glen, but much of the history of these pine woodlands is inferred from historical records and from modern field observations. It is these that led Steven and Carlisle (1959) to claim that the relict pine woodland has a structure that provides a link with the past. It is, however, impossible without recourse to proxy histories such as those obtained via pollen analysis, to confirm relict status or to see the detailed past structure and dynamics of the woodlands.

The development of understanding of the long-term ecological history of Glen Affric has been facilitated by ownership by the National Trust for Scotland in the west and the Forestry Commission in the east. The aims for the site as an area where re-wilding and restoration were possible (Steven and Carlisle, 1959), and the interest in and adoption of the area by the organisation Trees for Life whose vision is to bring the Caledonian forests 'back from the brink' (Trees for Life, n.d.), spurred three major palaeoecological studies and it is through these that we can begin to utilise the full power of palaeoecological analysis.

Pollen analysis in the east of the glen from a single lochan, Loch an Amair, showed that the woodland was indeed long-lived, maintaining a continuous cover throughout the last c.9,000 years (Froyd and Bennett, 2006). The assumption of Steven and Carlisle is supported – the woodland in east Affric is a relict population with links to the original post-glacial woodland. However, this analysis, from a single site, does not provide spatially resolved information on variation in vegetation cover at the stand scale. Furthermore, pollen analyses by Davies further to the west of the Glen showed a long-lived open landscape almost devoid of trees after the mid-Holocene pine decline. The woodland to the west was of also mixed in species composition (Tipping *et al.*, 2006). Given this evidence it was clear that the current structure of the relict pine stands may not reflect that of the past and that the recent degradation of the woodland to the west was assumed rather than confirmed.

A further eight pollen diagrams from Glen Affric formed a study to examine the temporal pattern of openness at the western edge of the extant woodland and the dynamics within the woodland (Shaw, 2006). This study showed both a complexity within the species composition of the relict stand and dynamism through time, with fluctuating periods of openness in some stand areas and shifts in the ratios of pine and birch through time in others (Figure 16.2). In ecological terms this should be expected; pine does not regenerate in its own shadow, but changes stance, regenerating in pulses lagging 20–30 years behind climatic shifts, with a cycling of as much as a hundred years between pulses (Zackrisson *et al.*, 1995).

The study also examined the woodland at the western edge of the current pine stand, thus providing information on longevity of pine presence between the western and eastern sites of Davies' study. Here the palaeoecological evidence showed a recent regenerative phase of pine woodland on to an area of heathland (Shaw and Tipping, 2006; Shaw, 2011), an expansion rather than retraction of the core area (Figure 16.2a). Furthermore this expansion looks, from available chronological controls on the core stratigraphy (Shaw, 2006), to have occurred during times that deer were present in high numbers in the glen (Shaw, 2011).

Pollen analysis can express the presence of woodland, and how it changed through time but drivers of these changes are not determined. The longevity of the east–west boundary of woodland, however, fits well with the theory that pine woodland was affected in the mid- and late Holocene by paludification (Crawford *et al.*, 2003) and thus implies a natural ecotone. The rainfall increases markedly to the west of the glen and the gentler slopes of the blanket peat landscape flanking Loch Affric are probably less conducive to pine growth. The native relict woodland in the east has a longevity that is remarkable, and leads us to question why it has survived the

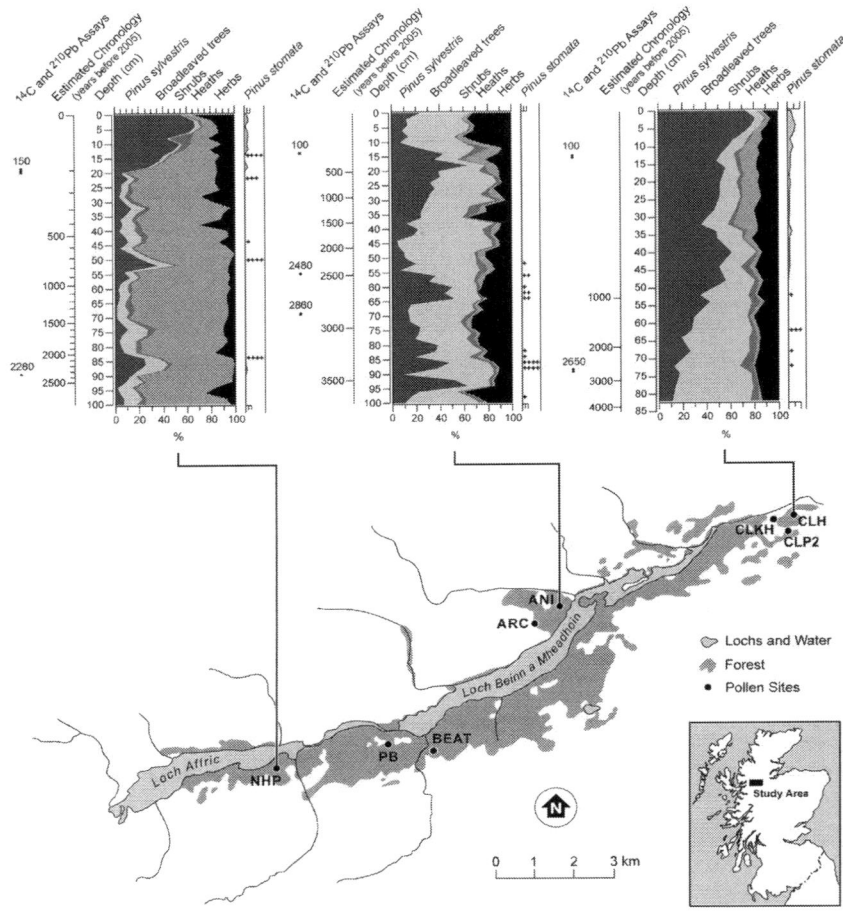

Figure 16.2 Map of Glen Affric with selected summary pollen diagrams from three selected sites in their locations along the glen, illustrating the differences along the east to west transect of sites
Note: Pollen is presented in key taxonomic groups as percentages of total land pollen. Full pollen diagrams are available in Shaw and Tipping (2006) and Shaw (2006).

onslaught of human occupation. Perhaps the area of pine woodland was always small, and therefore formed a scarce, valued and managed (or at least preserved) resource through prehistory (Shaw, 2011)?

Given the resilience of this woodland area, how should we manage it for the future? It is difficult to reconcile the need for planting within the relict woodland area given that it has survived so long. Steven and Carlisle and many others have been alarmed by a lack of regeneration, but given the findings of Zackrisson *et al.* (1995) it is likely that our understanding of the timing and density of regeneration in natural pine stands is still lacking. Indeed Edwards and Mason (2006), from a dendrochronological study of four Scottish pine woods including Glen Affric, highlight the need to examine pine woodland regeneration dynamics in terms of 300-year cycles. Additionally while deer numbers are known to adversely affect regeneration, in reality deer grazing forms part of a suite of complex ecological factors including competition with *Calluna vulgaris* that is not fully understood (Brooker *et al.*, 2006; Palmer and Truscott, 2003). A vast fenced exclosure and deer culling in the glen over recent decades may have boosted regeneration. However, planting has also been carried out; the success of regeneration in some areas has led now to concern that sufficient open space is maintained for valuable ground flora (Humphrey *et al.*, 2004).

The pollen diagrams from the core pinewood zone show that the forest was always mixed, with broadleaved trees such as birch, alder, hazel and oak as well as pine. This supports the diversity of species in the planting strategy of Trees for Life. However, the main influence of comprehensive palaeoecological analysis should, perhaps, be to highlight and confirm the Glen Affric woodland as one of our last true wild patches with a dynamic structure and function that has survived climatic shifts and maintained a mosaic of open space and closed woodland. The removal of forestry in the surrounding areas should be welcomed, but we should also caution against too much interference within this very special area so that we have a unique opportunity to observe natural processes and guard against an over-dense canopy.

Ribblesdale: The palaeohistory of an upland landscape

A background to the English uplands

The landscape of the English uplands is very different to that of Highland Scotland. Given the continued use of English uplands as productive pastoral ecosystems the large-scale re-wilding possible in areas of Scotland is not mirrored here. The English uplands also differ in recent history to other mountain areas of Europe that have suffered extensive land abandonment (MacDonald *et al.*, 2000). Here the long-term influence of land use is definite and persistent. However, much of the story of land degradation due to overgrazing is assumed to be similarly recent, driven by changes in land management and the intensification of grazing over the last c.50 years under the headage payments of the Common Agricultural Policy (CAP), by sheep in this case rather than deer. The early years of the CAP are seen as part of a system of damage inflicted by governance from a distance via a misunderstanding of the finely tuned more sustainable approach of commons management that had previously been in place for centuries.

This vision of a previous idyllic rural landscape has driven recent rural reform. Recent adjustments to the CAP include a shift to payments for environmental services rather than production, while traditional commons management is set up as an alternative regime for a biodiverse future. The cultural aspects of upland landscape management are increasingly boosted by tourism, and an integrated social-ecological approach to land management (CBD, 1992; Defra, 2007), which sees ecological agencies such as Natural England incorporating cultural concerns into ecological management.

However, traditional management may suffer similar deterministic and static views to that of twentieth-century conservation ecology. Evidence shows that traditional land management in the UK uplands has altered through time, influenced heavily by outside economies. Bronze and Iron Age settlement and cultivation, mining and quarrying, the medieval wool trade, famine and epidemic disease, post-medieval cattle droving, the development of local markets with the Industrial Revolution, the coming of the railways, the influence of war, agricultural improvement and drainage: all these influenced land management before the economic drivers of the CAP. Given this background it is relevant to ask what sustainable traditional management is, and is there a stable baseline management condition that can deliver currently desired biodiversity outputs? Despite two decades of environmental CAP payments, it is unclear whether the new reforms are delivering real environmental benefit.

Knowledge of the recent centuries of environmental and human history is, arguably, more important to the management of these landscapes than understanding the natural landscape state. Historical knowledge of management is vital. However, we have a surprisingly limited understanding of the detailed ecological changes that were linked to variations in land management through recent centuries. Detailed ecological and land management information is simply not widely available in the documentary record. A local-scale combined palaeohistorical approach can therefore be enlightening.

Ribblesdale: Centuries of traditional management and sustainability in a cultural landscape?

This local-scale approach is illustrated by a research in the open heavily grazed limestone pastures in upper Ribblesdale, North Yorkshire (Figure 16.3). A local history of vegetation change is revealed in a pollen diagram from Wife Park (Shaw and Whyte, 2010). Pollen evidence (Figure 16.4) shows that major woodland loss occurred in prehistory (Swales, 1987); however, patches of woodland, boundary trees and scattered scrub have previously created structural diversity in the cultural landscape. Documentary and palaeoecological evidence reveals the loss of this wood as a resource due to misuse. Although sheep and cattle have been at the core of the upland economy, a diversity of farm enterprise land uses and productive ecosystem services have been lost over recent centuries.

Woodland is one of these, with anthropogenic communities adapting to the loss via use of external resources for construction timber and local alternatives such as peat for fuel. Rather than extensive woodland, the pollen evidence suggests limited numbers of trees within the pastoral landscape, perhaps small woodland areas and scattered field-boundary trees and shrubs; less dense than Vera's parkland landscape, but trees in the landscape nonetheless. Documentary evidence demonstrates that, by the end of the seventeenth century and probably earlier, there was little woodland in the landscape, apart from some scrub on the limestone pavements and in steep-sided gills or ravines, and small stands of mature timber around the farmsteads. This was due in part to slack management by absentee manorial lords that allowed the customary tenants exceptional freedom to fell mature timber without permission from the lord of the manor or his bailiffs. In addition to the marked decline in arboreal pollen, the diagram from Wife Park shows a gradual decline in structural diversity from the fourteenth to the twentieth centuries. The investigation of the Ribblesdale area through palaeohistorical analysis raises questions about our understanding of the sustainability of traditional land management. It illustrates, in the case of the loss of the woodland resource, that a reduction of variability in the landscape can drive adaptation following a period of unsustainable resource use (e.g. Shaw and Whyte, 2010), however, it also illustrates that while several shifts in land management occurred, there was a gradual decline in structural diversity or heterogeneity, of which the woodland component was an important part.

Current post-production farm payments through the CAP favour a return to a more traditional grazing regime via an increased use of cattle. These animals graze selectively and create micro-regeneration sites for herbaceous species through trampling of the grass sward. However, although there is undoubtedly merit in this approach, the traditional management system that created biodiversity within the cattle-grazing system involved the extensive use of hay meadows and these are likely to have been the source of the vegetation diversity illustrated from the seventeenth century by pollen analysis. Although hay-meadow restoration is occurring today, it is not widespread enough to replace the previous system. In addition, this period is shown in the pollen record to be one where structural diversity had already declined due to poor management. Productive ecosystem services used by the traditional upland management system have, throughout post-medieval times, become less diverse. This is influenced by external economic pressures rather than a traditional indigenous subsistence community. A loss of spatial heterogeneity is likely to have led to losses in biodiversity (e.g. Benton *et al.*, 2003). This downward trend in structural diversity is likely to continue as the remaining scattered trees and shrubs within the current landscape become moribund.

Policy approaches to prescribe sustainable pastoral farming systems have tended to narrow the focus of the previous production system to the one

Figure 16.3 Open grazed landscape of upper Ribblesdale, North Yorkshire
Source: Helen Shaw

that is in place now (*sensu* Baskerville, 1997). There is a tension between the rhetoric of traditional management and the reality of implementing multi-functional approaches. Recreating past diversity associated with traditional management may involve the widespread reinstatement of hay meadows and the incorporation of trees into the landscape, in addition to maintaining and/or adjusting the pastoral elements of the system.

Discussion

How do these case studies contribute to thinking about woodland management and about 'Animals, Man and Treescapes', the theme of the 2011 Sheffield conference? In the case of Glen Affric, although the relict status of the woodland to the east of the glen is confirmed, the recent regeneration of woodland in the west is surprising. Advice may be to guard against too much woodland planting, in case the valuable open space between the trees creating a myriad of edge effects is lost. Woodland of such longevity and resilience should be able, given space, to maintain itself. In addition, species shifts should be expected and mixed woodland may provide a glimpse of the natural condition of the wooded Caledon. Contrastingly, in Ribblesdale, where an open pastoral landscape is the goal of reinstated

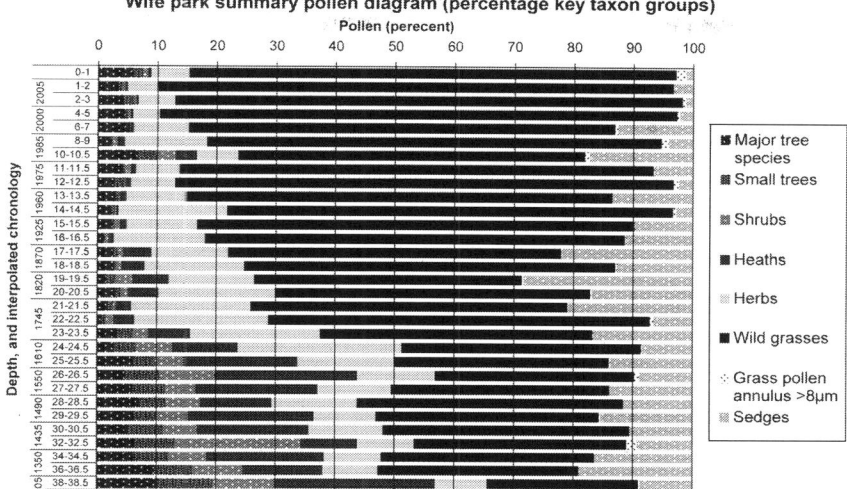

Figure 16.4 Summary pollen diagram from Wife Park, Ribblesdale
Note: Pollen is presented in key taxonomic groups as a percentage of total land pollen – excluding pine pollen assumed to be recently planted.
Source: Shaw and Whyte, 2013

traditional management regimes, more attention should be given to scattered stands of trees, lightly re-wooding some of the limestone pavements and adding tree and shrub species scattered within pastures to increase the diversity of vegetation without losing the pastoral elements. It may be too much to expect these wooded elements to return naturally in the pastoral landscape; planting and protection of saplings may be required to maintain heterogeneity.

Dynamic pasts mean that there are difficulties with implementing a baseline approach for biodiversity and conservation. Myllyntaus (2010) discusses the need for conservationists to make choices about which historical environment to recapture; however, historical environments, like natural ones, are created through process. Without a full understanding of process, idealised landscapes may be envisioned that may never have been sustainable – or even have existed at all. A more dynamic future needs to be planned for, where regeneration of trees is encouraged both within existing wood-pasture and within open pastoral systems, while, in woodlands, open space needs to grade to closed stands and in the last valuable relicts of wildwood we must be cautious in our approach to planting.

The results discussed illustrate the power of palaeoecological and combined palaeohistorical analysis to elucidate sometimes surprising, unexpected histories. This is by no means new for palaeoecology, which has provided many site-specific environmental histories where past dynamics can directly inform future management (e.g. Oldfield, 1970; Segerström et al., 1994).

Within Vera's woodland paradigm there is some critique of pollen analysis that has been variously reviewed, debated and refuted. However, it is important to recognise the widespread and meticulous work carried out by palaeoecologists over recent decades as well as ongoing research, in order to understand better the pollen–vegetation relationship. There continues to be a drive for scaled-up models to validate climate, elucidate regional and global-scale environmental change and contribute to overarching policy direction. However, to ensure high-quality conservation management, generalisations must be tested at the site scale. At this level modern local-scale palaeoecological and historical data can provide useful information. There may be no baseline but information on ecological processes and drivers of change and adaptation in the shifting systems; natural and semi-natural systems can be enlightening and provide management direction.

Conclusions

The findings discussed above suggest a need to challenge perceptions based upon the rural idyll of traditional management as well as myths and value judgements surrounding past natural landscapes and the drivers of their destruction. If we accept that a closed forest system is no longer relevant (Smout, 2010), even if it did once exist briefly in the mid-Holocene, and if we accept from ecology and palaeohistorical analyses, that some form of past openness created biodiversity through a myriad of shifting mosaics between grassland, heathland, bog and woodland, it is the scale and timing of the dynamic processes of ecosystems that are important for their management. We need to move beyond a baseline for cultural landscapes and guard against inflexible or separated woodland and grazing policies so that treescapes can be part of the wider landscape rather than confined to forestry, ancient woodland and amenity areas. This is especially so within the grazed landscape of the English uplands, where policy needs to reflect the very distinctive management histories. A process-based approach means accepting that transitions occur between communities, including wooded and non-wooded landscapes. Since historical landscapes leave a legacy more powerful for creating diversity than structure alone (Vellend et al., 2007), it will be important to build upon historical knowledge and to reinstate trees and woodland in areas of recent loss.

Developing open areas in woodlands and treescapes in pastoral areas forms challenges to thinking from a cultural perspective and from within the narrower confines of the biodiversity conservation movement. The development of a dialogue between social and ecological aspects of woodland futures is essential; however, we must be careful to identify and separate evidence-based knowledge of process in ecology from 'myth' and to ensure that current science knowledge is widely disseminated and accurately used to form the basis for discussion and implementation of ecological conversation.

Acknowledgements

The Glen Affric research was undertaken for a PhD at the University of Stirling, funded by Stirling University and Forest Research (Jonathan Humphrey) and supervised by Richard Tipping, Andrew Tyler and Philip Wookey. AMS radiocarbon dates mentioned in the text were funded by an application to the NERC Radiocarbon Laboratory. The Ribblesdale research was funded by the Leverhulme Trust (Research Grant no F/100 185/V). Radiocarbon dates were supplied by the NERC SUERC Radiocarbon facility and funded by the Leverhulme Trust. The authors are grateful to farmers and landowners in Ribblesdale for land access.

References

Adams, W.M. (1997) 'Rationalization and conservation: Ecology and the management of nature in the United Kingdom', *Transactions of the Institute of British Geographers, New Series*, 22(3): 277–91.
Adams, W.M. (2003) *Future Nature: A Vision for Conservation*, Earthscan, London.
Aldhous, J.R. (ed.) (1995) *Our Pinewood Heritage*, Conference Proceedings, 20–22 October 1994, Inverness, Forestry Authority, Edinburgh.
Baskerville, G.L. (1997) 'Advocacy, science, policy, and life in the real world', *Conservation Ecology*, 1(1): 9.
Bennett, K.D. (1984) 'The post-glacial history of *Pinus sylvestris* in the British Isles', *Quaternary Science Reviews*, 3: 133–55.
Bennett, K.D. (1989) 'A provisional map of forest types for the British Isles 5000 years ago', *Journal of Quaternary Science*, 4: 141–44.
Bennett, K.D. (1995) 'Post-glacial dynamics of Pine (*Pinus sylvestris* L.) and pinewoods in Scotland', in Aldhous, J.R. (ed.) *Our Pinewood Heritage: Proceedings of a Conference at Culloden Academy 20–22 Oct 1994*, Inverness, Forestry Commission, The Royal Society for the Protection of Birds, Scottish Natural Heritage.
Benton, T.G., Vickery, J.A. and Wilson, J.D. (2003) 'Farmland biodiversity: Is habitat heterogeneity the key?', *Trends in Ecology and Evolution*, 18: 182–8.
Bignal, E. and McCracken, D. (2009) 'Herbivores in space: Extensive grazing systems in Europe', *British Wildlife*, 20(5) (Special supplement) June: 44–9.
Birks, H.J.B. (1989) 'Holocene isochrone maps and patterns of tree-spreading in the British-Isles', *Journal of Biogeography*, 16: 503–40.
Birks, H.J.B. (1997) 'Loch Sionascaig', Chapter 6 in 'North-west Highlands', *Geological Conservation Review*, Vol. 6, Quaternary of Scotland: extract at http://jncc.defra.gov.uk/pdf/gcrdb/GCRsiteaccount2747.pdf accessed 23/02/2012
Botkin, D.B. (1992) *Discordant Harmonies: A New Ecology for the 21st Century*, Oxford University Press, USA, new edition.
Breeze, D. (1992) 'The great myth of Caledon', *Scottish Forestry*, 46: 331–5.

Bridge, M.C., Haggart, B.A. and Lowe, J.J. (1990) 'The history and paleoclimatic significance of subfossil remains of *Pinus sylvestris* in blanket peats from Scotland', *Journal of Ecology*, 78: 77–99.

Brooker, R.W., Scott, D.D., Palmer, S.F. and Swaine, E.E. (2006) 'Transient facilitative effects of heather on Scots pine along a grazing disturbance gradient in Scottish moorland', *Journal of Ecology*, 94:(3): 637–45.

Broström, A., Sugita, S. and Gaillard, M.-J. (2004) 'Pollen productivity estimates for reconstruction of past vegetation cover in the cultural landscape of Southern Sweden', *The Holocene*, 14: 371–84.

Broström, A., Nielsen, A.-B., Gaillard, M.-J., Hjelle, K., Mazier, F., Binney, H., Bunting, M.-J., Fyfe, R., Meltsov, V., Poska, A., Rasanen, S., Soepboer, W., von Stedingk, H., Suutari, H. and Sugita, S. (2008) 'Pollen productivity estimates of key European plant taxa for quantitative reconstruction of past vegetation: A review', *Vegetation History and Archaeobotany*, 17(5): 461–78.

Bunce, R.G.H. and Jeffers, J.N.R. (eds) (1977) *Native pinewoods of Scotland*, Institute of Terrestrial Ecology, Cambridge.

CBD (Convention on Biological Diversity) (1992) *Convention on Biological Diversity*, UNEP, www.cbd.int/doc/legal/cbd-en.pdf, accessed 1 March 2012.

COP2 (Conference of the Parties 2) (1995) *COP 2 Decisions. Second Ordinary Meeting of the Conference of the Parties to the Convention on Biological Diversity*, 6–17 November 1995, Jakarta, Indonesia, decision II/8 http://www.cbd.int/decision/cop/?id=7081, accessed 1 March 2012.

Crawford R.M.M. (2000) 'Ecological hazards of oceanic environments', *New Phytologist*, 147: 257–81.

Crawford, R.M.M., Jeferee, C.E. and Rees, W.G. (2003) 'Paludification and forest retreat in northern oceanic environments', *Annals of Botany*, 91: 213–26.

Defra (Department for Environment, Food and Rural Affairs) (2007) 'Securing a healthy natural environment: An action plan for embedding an ecosystems approach', Department for Environment, Food and Rural Affairs, London.

Dubois, A.D. and Ferguson, D.K. (1985) 'The climatic history of pine in the Cairngorms based on radiocarbon dates and stable isotope analysis, with an account of the events leading up to its colonization', *Review of Palaeobotany and Palynology*, 46: 55–80.

Edwards, C. and Mason, W.L. (2006) 'Stand structure and dynamics of four native Scots pine (Pinus sylvestris L.) woodlands in northern Scotland', *Forestry*, 79: 261–77.

Fenton J. (1997) 'Native woods in the Highlands: Thoughts and observations', *Scottish Forestry*, 51: 160–64.

Forestry Commission (1998) Caledonian Pinewood Inventory 1998, Forestry Commission, Edinburgh. Froyd, C. and Bennett, K. (2006) 'Long-term ecology of native pinewood communities in East Glen Affric, Scotland', *Forestry*, 79: 279–91

Gaillard, M.-J., Sugita, S., Bunting, M. J., Middleton, R., Brostrom, A., Caseldine, C., Giesecke, T., Hellman, S., Hicks, S., Hjelle, K., Langdon, C., Nielsen, A.-B., Poska, A., von Stedingk, H., Veski, S. and POLLANDCAL members (2008) 'The use of modelling and simulation approach in reconstructing past landscapes from fossil pollen data: A review and results from the POLLANDCAL network', *Vegetation History Archaeobotany*, 17(5): 419–43.

Gear, A.J. and Huntley, B. (1991) 'Rapid changes in the range limits of Scots Pine 4,000 years ago', *Science*, 251: 544–7.

Gleason, H.A. (1926) 'The individualistic concept of the plant association', *Bulletin of the Torrey Botanical Club*, 53: 7–26.
Hellman, S., Gaillard, M.-J., Brostrom, A. and Sugita, S. (2008) 'The REVEALS model, a new tool to estimate past regional plant abundance from pollen data in large lakes: Validation in southern Sweden', *Journal of Quaternary Science*, 23(1): 21–42.
Hodder, K.H., Buckland, P.C., Kirby, K.J. and Bullock, J.M. (2009) 'Can the pre-Neolithic provide suitable models for re-wilding the landscape in Britain?', *British Wildlife*, 20(5) (Special supplement), June.
Humphrey, J., Ray, D., Watts, K., Brown, C., Poulsom, L., Griffiths, M. and Broome, A. (2004) *Balancing Upland and Woodland Strategic Priorities*, Scottish Natural Heritage Commissioned Report No. 037 (ROAME No. F02AA101).
Humphrey, J.W. (2006) Ecology and management of native pinewoods: Overview of special issue', *Forestry*, 79(3): 245–7.
Jacobson, G.L. Jr. and Bradshaw, R. (1981) 'The selection of sites for paleovegetational studies', *Quaternary Research*, 16: 80–96.
Kirby, K.J. (2003) *What Might a British Forest Landscape Driven by Large Herbivores Look Like?*, English Nature Research Report 530, Peterborough.
Kirby, K.J. (2004) 'A model of a natural wooded landscape in Britain as influenced by large herbivore activity', *Forestry*, 77(5): 405–20.
Kreuz, A. (2008) 'Closed forest or open woodland as natural vegetation in the surroundings of Linearbandkeramik settlements?', *Vegetation History and Archaeobotany*, 17(1): 51–64.
MacDonald, D., Crabtree, J.R., Wiesinger, G., Dax, T., Stamou, N., Fleury, P., Gutierrez Lazpita, J. and Gibon, A. (2000) 'Agricultural abandonment in mountain areas of Europe: Environmental consequences and policy response', *Journal of Environmental Management*, 59(1): 47–69.
McLauchlan, K.K., Elmore, A.J., Oswald, W.W. and Sugita, S. (2007) 'Detecting open vegetation in a forested landscape: Pollen and remote sensing data from New England, USA', *Holocene*, 17(8): 1233–43.
Miles, H. and Jackman, B. (1991) *The Great Wood of Caledon*, Seven Hills Books, Lanark.
Mitchell, F.J.G. (2005) 'How open were European primeval forests? Hypothesis testing using palaeoecological data', *Journal of Ecology*, 93: 168–77.
Myllyntaus, T. (2010) 'Changing Forests, moving targets in Finland', in Hall, M. (ed.) *Restoration and History: The Search for a Useable Environmental Past*, Routledge, London, pp46–57.
Oldfield, F. (1970) 'The ecological history of Blelham Bog National Nature Reserve', in Walker, D. and West R.G. (eds) *Studies in the Vegetational History of the British Isles*, Cambridge University Press, Cambridge, pp141–57.
Pauly, D. (1995) 'Anecdotes and the shifting baseline syndrome', *Trends in Ecology & Environment*, 10: 430.
Palmer, S.C.F. and Truscott, A.-M. (2003) 'Browsing by deer on naturally regenerating Scots pine (Pinus sylvestris L.) and its effects on sapling growth', *Forest Ecology and Management*, 182(1–3): 31–47.
Scottish Executive (2006) *The Scottish Forestry Strategy*, Forestry Commission Scotland, Edinburgh.
Segerström, U., Bradshaw, R.H.W., Hörnberg, G. and Bohlin E. (1994) 'Disturbance history of a swamp-forest refuge in northern Sweden', *Biological Conservation*, 68(2): 189–96.

Shaw, H. (2006) 'A palaeoecological investigation of stand-scale ecological dynamics in semi-open native pine woods', unpublished PhD thesis, School of Biological and Environmental Science, University of Stirling, Stirling.

Shaw, H. (2011) 'The history of woodland dynamics in relict east Glen Affric pinewoods: A lack of human impact?', in *Woods as Working and Cultural Landscapes – Past and Present*, Scottish Woodland History Discussion Group Notes, XV, 2011.

Shaw, H. and Tipping, R. (2006) 'Recent pine woodland dynamics in east Glen Affric, northern Scotland, from highly resolved palaeoecological analyses', *Forestry*, 79: 331–40.

Shaw, H. and Whyte, I. (2008) 'Shifting ecosystem services through time in the North West uplands and implications for planning adaptation in the future', *Aspects of Applied Biology*, 'Shaping a vision for the Uplands', 85: 99–106.

Shaw, H. and Whyte, I. (2010) 'Land management and biodiversity through time in upper Ribblesdale, North Yorkshire, UK: Understanding the impact of traditional management', *Landscape Archaeology and Ecology – End of Tradition*, 8(1–2): 165–75.

Shaw, H. and Whyte, I. (2013) 'Land management and biodiversity through time in Upper Ribblesdale, North Yorkshire, UK: Understanding the impact of traditional management', in Rotherham, I.D. (ed.) *Cultural Severance and the Environment: The Ending of Traditional and Customary Practice on Commons and Landscapes Managed in Common*, Springer, Dordrecht, The Netherlands.

Sheppard, C. (1995) 'Shifting baseline syndrome', *Marine Pollution Bulletin*, 30: 766–7.

Smout, T.C. (2000) *Nature Contested: Environmental History in Scotland and Northern England since 1600*, Edinburgh University Press, Edinburgh.

Smout, T.C. (2010) 'Regardening and the rest', in Hall, M. (ed.) *Restoration and History: The Search for a Useable Environmental Past*, Routledge, New York, pp111–24.

Staines, B.W., Balharry, R. and Welch, D. (1995) 'The impact of red deer and their management on the natural heritage in the uplands', in Thompson, D.B.A. and Usher, M.B. (eds) *Heaths and Moorland: Cultural Landscapes*, SNH/HMSO, Edinburgh, pp294–305.

Steven, H.M. and Carlisle, A. (1959) *The Native Pinewoods of Scotland*, Oliver and Boyd, Edinburgh.

Stewart, M.J. (2010) 'Does the past matter in Scottish woodland restoration?', in Hall, M. (ed.) *Restoration and History: The Search for a Useable Environmental Past*, Routledge, New York, pp63–73.

Sugita, S. (2007a) 'Theory of quantitative reconstruction of vegetation. I. Pollen from large sites REVEALS regional vegetation', *Holocene*, 17: 229–41.

Sugita, S. (2007b) 'Theory of quantitative reconstruction of vegetation. II. All you need is love', *Holocene*, 17: 243–57.

Svenning, J-C. (2002) 'A review of natural vegetation openness in north-western Europe', *Biological Conservation*, 104: 133–48.

Swales, S. (1987) 'The vegetational and archaeological history of the Ingleborough Massif, North Yorkshire', unpublished PhD Thesis, University of Leeds, Leeds.

Tipping, R. (1994) 'The form and fate of Scotland's woodlands', *Proceedings of the Society of Antiquaries in Scotland*, 124, pp1–54.

Tipping, R. (2003) 'Living in the past: Woods people and prehistory to 1000 BC', in Smout, T.C. (ed.) *People and Woods in Scotland*, Edinburgh University Press, Edinburgh, pp14–39.

Tipping, R., Davies, A. and Tisdall, E. (2006) 'Long-term woodland dynamics in West Glen Affric, northern Scotland', *Forestry*, 79: 351–9.

Trees for Life (n.d.) www.treesforlife.org.uk/, accessed 25 February 2012.

Vellend, M., Verheyen, K., Flinn, K., Jacquemyn, H., Kolb, A., Van Calster, H., Peterken, G., Graae, B., Bellemare, J., Honnay, O., Brunet, J., Wulf, M., Gerhardt, F. and Hermy, M. (2007) 'Homogenization of forest plant communities and weakening of species–environment relationships via agricultural land use', *Journal of Ecology*, 95(3): 565–73.

Vera, F.W.M. (2000) *Grazing Ecology and Forest History*, CABI Publishing, Wallingford, Oxon.

Vera, F.W.M (2009) 'Large-scale nature development – the Oostvaardersplassen', *British Wildlife*, 20(5) (Special supplement): 28–36.

Zackrisson, O., Nilsson, M-C., Steijlen, I. and Hornberg, G. (1995) 'Regeneration pulses and climate–vegetation interactions in non-pyrogenic boreal Scots pine stands', *Journal of Ecology*, 83: 469–83.

17 Integrated conservation of a park and its associated cattle herd

Chillingham Park, northern England

Stephen J.G. Hall

Introduction

Chillingham Park in Northumberland is one of the most famous English parklands. The chapter considers how the landscape has been managed in the past and what the future might hold. The Chillingham cattle breed and their habitat are placed in a wider context.

Chillingham Park, Northumberland has, at least since 1646 and perhaps for much longer, been the home of the Chillingham wild white cattle (Figure 17.1). The cattle are of zoological interest because of their relatively unmanaged state, and because of their distinct history and isolation they are a breed in their own right. Apart from a reserve herd in north-east Scotland they are found nowhere else. Chillingham Park is relatively well-documented from the eighteenth century onwards but its medieval history is almost unknown.

The cattle through the works of artists notably Bewick and Landseer have played a part in developing concepts of the wild animal kingdom; making the Chillingham wild white bull, according to Schama (2002: 126), 'an image of massive power{...}perhaps the greatest icon of British natural history'. Their charisma can be partly attributed to the mystery that has shrouded their origins, which has enabled legends to arise, to be fostered and to flourish around them (Hemming, 2002; Ritvo, 1992).

Some aspects at least of their history and biology are accessible to research. While in very many ways the cattle and their habitat are unique, experience with large herbivores elsewhere will be of benefit to their management, and they can also impart lessons that may also benefit conservation generally. Understanding the history of the park requires it to be placed in several contexts; its relationship to Chillingham Castle, the parish and its ancient church (Heslop and Harbottle, 1999; Pevsner *et al.*, 2002), indeed the county of Northumberland as a whole. A start has been made on this contextualisation (McGowan, 2010), and here I focus on the evident functional relationships between the cattle, their pastures and the trees.

The park is a Georgian design, originally of 610ha encompassed by a stone wall 9.2km in length, imposed on an earlier layout (Hall, 2007 and 2010).

Figure 17.1 Members of the herd on pasture near Chillingham Burn, extreme east of the park
Source: Stephen J.G. Hall

A licence to enclose had been issued to Sir Ralph Grey of Chillingham in 1626 and in 1634 tenants in the neighbouring parish of Chatton complained that he had enclosed land from Chatton common (Dodds, 1935). The earliest known map, the estate terrier of 1711, reproduced by Hall (2010), depicts the Great Wood of Chillingham occupying the lower relatively flat ground, with open land with streamside woodland and scattered trees (presumably a wood-pasture) around it. However, the subsequent re-modellings, in the late 1750s and between 1799 and 1808 involved, initially, the clearing of the Great Wood and felling of almost all trees, and later, the grubbing up of the remaining stumps. Most of the Great Wood area was converted into pasture and (continuing into the nineteenth century) new woods were established on the sloping ground.

Under economic pressure, and over several decades, land was removed from the park during the twentieth century as from the rest of the Tankerville estates (Hodgson, 1916). The range historically occupied year-round by the cattle remained unchanged and, with the surrounding woodlands, is now owned by the Chillingham Wild Cattle Association (CWCA), a registered charity. Other woodlands and former hayfields, to which the cattle probably had access for limited periods each year, are now under separate ownerships. This account focuses on the CWCA holding.

Background to the Chillingham wild cattle

The Chillingham wild cattle are of unique significance as a breed (Hall and Clutton-Brock, 1988). This is partly because, unique among British cattle, they were not selectively bred, except for, reportedly, culling of wrongly coloured calves during the eighteenth century (Whitehead, 1953) and, perhaps, a degree of selective castration during the nineteenth century (Hall and Hall, 1988). They have contributed genetic material to the White Park cattle, but the latter is a composite breed primarily derived from the Cadzow, Vaynol, Dynevor and Chartley herds and with contributions from the Longhorn, Highland and Welsh Black. These four principal herds were associated with, respectively, Cadzow (Lanarkshire), Glan Faenol (Vaynol; Carnarvonshire) Dinefwr (Dynevor; Carmarthenshire) and Chartley (Staffordshire). They were crossed either with each other or with other breeds (for details see Whitehead, 1953); the Vaynol herd has only been there since 1872 and the continuity of occupancy was lost for this herd and for Chartley; and for all a more intensive form of husbandry (planned matings, castration and culling) was practised than has been the case at Chillingham.

The Chillingham herd is unique in having a long documented history (back to 1645; the histories of the Cadzow and Dynevor herds are equivocal before the nineteenth century), and in having an undeniable association with the park they inhabit. They have inspired a considerable amount of scientific work that has contributed substantially to the development of conservation biology (see for example Burthe et al., 2011; Visscher et al., 2001).

The cattle

Improved management and the extinction of a sheep-grazing tenancy are credited with a rise in total herd size from an average of 51 head in the last decade of the twentieth century to 108 in October 2012 (C.J. Leyland, pers. comm.). Although animals are culled on welfare grounds (Hall et al., 2005) there is no castration and winter hay feeding is the only human intervention. Local experience suggests 120 is the manageable herd size for the area; but as males slightly outnumber females and because the cows tend to be late maturing, a herd of that size is unlikely to include more than 30 breeding females.

In contrast, between 1862 and 1899, the herd was maintained at an average size of 61 (17 entire males, 13 castrated males, 31 females) with animals occasionally being taken for feasts and trophies (Hall and Hall, 1988) Data on numbers of calves born in relation to cows in the herd can be compared between Victorian times and today and there is no apparent diminution in herd fertility, which argues for the long period of inbreeding (there being no record of introduction from outside at least since the mid-1700s) having acted, perhaps with a degree of natural selection, to purge the population of harmful recessive genes (Visscher *et al.*, 2001). Generally, mortality patterns and population dynamics appear similar to those of other large, northern herbivores, and this is the subject of current research.

The pasture

The cattle park comprises 142.61ha, of which the uppermost 10.70ha is heathland and 18.84ha is mature unfenced woodland. The park is not really to be seen as a wood-pasture; it is a set of unenclosed pastures (Figure 17.2) with extensive streamside woodlands and scattered plantations (Hall and Bunce, 2011). From 1977 to 1982, behavioural studies on the herd showed sex differences in ranging and grazing behaviour with seasonal variation in the use of different vegetation types (Hall, 1988). Underpinning this work, the vegetation of the park was mapped in 1979 (Hall and Bunce, 1984).

Historically, pasture management had been limited to opportunistic cutting of bracken (Bennet, 1979) though there had been sporadic fertilisation and under-draining of certain areas before about 1970. In early 1980, six lactating cows died apparently of magnesium deficiency and a rotational liming programme was initiated to improve herd nutrition through increasing herbage quality and yield. This continued until 2005, and was successful in that mortality of this kind has not subsequently been observed.

In 2008, the 1979 vegetation survey was repeated (Bunce and Hall, in press). Elsewhere in upland Britain, over the intervening period substantial reductions in plant species diversity had been observed (Carey *et al.*, 2008); the reduction in species richness of sampled areas by 23 per cent was therefore not surprising. This is thought to be due to a combination of the liming programme, inputs of anthropogenic fixed nitrogen and the redistribution of nutrients by grazing sheep.

The trees

The oldest tree in Chillingham is the Dwarf Oak (Anon., 1906), a lapsed pollard in the grounds of Chillingham Castle. Trees within Chillingham but outside the park have not otherwise been formally described. Trees within the park were surveyed in 2008 (Hall and Bunce, 2011) and can be described, in descending order of biological interest, as follows:

1   Stem apparently untouched in the fellings of the 1750s – one oak (Figure 17.3), perhaps marking the eastern extremity of the Great Wood;
2   Stems representing re-growth from stumps since the 1750s fellings – alder and ash along the streamsides (Figure 17.3) and in the wet ground known as the Allers, which appears to be the only part of the Great Wood that was not grubbed up;

3   Small woodlands and open-grown trees planted from 1789 onwards, primarily of beech and oak, but also Scots pine and other species;
4   Individual exotic species including deodar, Norway maple, larch, Douglas fir, in very small numbers, after about 1860, though documentation is lacking;
5   In the 1960s, commercial conifers (under dedication schemes) in the woodlands around the park, which comprise the scenic backdrop. These replaced early nineteenth century plantings, primarily of oak, which had been felled during World War I. Some of the boundaries between the park and the former hayfields were also demarcated by strips of new woodland, mainly of conifers;
6   From 1991 (Hall, 1992) replacement trees were planted to fill gaps in the woods within the park; these were primarily oak and beech with Scots pine nurses.

From 2005, the current programme of tree planting has been in effect with three components:

1   Replacement of the Scots and lodgepole pine plantation on the skyline to the east of the park, with juniper and heather regeneration;
2   Installation of circular cattle-proof roundels of approximately 200m^2 in appropriate areas to protect oak and beech regeneration;
3   Protection of individual self-seeded alders, with some transplantation.

Figure 17.2 Two bulls on the Sandy Banks at the north-west of the park
Note: The southeast boundary of the Park is visible, with tree re-growth beyond and the heather moorland of Hepburn Moor on the skyline
Source: Stephen J.G. Hall

Figure 17.3 Winter at Chillingham
Note: The oldest oak (estimated planting/seeding date 1620) is in the foreground, surrounded by fallen dead wood. Ash/alder streamside vegetation is on the left-hand side (last coppiced around 1754–60); beech plantation (planted 1789) on the right.
Source: Stephen J.G. Hall

Integrated conservation at Chillingham

CWCA was founded in 1939 by the 8th Earl of Tankerville and the survival of the herd thus far had been due to the commitment of the earl and his forebears. His widow, Violet, Dowager Countess of Tankerville died in November 2003, pre-deceased by her son, the Hon. Ian Bennet five years earlier, and between them they had effectively taken responsibility for the herd since the 8th earl died in 1973. Latterly, financial and other support from the Knott Trust with donations from the Duke of Northumberland and others enabled CWCA to purchase the freehold of the park and surrounding woodlands, and these were reunited under the same ownership with the cattle in 2005.

In 2005, Higher Level Stewardship support was obtained but unlike many beneficiaries of the scheme, Chillingham Park did not include any sites of special scientific interest (SSSIs). Generous and indispensable funding and in-kind contributions have come from other individuals, trusts and organisations, a key grant at present being from the Tubney Trust for environmental remediation.

Figure 17.4 The herd during winter at Chillingham
Note: The herd is being fed hay in the distance; a small subgroup of bulls is in the foreground. Between the cattle groups can be seen the historic boundary of the former Great Wood of Chillingham (cleared and boundary suppressed 1754–60). Just behind the main cattle group is Foxes Knowes, felled probably in 1914–18 with outermost trees retained, fenced and restocked in 1991. In the left background is Ros Hill (not accessible to the cattle), where the 1960s pine plantation is currently being cleared and replaced with juniper and heather.
Source: Stephen J.G. Hall

It is relatively unusual for a park to have both trees and pasture in relatively good conservation condition (Cox and Sanderson, 2001), and Chillingham is unique in the unbroken association with its indigenous breed. While the maintenance of the herd is the primary aim of CWCA, conservation of its environment and landscape (Figure 17.4) are now also acknowledged as within its remit.

Implications for elsewhere

Growing interest in the conservation of wood-pastures and other pastoral systems can benefit from experience at Chillingham. A further topicality is added through the ancient association of the park with a specific cattle breed. In continental Europe, many parks and designed landscapes survive (Hunt, 2003) but are appreciated as cultural landscapes rather than refuges for biodiversity. Most northern European wood-pastures have been lost (but see Buttler *et al.*, 2008; Finck *et al.*, 2005; Luoto *et al.*, 2003; Mitlacher *et al.*, 2002; Pott, 1999; Sammul, 2008), while in southern Europe (Milán *et al.*, 2006; Vieira and Eden, 1995), they are still extensive but under considerable threat. The remaining fragments of continental wood-pasture have proved difficult to conserve, apparently because this habitat type is only 'inconsistently' recognised in the EU Habitats Directive (Bergmeier et al., 2010) and risks are being posed to high nature value farming (Jones, 2011). Their conservation and restoration as functional landscapes will require an understanding of the interrelationships of the livestock, pasture and trees.

The main focus of wood-pasture conservation has been the maintenance of floral and faunal biodiversity, and little attention has been paid to the opportunities for conserving the local, adapted livestock breeds of these land classes (but see Rois-Días et al., 2006). Admittedly, management of wood-pastures for floral and faunal biodiversity would be simpler if genetic conservation of the grazing livestock were not a concern, as herd managers and conservation managers will usually have different goals (Small, 2010). However, wood-pasture conservationists might care to note the large sums of money made available under agri-environmental schemes for local traditional livestock in general and the clear political will to support these breeds. In 2001, in the EU there were 8,442 agri-environment contracts covering 60,568 livestock units (Mills et al., 2007). Considering England alone, in 2010 there were 616 current contracts for the HR2 Native Breeds at Risk supplement in the English Higher Level Stewardship scheme covering 30,635ha at a total cost of £18.3 million (J. Hosking, pers. comm.).

Wood-pastures and their traditional livestock

Local hardy breeds of grazing livestock traditionally associated with wood-pastures include the New Forest pony (England: (Tubbs, 1997), Iberian pig and Retinta, Morucha and Avileña-Black Iberian cattle (dehesas, Spain: (Milán et al., 2006), and Mirandesa, Arouquesa, Barrosã, Marinhoa, Maronesa, Alentejana, Mertolenga and Raça Preta cattle (montados, Portugal: Coelho et al., 1999; Rodrígues et al., 1999).

Traditionally, wood-pasture was also used for the making of hay and tree fodder, and livestock was often housed in winter or depended on seasonal transhumance (Bunce et al., 2008; Clément, 2008). Presumably as a consequence of low-nutrient status, wood-pasture has been insufficiently productive to fatten livestock, as attested by the historic importance of cattle droving with cattle being bred in extensive systems and then brought to slaughter condition near the cities (UK example: Haldane, 1952). Similarly, today most Iberian dehesa beef cattle are fattened on the farm and not on pasture (Milán et al., 2006) and acquire 34 per cent of their dietary energy as supplementation (Polo et al., 2004). In step with adoption of higher-input management, more productive livestock genotypes are being used (Plieninger and Wilbrand, 2001). Clearly, more information is needed on how resilient north European wood-pastures are to overexploitation and intensification of grazing.

A compelling reason why genetic conservation of livestock breeds should be considered in parallel with wood-pasture biodiversity conservation is the difficulty of conserving, *ex situ*, the adaptations of breeds. In principle, embryo and semen banks (Simm *et al.*, 2004) could conserve these traits, but with an *in situ* programme the traits will become better understood and characterised, and the cultural benefits will be maintained. The main factors in limiting loss of genetic variation are the maximising of effective population size (adopted as an indicator by Defra, 2010) and the equalising of family size, both of which are promoted by having many breeding males all of approximately equal influence (Falconer and Mackay, 1996). Large numbers of small herds would be an effective way of maintaining traits for local adaptation.

Ideally a conservation herd would be self-contained, and for this a minimum size of herd and therefore of range is required. Guidance as to the optimum size of such a herd might come from general experience of keepers of rare and minority breeds. For example in the UK, the median number of adult females in Lincoln Red cattle herds in 2008 was 16, in Longhorn cattle herds in 2009 it was 11 (means were 30 and 18 respectively) (Hall, 2011). Presuming therefore that a herd of about 20 cows plus followers is optimal for management, and a stocking rate similar to those recorded by Hearn (1995) of 360–800kg ha^{-1} is appropriate, coordinated conservation of a holarctic wood-pasture and a herd of the local breed of cattle would require around 50ha.

Decoupling of CAP farm income support from production is intended to benefit biodiversity but may threaten genetic conservation of farm animals by reducing numbers of extensively farmed livestock (Acs *et al.*, 2010; Osterburg and von Horn, 2006; Uthes *et al.*, 2008). In this connection, Small and Hosking (2010) reviewed modes of support for farm animal genetic resources in 25 European countries. England and Scotland link support to agri-environment schemes; all others support eligible breeds on a per head basis. English and Scottish participants in certain schemes, who keep specified breeds, can receive support proportional to the area occupied by the livestock, and there is no stipulation that the breeds kept should be native to the particular area. However, the use of non-priority breeds would disqualify. The other European schemes would tend to favour traditional systems but this is not necessarily associated with specific faunal and plant biodiversity outcomes.

Support allocated on a per head basis might risk overstocking and overproduction and (although authorised under EU Regulation 1698/2005) would appear to be at variance with the concept of decoupling. Area-based support systems avoid this, but breed support is not accessible to all keepers of relevant breeds because the breed component is only available as a supplement to the main conservation activities. Thus both these contrasting approaches are open to criticism, and experience in the maintenance of breeds within traditional systems will be valuable should policy options in this area be reviewed.

This is the first account bearing explicitly on the parallel conservation of a grazed ecosystem and a livestock breed. Although the cattle still have priority at Chillingham, the trees have probably been resilient and the sward has retained much of its interest and it will be possible to assess in due course, whether the more sympathetic pasture-management regime will have retrieved the situation further. This study also draws attention to the previously neglected notion that the conservation of an indigenous, locally adapted breed could, and should, be planned and conducted in partnership with that of its environment.

References

Acs, S., Hanley, N., Dallimer, M., Gaston, K.J., Robertson, P., Wilson, P. and Armsworth, P.R. (2010) 'The effect of decoupling on marginal agricultural systems: Implications for farm incomes, land use and upland ecology', *Land Use Policy*, 27: 550–563.

Anon. (1906) 'Report of meeting at Chillingham', *History of the Berwickshire Naturalists' Club*, 20(2): 11–19.

Bennet, I. (1979) 'The challenge of Chillingham Park', *Roebuck. Journal of the Northumberland Wildlife Trust*, 26: 13–18.

Bergmeier, E., Petermann, J. and Schröder, E. (2010) 'Geobotanical survey of wood-pasture habitats in Europe: diversity, threats and conservation', *Biodiversity and Conservation*, 19: 2995–3014.

Bunce, R.G.H. and Hall, S.J.G. (in press) 'Vegetation change from 1979 to 2008 at Chillingham Park in relation to conservation of the Chillingham Wild Cattle', *Northumbrian Naturalist*.

Bunce, R.G.H., Pérez-Soba, M. and Smith, M. (2008) 'Assessment of the extent of agroforestry systems in Europe and their role within transhumance systems', in A. Rigueiro-Rodríguez, J. McAdam and M. Mosquera-Losada (eds) *Agroforestry in Europe: Current Status and Future Prospects. Advances in Agroforestry*, Vol. 6, Springer, Netherlands, pp321–9.

Burthe, S., Butler, A., Searle, K.R., Hall, S.J.G. and Thackeray, S.J. (2011) 'Demographic consequences of increased winter births in a large aseasonally breeding mammal (*Bos taurus* L.) in response to climate change', *Journal of Animal Ecology*, 40: 1134–44.

Buttler, A., Kohler, F. and Gillet, F. (2008) 'The Swiss mountain wooded pastures: Patterns and processes', *Advances in Agroforestry*, 6: 377–96.

Carey, P.D., Wallis, S.M., Emmett, B.E., Maskell, L.C., Murphy, J., Norton, L.R., Simpson, I.C. and Smart, S.S. (2008) *Countryside Survey: UK Headline Messages from 2007*, NERC Centre for Ecology and Hydrology, Wallingford, Oxon.

Clément, V. (2008) 'Spanish wood pasture: Origin and durability of an historical wooded landscape in Mediterranean Europe', *Environment and History*, 14: 67–87.

Coelho, I., Fragata, A., Galvao-Teles, C. and Simoes, J. (1999) '"Raça Preta", a Portuguese beef breed: Economic and environmental objectives in natural resources management', in J.P. Laker and J.A. Milne (eds) *Livestock Production in the European Less Favoured Areas: Meeting Future Economic, Environmental, and Policy Objectives through Integrated Research. Proceedings of the 2nd Conference of the LSIRD Network, Bray, Dublin, Ireland*, Macaulay Land Use Research Institute, Craigiebuckler, Scotland, pp189–91.

Cox, J. and Sanderson, N. (2001) 'Livestock grazing in National Trust parklands – its impact on tree health and habitat', report commissioned by the National Trust Estates Department, Cirencester, Jonathan Cox Associates, Lymington, Hampshire, UK.

Defra (Department for Environment, Food and Rural Affairs) (2010) 'UK biodiversity indicators in your pocket', http://jncc.defra.gov.uk/pdf/BIYP_2010.pdf

Dodds, M.H. (1935) 'Chillingham parish. Area and population', in M.H.Dodds (ed.) *A history of Northumberland (Victoria County History) Vol. XIV*, Andrew Reid and Co., Newcastle upon Tyne, pp301–6.

Falconer, D.S. and Mackay, T.F.C. (1996) *Introduction to Quantitative Genetics*, 4th edition, Longman, Harlow, UK.

Finck, P., Riecken, U. and Glaser, F. (2005) 'Situation and perspectives of silvopastoral systems in Germany', in M. Mosquera-Losada, J. McAdam and A. Rigueiro-Rodríguez (eds) *Silvopastoralism and Sustainable Land Management. Proceedings of an International Congress on Silvopastoralism and Sustainable Management held in Lugo, Spain, in April 2004*, CABI Publishing, Wallingford, pp397–9.

Haldane, A.R.B. (1952) *The Drove Roads of Scotland*, Nelson, London.

Hall, S.J.G. (1988) 'Chillingham Park and its herd of white cattle: Relationships between vegetation classes and patterns of range use', *Journal of Applied Ecology*, 25: 777–89.

Hall, S.J.G. (1992) 'Caring for Chillingham Park', *Ark*, 19: 369.

Hall, S.J.G. (2007) 'Chillingham wild cattle park, Northumberland', *Landscape Archaeology and Ecology*, 6: 53–7.

Hall, S.J.G. (2010) 'Caring for the legend of the wild bull: An interpretation of the Georgian landscape of Chillingham Park, Northumberland', *Garden History*, 38: 213–30.

Hall, S.J.G. (2011) 'Number of females in cattle, sheep, pig, goat and horse breeds predicted from a single year's registration data', *Animal*, 5: 980–5.

Hall, S.J.G. and Bunce, R.G.H. (1984) 'Vegetation survey of Chillingham Park, Northumberland', *Transactions of the Natural History Society of Northumbria*, 52: 5–14.

Hall, S.J.G. and Bunce, R.G.H. (2011) 'Mature trees as keystone structures in Holarctic ecosystems – a quantitative species comparison in a northern English park', *Plant Ecology and Diversity*, 4: 243–50.

Hall, S.J.G. and Clutton-Brock, J. (1988) *Two Hundred Years of British Farm Livestock*, British Museum (Natural History), London.

Hall, S.J.G. and Hall, J.G. (1988) 'Inbreeding and population dynamics of the Chillingham cattle (*Bos taurus*)', *Journal of Zoology*, 216: 479–93.

Hall, S.J.G., Fletcher, T.J., Gidlow, J.R., Ingham, B., Shepherd, A., Smith, A. and Widdows, A. (2005) 'Management of the Chillingham wild cattle', *Government Veterinary Journal*, 15: 4–11.

Hearn, K.A. (1995) 'Stock grazing of semi-natural habitats on National Trust land', *Biological Journal of the Linnaean Society*, 56 (Suppl.): 25–37.

Hemming, J. (2002) '*Bos primigenius* in Britain: Or, why do fairy cows have red ears?', *Folklore*, 113: 71–82.

Heslop, D. and Harbottle, B. (1999) 'Chillingham Church, Northumberland: The south chapel and the Grey Tomb', *Archaeologia Aeliana*, fifth series, 27: 123–34.

Hodgson, J.C. (1916) 'The dismemberment of the Tankerville estates', *History of the Berwickshire Naturalists' Club*, 22: 303–13.

Hunt, J.D. (2003) *The Picturesque Garden in Europe*, Thames and Hudson, London.
Jones, G. (2011) 'HNV farming and permanent pasture – the gap between EU rules and reality', *La Cañada. Newsletter of the Euroean Forum on Nature Conservation and Pastoralism*: 1–3.
Luoto, M., Rekolainen, S., Aakkula, J. and Pykälä, J. (2003) 'Loss of plant species richness and habitat connectivity in grasslands associated with agricultural change in Finland', *Ambio*, 32: 447–52.
McGowan, P. (2010) 'Chillingham parkland plan. Final report, May 2010', Commissioned by Chillingham Wild Cattle Association and Natural England, Chillingham, Northumberland.
Mills, J., Rook, A.J., Dumont, B., Isselstein, J., Scimone, M. and Wallis de Vries, M.F. (2007) 'Effect of livestock breed and grazing intensity on grazing systems: 5. Management and policy implications', *Grass and Forage Science*, 62: 429–36.
Milán, M.J., Bartolomé, J., Quintanilla, R., Garcia-Cachán, M.D., Espejo, M., Herráiz, P.L., Sánchez-Recio, J.M. and Piedrafita, J. (2006) 'Structural characterisation and typology of beef cattle farms of Spanish wooded rangelands (dehesas)', *Livestock Science*, 99: 197–209.
Mitlacher, K., Poschlod, P., Rosén, E. and Bakker, J.P. (2002) 'Restoration of wooded meadows – a comparative analysis along a chronosequence on Öland (Sweden)', *Applied Vegetation Science*, 5: 63–73.
Osterburg, B. and von Horn, L. (2006) 'Assessing the impacts of decoupling EU direct payments from agricultural production and the potential for "recoupling"', *Outlook on Agriculture*, 35: 107–113.
Pevsner, N., Richmond, I., Grundy, J., McCombie, G., Ryder, P. and Welfare, H. (2002) *The Buildings of England. Northumberland*, reprint of 2nd edition, Yale University Press, New Haven and London.
Plieninger, T. and Wilbrand, C. (2001) 'Land use, biodiversity conservation, and rural development in the dehesas of Cuatro Lugares, Spain', *Agroforestry Systems*, 51: 23–34.
Polo, J.L.M., Bellido, I.G. and Rodríguez, M.E.S. (2004) 'Meat production on savannah-like grasslands (dehesas) in semi-arid zones of the province of Salamanca', *Spanish Journal of Agricultural Research*, 2: 107–113.
Pott, R. (1999) 'Diversity of pasture-woodlands on north-western Germany', in A. Kratochwil (ed.) *Biodiversity in Ecosystems*, Kluwer, Netherlands, pp107–32.
Ritvo, H. (1992) 'Race, breed and myths of origin: Chillingham cattle as ancient Britons', *Representations*, 39: 1–22.
Rodrígues, A.M., Pinto de Andrade, L. and Várzea Rodrigues, J. (1999) 'Extensive beef cattle production in Portugal: the added value of indigenous breeds in the beef market', in J.P. Laker and J.A. Milne (eds) *Livestock Production in the European Less Favoured Areas: Meeting Future Economic, Environmental, and Policy Objectives through Integrated Research. Proceedings of the 2nd Conference of the LSIRD Network, Bray, Dublin, Ireland*, Macaulay Land Use Research Institute, Craigiebuckler, Scotland, pp61–9.
Rois-Días, M., Mosquera-Losada, R. and Rigueiro-Rodríguez, A. (2006) *Biodiversity indicators on Silvopastoralism Across Europe. EFI technical report 21*, European Forest Institute, Joensuu, Finland.
Sammul, M. (2008) 'Wooded meadows of Estonia: Conservation efforts for a traditional habitat', *Agricultural and Food Science*, 17: 413–29.

Schama, S. (2002) *A History of Britain: The Fate of Empire 1776–2000*, BBC Worldwide, London.

Simm, G., Villanueva, B., Sinclair, K.D. and Townsend, S.J. (2004) *Farm Animal Genetic Resources. BSAS publication 30*, Nottingham University Press/British Society of Animal Science, Nottingham, UK.

Small, R. and Hosking, J. (2010) 'Rural development programme funding for farm animal genetic resources: A questionnaire survey', unpublished report to Defra, *http://www.defra.gov.uk/fangr/documents/nsc-survey.pdf*

Small, R.W. (2010) 'Conservation grazing: Delivering habitat management for conservation with livestock', *Journal of the Royal Agricultural Society of England*, 171: 38–44.

Tubbs, C.R. (1997) 'The ecology of pastoralism in the New Forest', *British Wildlife*, 9(1): 7–16.

Uthes, S., Sattler, C., Reinhardt, F.-J., Piorr, A., Zander, P., Happe, K., Damgaard, M. and Osuch, A. (2008) 'Ecological effects of payment decoupling in a case study region in Germany', *Journal of Farm Management*, 13: 219–30.

Vieira, M. and Eden, P. (1995) 'Portuguese montados', *La Cañada. Newsletter of the European Forum on Nature Conservation and Pastoralism*, 3, May: 5.

Visscher, P.M., Smith, D., Hall, S.J.G. and Williams, J.L. (2001) 'A viable herd of genetically uniform cattle', *Nature*, 409: 303.

Whitehead, G.K. (1953) *The Ancient White Cattle of Britain and their Descendants*, Faber and Faber, London.

Part V
Conservation, management and wildscapes

18 The impacts of the reintroduction of wild boar in the Forest of Dean, Great Britain

Martin Goulding

Introduction

Free-living wild boar in Britain became extinct 700 years ago through loss of woodland habitat, over-hunting and out-breeding with domestic pigs (Clutton-Brock, 1999; Yalden, 1986). Since the early 1990s, escaped or deliberately released farmed wild boar, from stock originally imported from continental Europe, have re-established free-living populations in several English counties, for example, Dorset, East Sussex, Gloucestershire, Herefordshire and Kent (Goulding *et al.*, 2003; Wilson, 2005). The impact of the wild boars' reintroduction into their ancestral habitat is uncertain because the countryside, the rural way of life and recreational use of the countryside have changed dramatically in the intervening years.

In the Forest of Dean, Gloucestershire, the day-to-day management of the wild boar is carried out by the Forestry Commission, the government department responsible for the management and protection of Britain's forests and woodlands. The Forestry Commission's Feral Wild Boar Management Plan for the Forest of Dean, 2011 to 2016 (Forestry Commission, 2011) informs that wild boar were unofficially reintroduced into the forest in 2004 when a group of 60 suddenly appeared. The tame demeanour and diurnal behaviour of the wild boar suggests they were farm-raised and had been deliberately released. However, because wild boar readily cross-breed with domestic pigs and some British farmers hybridise their wild boar stock with domestic pigs to encourage more frequent farrowing and increased litter size, uncertainty exists over the genetic purity of the reintroduced stock (Goulding, 2001). Genetic testing to date has proved inconclusive (Frantz *et al.*, 2010) although the animals have the appearance of wild boar, behave like wild boar, and appear to fulfil the ecological niche of wild boar. The Forestry Commission refer to the animals as 'feral wild boar' while the Forest of Dean District Council, after considerable debate, opted for the more diplomatic 'feral wild boar or wild boar like pigs' (Goulding, 2009).

Wild boar

Wild boar live in highly organised social groups typically consisting of two to five reproductive females, their most recent litters and surviving young and sub-adults from previous litters. A matriarchal female (Figure 18.1) leads the group and group size varies between 6 and 30 animals. Wild boar breed at approximately one year of age and, unusually for an ungulate species, have large litter sizes (four to six piglets) once a year, occasionally twice.

Wild boar are typically associated with deciduous woodlands but the species can occupy a wide range of habitats, including pine woodlands (Leaper *et al.*, 1999), wetlands (Dardaillon, 1987) and open habitats such as heathland and grassland (Groot Bruinderink and Hazebroek, 1995). Such diversity in habitats is enabled in part by the wild boars' omnivorous diet, approximately 95 per cent of which consists of a wide variety of plant material including fruit and agricultural crops, forest fruits (particularly acorns, beech mast and chestnuts) supplemented with 5 per cent animal species such as worms, insects and insect larvae, occasional small animals such as mice, and carrion (Dardaillon, 1987; Groot Bruinderink *et al.*, 1994; Schley and Roper, 2003).

Figure 18.1 Wild boar sow in woodland
Source: Martin Goulding

Wild boar to date have limited distribution in Britain. The Forest of Dean is thought to house one of the largest British populations, although the actual number of wild boar within the forest is unknown. Wild boar is a difficult species to census as they are secretive, wary of people and mainly nocturnal. They are though more diurnal in times of food shortage, in areas where they are not hunted, or where they have become accustomed to being fed by people.

Forest of Dean

The Forest of Dean is a 110km^2 public estate situated predominantly in the county of Gloucestershire and bounded by the rivers Severn and Wye. In Norman times (1066–1272), the forest was a protected game reserve where red, roe and fallow deer, and wild boar were hunted. Historical records show that in 1251 King Henry III (reigned 1216–72) ordered 200 wild boar to be caught in the Forest of Dean for a royal Christmas feast (Rackham, 1997). Records also show that an undated order for wild boar from the forest placed by King Edward II (reigned 1307–27) could not be fulfilled (Yalden, 1999), possibly signalling the final demise of wild boar in the Forest of Dean was at the turn of the thirteenth century.

The Forestry Commission have managed the Forest of Dean since 1924, primarily for timber production (Currie *et al.*, 1996) and more recently as an area for an increasing array of outdoor recreational activities, such as walking, cycling, horse riding, high-wire rope courses, canoeing, caving and abseiling. Certain ancient forest traditions do still remain alive today, for example, domestic sheep are still grazed unrestricted through the forest in accordance with rights bestowed upon foresters since Norman times. The right of pannage (allowing the grazing of domestic pigs in the forest during the autumn months to feed on acorns) is also still permitted.

The landscape of the Dean is a varied mixture of woodland, pasture land, arable land, hills, escarpments and riparian environments. The diverse array of flora is a result of the limestone and sandstone geology of the forest. Areas of remnant ancient woodlands house species such as herb paris (*Paris quadrifolia*), sanicle (*Sanicula europaea*), sweet woodruff (*Galium odoratum*) and yellow archangel (*Lamium galeobdolon*). In the central areas of the forest, swathes of nationally important bluebells (*Hyacinthoides non-scripta*) dominate and in wetter shaded areas numerous species of ferns are found, such as broad buckler (*Dryopteris dilatata*), harts tongue (*Asplenium scolopendrium*) and various spleenworts (*Asplenium* sp. and polypody *Polypodium* sp.). Examples of nationally rare avian species include northern goshawks (*Accipiter gentilis*) and peregrine falcons (*Falco peregrinus*), ravens (*Corvus corax*), hawfinches (*Coccothraustes coccothraustes*) and pied flycatchers (*Ficedula hypoleuca*). Mammalian fauna includes badgers (*Meles meles*), red foxes (*Vulpes vulpes*), grey squirrels (*Sciurus caroliniensis*), voles (*Microtus* sp.), dormice (*Muscardinus avellanarius*) and several protected species of bats, of which the pipistrelle (*Pipistrelle pipistrelle*) is the most common. The larger mammals are predominantly fallow deer (*Dama dama*), with occasional roe deer (*Capreolus capreolus*) and muntjac (*Muntiacus reevesi*), and the reintroduced wild boar.

Managing wild boar in the Forest of Dean

Worldwide, the reasons for managing wild boar populations are varied and include the protection of sensitive habitats such as recreational or agricultural pastureland, agricultural crops, or areas valued for their sensitive or rare species, or where the wild boars' rooting is generally unwanted. Various techniques are employed to reduce or eradicate wild boar populations, such as poisoning, shooting, trapping, exclusion fencing and supplementary feeding. Deterrent systems involving loud sounds, bright light or unpleasant odours only have a brief effect, quickly losing their effectiveness (Schlageter and Haag-Wackernagel, 2010). The lack of any significant natural predation on the wild boar in Britain, where the wolf (*Canis lupus*) and lynx (*Lynx lynx*) are both extinct, suggests that wild boar numbers will increase to the extent whereby population management will become inevitable.

In Britain, individual landowners and local communities are responsible for managing wild boar populations (Defra, 2008). Management goals differ from encouraging wild boar numbers (for example, in areas where there are commercial shooting interests), maintaining consistent population numbers or eradication. In the Forest of Dean, the Forestry Commission's stated reasons for managing the wild boar are to: minimise the visual and physical damage to amenity grasslands; maintain the population at a manageable size so numbers can continue to be controlled in the future; and minimise the risk of adverse interaction between people, dogs, horses and the wild boar by keeping population densities low (Forestry Commission, 2011).

The Forestry Commission's chosen control method is shooting, either by stalking on foot, shooting from high seats, shooting over bait or shooting in cage traps. Their objective is to reduce, but not eradicate, wild boar populations in the forest (Forestry Commission, 2011). However, the culling of wild boar in the forest is not universally supported and in response wild boar support groups have started to form. Friends of the Boar (FOB, 2012), for example, are a group of local residents who campaign for more resource to be put into educating the public how to live safely alongside the wild boar, their reasoning being that a better-informed public may lead to a reduction in the number of conflict situations, thus reducing the need for culling. It is in the interest of both parties to find common ground because a conflict of opinions between organisations tasked to carry out culling and those opposed to it can add considerable extra challenges to wildlife management programmes (Goulding and Roper, 2002).

Public safety

The Forestry Commission's management objective of minimising the risk of conflict between people, dogs, horses and the wild boar is understandable because wild boar are potentially very dangerous. Wild boar farmed in Britain are covered by the Dangerous Wild Animals Act 1976, as amended in 1984, and adult males can exceed 150kg in weight and possess tusks that are sharp and long enough to cut with ease through human flesh. Fortunately, wild boar tend to avoid human contact and move away when they detect the presence of people. However, in common with other large woodland animals, such as species of deer, females can become aggressive just after farrowing if they feel their piglets are threatened, and male aggression can increase during the mating rut. Although there are very few documented cases of wild boar injuring people in Europe or elsewhere, wild boar and feral pig (escaped domestic pigs) attacks have been reported and although rare, the consequences can be fatal. Mayer (2008) reports that an analysis of 330 worldwide wild boar and feral pig attacks on humans showed that legs and feet are the body parts most frequently injured. Injuries were primarily lacerations and punctures, and fatalities were usually due to blood loss. In some cases, serious infections or toxaemia resulted from the injuries. In the Forest of Dean, the Forestry Commission reported only one recorded instance of a person being injured by a wild boar when the animal, which had become habitualised to people feeding it, reacted to the actions of a second bystander who, with a frightening lack of awareness of the species' temperament, was 'teasing the animal with a stick' as his companion was feeding it (Forestry Commission, 2011).

The Forestry Commission comment that the public's feelings toward the wild boar range from people who say they are now too afraid to walk in the forest to those who walk daily in the woods with no apprehension. To reduce the risk of a conflict situation with the wild boar, the Forestry Commission advise the general public:

- Do not walk through dense undergrowth where wild boar may be encountered at close quarters, such areas are favoured as breeding and resting sites. Wild boar have a long breeding season but most litters are born in the spring (February to May) when there may be potentially dangerous defensive sows with young piglets.
- Should you encounter wild boar in the Forest, do not approach them – if possible leave the area by the same route you approached by, or make a detour giving the animals a wide berth.
- Keep your dog under close control – a number of dogs have been seriously injured by wild boar and it is best to avoid the interaction if at all possible, this will also help reduce disturbance to other wildlife too.

The advice concerning domestic dogs is also understandable. Dogs running off the lead have been attacked by wild boar in the Forest of Dean and the resulting injuries have led to at least one dog being reported in the local media as being 'put to sleep' (BBC News, 2010). The Forestry Commission though suggests that from the direct reports they have received, the number of domestic dogs dying following a confrontation with a wild boar is probably around five or six, and that figure they state is likely to be an underestimate (Forestry Commission, 2011). The Forestry Commission note that although media reports may sensationalise the dangers of wild boar to people and domestic animals, they have the advantage of raising awareness among the local community that wild boar are present in the forest and can cause injury to dogs and people. To reduce the risk of confrontation between dogs and wild boar, the Forestry Commission advise owners through strategically placed warning notices that 'If your dog chases a boar, stay at a safe distance and continue to call the dog back – do not approach the boar or interfere with the wild boar'. Notices also advise owners to keep their dogs under close control or on a lead, however, personal observations suggests that this by-law requirement is not often adhered to.

The forest is criss-crossed by bridleways and horses are known to react unpredictably to the scent of wild boar. The Forestry Commission state they are aware of anecdotal reports of riders being thrown or wild boar charging a horse, although 'no direct reports have been received' (Forestry Commission, 2011).

Road Traffic Accidents

Throughout Europe, wild boar are recognised to be a significant cause of Road Traffic Accidents (RTAs) because under the cover of darkness wild boar will move from one feeding area to another, which may involve frequent crossing of roads (Groot Bruinderink and Hazebroek, 1996). In mitigation, perimeter fencing has been installed in some European countries and also fauna passages in certain high risk areas. Wild boar will use fauna passages once they have become accustomed to them, and their location is reported to be the key factor determining use (Rosell *et al.*, 2010).

RTAs have injured or claimed the lives of several wild boar in Britain to date and it is probably only through good fortune that no human fatalities have been reported; unfortunately European experiences suggest it is may only be a matter of time. An estimation made in 2005, using data from continental populations and then current British population estimates, suggested that Britain could expect about six wild boar-related RTAs annually (Wilson, 2005). However, between the 1 April 2010 and 31 March 2011, the Forestry Commission reported that 22 wild boar were recorded as being killed in road traffic accidents in the Forest of Dean alone (Forestry Commission, 2011). A continually increasing wild boar population will bring a concurrent increase in wild boar-related RTAs.

Ecological impact

Wild boar rooting disturbs the soil to such an extent that plant cover is almost completely removed, leaving bare patches of earth that are exposed to subsequent plant colonisation. Bratton (1975) found that in certain areas of the Great Smoky National Park, USA, rooting reduced the forest understorey from 80–100 per cent to as little as 2–15 per cent, with a significant reduction in plant species number. Bialy (1996) also suggests that rooting wild boar decrease the growth of wood anemones (*Anemone nemorosa*) by feeding on the rhizomes. Conversely, rooting among set-aside land in Germany was found to increase species diversity (Milton *et al.*, 1997) and an increase in plant species diversity was also recorded in areas of a wetland environment rooted by feral pigs (Arrington *et al.*, 1999). Welander (2000) in a study of Swedish wild boar, which in common with the British situation had been reintroduced in habitat from which they had previously been driven extinct, also found that rooting increased plant species diversity.

The rooting activities of wild boar in British woodlands will create a disturbance regime that, with the occasional exception of autumnal pannaged domestic pigs, the woodland has not experienced for hundreds of years. Predicting the ecological impact of rooting is fraught with difficulty because rooting intensity varies from year to year. This is due to fluctuating wild boar numbers, unpredictable natural food supplies and weather conditions. Wild boar will also root up areas that have previously been rooted. In Britain, a short-term study of wild boar rooting in three habitats in rural East Sussex, woodland, grassland and woodland ride, suggested rooting could be beneficial to woodland habitats through observed increases in plant cover and diversity, increased emergence of viable seeds from the seed bank and increased decomposition of leaf litter (Sims, 2006). Concerns have been raised that wild boar rooting may destroy aesthetically pleasing monocultures of bluebells (*Hyacinthoides non-scripta*) that characterise numerous British woodlands. Although longer-term studies are needed, Sims' (2006) preliminary research suggested the impact on bluebell plant density was localised and short-lived.

Further ecological impacts of the wild boar involve the species' role in plant dispersal. Wild boar deposit the seeds caught up in their bristly coats through their grooming behaviour of wallowing in mud followed by a rubbing of their flanks against tree bark to remove parasites or loose hair during the moult. Wallowing also creates ephemeral pools for aquatic insects, dragonflies and amphibians.

The Forestry Commission recognise that understanding the ecological impacts of wild boar in the forest has 'hardly begun although the consensus of opinion amongst researchers is that ecologically, wild boar (at current density levels) are unlikely to have a negative impact upon the Forest' and 'there is likely to be a locally positive impact on the woodland ecosystem'

(Forestry Commission, 2011). The Commission also report that wild boar are not negatively impinging on timber production in the forest because wild boar do not graze on the trees as deer do, and do not significantly damage tree bark, unlike deer and squirrels. There has been damage by wild boar to fences erected to protect young trees from grazing by sheep and deer, although this is reported not to be significantly different to damage caused by other animals, such as badgers (Forestry Commission, 2011). However, the Commission note that during the autumn and winter of 2010 to 2011 the amount of rooting damage to amenity grassland and road and track verges (Figure 18.2) was the worst observed so far, with some grass verges repeatedly overturned during the winter months. The reasons for this were not clear but local butterfly conservation groups have expressed concern due to the number of locally and nationally rare butterfly species that overwinter as pupae in the verge grasses (Forestry Commission, 2011).

Agricultural impact

Wild boar are an important agricultural pest in many regions of the world. Their impact on agriculture includes rooting and trampling damage to agricultural crops and pasture land, breaching stock fencing, predation of domestic livestock, interbreeding with domestic pigs and acting as a vector

Figure 18.2 Wild boar rooting at a roadside
Source: Martin Goulding

of disease to domestic livestock. Over 40 different crop plant species are consumed by wild boar (Schley and Roper, 2003) and crop losses attributed to wild boar are substantial enough for some countries, for example Poland, Italy and France, to adopt compensation schemes to reimburse farmers for their economic losses.

The most evident form of damage attributable to the returned British wild boar is rooting of pastureland, although damage to cereal crops has also been recorded (Goulding, 2003; Wilson, 2005). To date, agricultural damage in Britain by wild boar is only minor and localised. Damage to agricultural fields has been shown to be related to shooting pressure, crop type, presence of livestock and proximity to woodland, but not by proximity to roads or field size (Goulding, 2003; Moore, 2004). In the Forest of Dean, wild boar have damaged growing crops and rooted up pasture in fields in close proximity to the forest, but to date no assessment of the scale of damage has been reported. Again, an increasing wild boar population will bring a concurrent increase in wild boar-related agricultural damage.

A significantly more serious to Britain's agricultural economy is that wild boar are susceptible to many infectious diseases of livestock such as classical swine fever, African swine fever, foot-and-mouth, Aujeszky's disease and bovine tuberculosis. Defra's wild boar action plan (Defra, 2008) encouragingly reports that the low likelihood of contact between wild boar and domestic pigs means that wild boar will not significantly impact on the countries' ability to control endemic diseases in the domestic pig stock. However, Defra recognises that wild boars' behavioural ecology may favour the spread of disease as wild boar are frequently in contact with one another, particularly when concentrated around food and water sources. Additionally home ranges can be large and overlap with other wild boar populations, providing the potential for mixing of animals. Wild boar can also disperse over considerable distances and their omnivorous, opportunistic, scavenging behaviour also increases the likelihood of wild boar becoming infected if there is infected wildlife or livestock in the proximity. Free-ranging wild boar are also drawn towards outdoor domestic pig units, enticed by the availability of food, social interaction and opportunities to mate. Defra notes that interactions with domestic pigs are particularly concerning as diseases may be transmitted in either direction. The greatest risks of exotic disease incursion into Britain are reportedly associated with disease entering through the consumption of infected pork meat or meat products by either wild boar or domestic swine. The diseases of highest risk are therefore classic swine fever, foot-and-mouth and *Trichinella* sp. (Defra, 2008).

In the Forest of Dean, livestock disease has not to date become an issue, but one wild boar in woodland close to the Forest of Dean had lesions consistent with bovine tuberculosis during post-mortem examination (Defra, 2012).

Wild boar and their relationship to other woodland species

Research into possible competition between wild boar and other woodland species is lacking and it is not known how the wild boar in the Forest of Dean will impact on the native species already resident. Research has shown that corvids will directly associate with wild boar that deliberately encourage the birds' attention. This is presumably to encourage the removal of ticks or parasites from the wild boars' coats (Massei and Genov, 1995). Hunting returns have though implicated wild boar in causing a fall in woodcock (*Scolopax rusticola*) numbers in some German hunting estates (Nyenhuis, 1991), and wild boar have reportedly predated the nests of red-legged partridges (*Alectoris rufa*) in Spain (Leaper *et al.*, 1999). Furthermore, in a Spanish wetland environment, stomach content analysis has shown that wild boar predate the eggs and chicks of several protected bird species, including those of the bittern (*Botaurus stellaris*) (Bertolino *et al.*, 2010). Wild boar may also compete with small mammals by digging up caches of buried mast and may opportunistically predate the young (Focardi *et al.*, 2000).

Next steps

Following the wild boar's accidental reintroduction into the Forest of Dean in 2004, the Forestry Commission stated in 2011 that 'at the current population densities' the wild boar are thought to not have a negative impact upon the ecology of the forest or on any rare or endangered species (Forestry Commission, 2011). The Forestry Commission also acknowledged that to improve their management of the wild boar a better understanding of the species' habitat requirements within the forest is required to enable better distribution management and reduce risk of damage to more sensitive areas. Furthermore, reliable methods of estimating wild boar population size are needed to allow accurate and defendable cull levels to be implemented.

As well as reducing conflict between wild boar and people, wildlife management programmes, such as that implemented in the Forest of Dean, also need to avoid conflict arising with the local community. A report on the public's perception of the wild boar in the Forest of Dean highlighted that a lack of awareness of a clear management strategy being implemented by the Forestry Commission had led initially to 'confusion and disquiet' (Dutton and Clayton, 2011). However, eight years down the road from the wild boar's unofficial reintroduction into the Forest of Dean, a wild boar management plan is now in place and future research needs are identified. With time, the impacts of the returning wild boar on the woodlands and associated habitats will become clearer still.

References

Arrington, D.A., Toth, L.A. and Koebel, J.W. (1999) 'Effects of rooting by feral hogs *Sus scrofa* L. on the structure of a floodplain vegetation assemblage', *Wetlands*, 19: 535–44.

BBC News (2010) 'Warning to Forest of Dean dog owners after boar attacks', http://news.bbc.co.uk/1/hi/england/gloucestershire/8711014.stm, accessed 1 June 2012.

Bertolino, S., Angeliciz, C., Scarfò, F. et al. (2010) 'Is the wild boar an important nest predator in wetland areas? An experiment with dummy nests', in *Book of Abstracts: 8th International Conference on Wild Boar and other Suids*, Food and Environment Research Agency (FERA), York, p54, https://secure.fera.defra.gov.uk/wildboar2010/documents/bookOfAbstractsWildBoarNov10.pdf, accessed 2 June 2012

Bialy, K. (1996) 'The effect of boar (*Sus scrofa*) rooting on the distribution of organic matter in soil profiles and the development of wood anemone (*Anemone nemorosa* L.) in the oak-hornbeam stand (*Tilio-carpinetum*) in the Białowieża primeval forest', *Folia Forestalia Polonica Series A – Forestry*, 38: 77–88.

Bratton, S.P. (1975) 'The effect of the European wild boar (Sus scrofa) on Gray beech forest in the Great Smokey Mountains', *Ecology*, 56: 1356–66.

Clutton-Brock, J. (1999) *A Natural History of Domesticated Animals*, Cambridge University Press, Cambridge.

Currie, C.R.J., Herbert, N.M., Baggs, A.P. et al. (1996) 'Forest of Dean: Introduction', in W. Page and N.M. Herbert (eds) *A History of the County of Gloucester: Volume 5: Bledisloe Hundred, St. Briavels Hundred, The Forest of Dean*, Oxford University Press, www.british-history.ac.uk/report.aspx?compid=23264, accessed 4 June 2012.

Dardaillon, M. (1987) 'Seasonal feeding habits of the wild boar in a Mediterranean wetland, the Camargue (southern France)', *Acta Theriol*, 32: 389–401.

Defra (Department for Environment, Food and Rural Affairs) (2008) *Feral Wild boar in England: An Action Plan*, www.naturalengland.org.uk/Images/feralwildboar_tcm6-4508.pdf, accessed 2 June 2012.

Defra (2012) 'Bovine TB: TB in other species', http://archive.defra.gov.uk/foodfarm/farmanimal/diseases/atoz/tb/abouttb/otherspecies.htm, accessed 2 June 2012.

Dutton, J. and Clayton, H. (2010) 'Public perception of wild boar in the Forest of Dean, England, potential implications for their future management', in *Book of Abstracts 8th International Conference on Wild Boar and other Suids*, Food and Environment Research Agency (FERA), York, p38, https://secure.fera.defra.gov.uk/wildboar2010/documents/bookOfAbstractsWildBoarNov10.pdf, accessed 2 June 2012.

FOB (Friends of the Boar) (2012) 'Friends of the Boar', www.friendsoftheboar.co.uk/, accessed 3 June 2012.

Forestry Commission (2011) *Feral Wild Boar Management Plan, Forest of Dean, 2011 to 2016*, Forestry Commission, England, www.britishwildboar.org.uk/Feral Wild Boar Management Plan Forest of Dean.pdf, accessed 2 June 2011.

Frantz, A.C., Massei, G. and Burkeet, T. (2010) 'How "wild" are British wild boar?', in *Book of Abstracts: 8th International Conference on Wild Boar and other Suids*, Food and Environment Research Agency (FERA), York, p70, https://secure.fera.

defra.gov.uk/wildboar2010/documents/bookOfAbstractsWildBoarNov10.pdf, accessed 2 June 2012.

Focardi, S., Capizzi, D. and Monetti, D. (2000) 'Competition for acorns among wild boar (Sus scrofa) and small mammals in a Mediterranean woodland', *Journal of Zoology, London*, 250: 329–34.

Fruzinski, B. (1995) 'Situation of wild boar populations in western Poland', *IBEX Journal of Mountain Ecology*, 3: 186–7.

Goulding, M.J. (2001) 'Possible genetic sources of free-living wild boar (*Sus scrofa*) in southern England', *Mammal Review*, 31: 245–8.

Goulding, M.J. (2003) 'Investigation of free-living wild boar (Sus scrofa) in southern England', unpublished PhD thesis, University of Sussex, Brighton.

Goulding, M.J. (2009) 'Wild boar issues and arguments – a case study', *Int. Urban Ecol. Rev.*, 4: 60–6.

Goulding, M.J. and Roper, T.J. (2002) 'Press responses to the presence of free-living wild boar in southern England', *Mammal Review*, 32: 272–82.

Goulding, M.J., Roper, T.J., Smith G.C. and Baker, S.J. (2003) 'Presence of free-living wild boar *Sus scrofa* in southern England', *Wildl. Biol.*, 9 (Suppl. 1): 15–20.

Groot Bruinderink, G.W.T.A. and Hazebroek, E. (1995) 'Modelling carrying capacity for wild boar *Sus scrofa* in a forest/heathland ecosystem', *Wildl. Biol.*, 1: 81–7.

Groot Bruinderink, G.W.T.A. and Hazebroek, E. (1996) 'Ungulate traffic collisions in Europe', *Conservation Biology*, 10: 1056–7.

Groot Bruinderink, G.W.T.A., Hazebroek, E. and van der Voet, H. (1994) 'Diet and condition of wild boar, *Sus scrofa scrofa*, without supplementary feeding', *Journal of Zoology*, 233: 631–48.

Leaper, R., Massei, G., Gorman, M.L. *et al.* (1999) 'The feasibility of reintroducing Wild Boar (*Sus scrofa*) to Scotland', *Mammal Review*, 29: 239–59.

Lever, C. (1994) *Naturalised Animals: The Ecology of Successfully Introduced Species*, T. & A.D. Poyser, London.

Massei, G. and Genov, P. (1995) 'Preliminary analysis of factors influencing habitat-use by the wild boar', *IBEX Journal of Mountain Ecology*, 3: 168–170.

Mayer, J. (2008) 'Wild pig attacks on humans', 2008 National Conference on Feral Hogs, 13–15 April, www.britishwildboar.org.uk/JohnMayer.pdf, accessed 2 June 2011.

Milton, S.J., Dean, W.R.J. and Klotz, S. (1997) 'Effects of small scale animal disturbances on plant assemblages of set-aside land in Germany', *Journal of Vegetation Science*, 8: 45–54.

Moore, N. (2004) *The Ecology and Management of Wild Boar in Southern England*, Defra Final Project Report, VC0325, http://sciencesearch.defra.gov.uk/Document.aspx?Document=VC0325_2113_FRP.doc, accessed 2 June 2012.

Nyenhuis, H. (1991) 'Predation between woodcock (Scolopax rusticola L.) game of prey and wild boar (*Sus scrofa* L.)', *Allgemeine Forst-u.J.-Ztg*, 162: 174–80 (English Summary).

Rackham O. (1997) *The Illustrated History of the Countryside*, Phoenix Illustrated, London.

Rosell, C., Navàsl, F., Carol, Q. *et al.*. (2010) 'Use of fauna passages by wild boar: Some change is observed', in *Book of Abstracts: 8th International Conference on Wild Boar and other Suids*, Food and Environment Research Agency (FERA), York, p35, https://secure.fera.defra.gov.uk/wildboar2010/documents/bookOfAbstractsWildBoarNov10.pdf, accessed 2 June 2012.

Schlageter, A. and Haag Wackernagel, D. (2010) 'Investigation of the effectiveness of deterrent systems against wild boar *Sus scrofa*', in *Book of Abstracts: 8th International Conference on Wild Boar and other Suids*, Food and Environment Research Agency (FERA), York, p34, https://secure.fera.defra.gov.uk/wildboar2010/documents/bookOfAbstractsWildBoarNov10.pdf, accessed 2 June 2012.

Schley, L. and Roper, T.J. (2003) 'Diet of wild boar *Sus scrofa* in Western Europe, with particular reference to consumption of agricultural crops', *Mammal Review*, 33: 43–56.

Sims, N.K.E. (2006) 'The ecological impacts of wild boar rooting in East Sussex', unpublished DPhil thesis, University of Sussex, Brighton.

Welander, J. (1995) 'Are wild boars a future threat to the Swedish flora?', *IBEX Journal of Mountain Ecology*, 3: 165–7.

Welander, J. (2000) 'Spatial and temporal dynamics of a disturbance regime: Wild boar (Sus scrofa L.) rooting and its effects on plant species diversity', PhD thesis, Swedish University of Agricultural Sciences, Utgivningsort.

Wilson, C.J. (2005) 'Feral wild boar in England. Status, impact and management. A Report on behalf of Defra European Wildlife Division', Defra, Bristol, www.naturalengland.org.uk/Images/wildboarstatusImpactmanagement_tcm6-4512.pdf, accessed 2 June 2009.

Yalden, D.W. (1986) 'Opportunities for Reintroducing British Mammals', *Mammal Review*, 16(2): 53–63.

Yalden, D.W. (1999) *The History of British Mammals*, T. and A.D. Poyser, London.

19 Wild cattle and the 'wilder valley' experiences

The introduction of extensive grazing with Galloway cattle in the Ennerdale Valley, England

Gareth Browning and John Gorst

Introduction

The Ennerdale Valley presents a dramatic picture in a remote position on the western fringe of the Lake District National Park. Extending to 14km long and almost 5km wide, at its widest it encloses an area of around 5,000ha. The valley narrows from west to east and is surrounded by dramatic mountain ridges that include some of Lakeland's highest summits such as Great Gable and Pillar, both over 890m high.

The large scale and diversity of its landscapes, incorporating farming, mixed forest, rivers, lake, open fell and mountains, combined with the significant lack of roads, traffic and buildings all contribute to enhance the sense of Ennerdale as a wild, tranquil and spiritually refreshing place. Over 40 per cent of the area is designated a SSSI and special area of conservation (SAC). The river Liza is a major feature, falling wild and unchecked down the valley and is one of few rivers in England to show such uncontrolled dynamism.

The whole valley is highly significant for its rich legacy of archaeological remains and diverse habitats for flora and fauna, all with features that range from regional to international importance. Over 3,000 years of human activity, stretching from the Bronze Age to present day, are etched into the landscape.

At the western end of the valley lies Ennerdale Water, which supplies over 60,000 customers with drinking water. The network of footpaths, tracks and open access both in the forest and on the open fells provides a wealth of opportunity for people to explore the valley with a sense of freedom, adventure and challenge.

Ennerdale is a place where people can feel humbled by their surroundings, where signs of human influence are less and where nature remains, to varying degrees, the dominating force.

The Wild Ennerdale Partnership

Wild Ennerdale is a partnership between people and organisations led by the National Trust (NT), the Forestry Commission (FC) and United Utilities (UU), which are the primary landowners in the Ennerdale Valley, and Natural England, the government's advisor on the environment.

The partnership has a vision to 'to allow the evolution of Ennerdale as a wild valley for the benefit of people, relying more on natural processes to shape its landscape and ecology' (Wild Ennerdale Partnership, 2006).

Our approach

Wild land is a relatively new concept for nature conservation in the UK and involves giving natural processes greater freedom to develop our future landscapes. Nature conservation in England is generally focused on small-scale interventions, whereas in Ennerdale more weight is given to the landscape scale leaving the detail to natural processes. Wild Ennerdale is one of the UK's longest running and largest wild land projects allowing ecosystems throughout the valley to evolve with greater freedom. Its experience in managing land through minimal human intervention is already widely recognised and shared by others.

In the UK, all our landscapes and ecosystems have to some degree been impacted by human influences and in Ennerdale (as elsewhere) any future landscapes and ecosystems will continue to be affected by past management along with influences, such as climate change and airborne pollution, which show no respect for boundaries. As a result using the words 'wild' and 'natural' can be contentious.

The Wild Ennerdale approach involves reducing the scale and altering the nature of human intervention in the valley, so that human processes (whether they be farming, forestry or recreation) do not come to dominate the wide variety of other processes that operate. Put simply, we are trying to place constraints on the way in which people operate in the valley so that they become part of a 'natural system'. This is an attempt to allow a 'wild' place to evolve in which people are and continue to be an important part.

In the context of Wild Ennerdale the words 'natural' and 'natural system' are not used in an ecologically pure way. We are not attempting to recreate a set of landscapes and ecosystems that might once have existed at a particular point in time. Rather, by acknowledging that natural systems are dynamic and constantly changing, we are using the present as a starting point – a starting point that includes three millennia of human activity and a variety of species that people have both eliminated (or at least seriously constrained) and introduced. Of these, Sitka spruce (*Picea sitchensis*) is perhaps the most obvious. Our concession to the 'unnaturalness' of this starting point is the management approach we have been undertaking over the last five few years and will continue into the near future. This involves introducing some of the more obvious and significant missing processes, such as extensive grazing by large herbivores and broadleaf tree planting, and providing some control on processes we have introduced in the past such as Sitka spruce regeneration. The intention is to create a more balanced starting point in which a broader array of processes has the opportunity to operate and influence the valley.

As the valley develops, it is hoped that there will be a series of naturally evolving and interacting ecosystems across the valley that are far more robust in the face of stresses such as climate change and that farming and forestry will maximise ecology and landscape value. It cannot be predicted exactly how biodiversity may develop as natural processes are given greater freedom. However, being able to observe these processes at work, over generations, will be one of the marvels of change in Ennerdale, and ensure that the lessons learnt will have a resonance far beyond the boundaries of the valley.

Why cattle?

When the Wild Ennerdale partners started to share their new vision with others the feedback from ecologists was unanimous in encouraging the partnership to introduce a large herbivore into the valley. This was said to be a key missing natural process from many of our modern-day forests. The partnership looked around for inspiration and discovered the nature reserve of Oostvaardersplassen in the Netherlands 'where cattle and horses in the Oostvaardersplassen had a completely free life with a natural social order and graze extensively without tending' (Vera, 2009). The partnership also visited a grazing scheme in south Cumbria where cattle (*B. taurus*) were rotated across a number of extensive sites.

As we found out more, we realised that perhaps cattle could help us achieve our aspiration to see the blurring of the traditional functional management of the valley and the removal of boundaries to natural processes (Figure 19.1). In the past, forestry and farming were kept separate, often divided by a fence or a wall leading to the development of a stark boundary where landscape texture, colour, look and feel change suddenly. Our vision for the valley challenges us to facilitate the development of more blurred boundaries where the extent of one habitat merging with another is difficult to define and perhaps new habitats develop that challenge our stereotypical understanding of habitat types such as forest and fell.

The introduction of cattle is principally about enabling a number of opportunistic processes where cattle disturb ground, creating niches into which the seed from different species can germinate and grow. Whether they make it to full height depends on the availability of nutrients, light and water and grazing by deer (*Cervidae*), sheep (*Ovis aries*) and the very cattle that provide the seedbed in the first place.

Figure 19.1 The Ennerdale Middle Valley
    Source: Gareth Browning and John Gorst

Large herbivores in Ennerdale: A historic context

As the often quoted verse from Ecclesiastes 1:9 says 'there is nothing new under the sun' and so it is with cattle in Ennerdale. Research by Oxford Archaeology North (Wild Ennerdale Partnership, 2003) identified two vaccary (cattle farms) in Ennerdale in 1334 and these great walled enclosures can still be explored today. Indeed, one of the vaccaries, near Woundell Beck, was the catalyst for the area known as Silvercove to be identified as the first extensive area for reintroducing cattle. The vaccary, however, has been given some additional protection with internal fences, ensuring the historic walls are not damaged by modern relatives of the original medieval users.

Wild cattle: A brief timeline of introduction

The introduction of cattle to the valley become reality in early 2006 when the first herd of animals was introduced to around 140ha of recently clear felled conifer forest, ancient woodland and heath at Silvercove. Silvercove was chosen as the first area because it had no recent history of grazing and no existing tenancies and so was relatively easy to establish.

In 2008, the renewal of a farm business tenancy provided the opportunity to introduce a second herd into the Middle Valley extending to around 250ha. This herd was initially excluded from an area of previously intensively sheep-grazed in-bye fields so that these would become rougher. The concern was that the cattle may decide to spend the whole time in the fields and not explore the valley bottom and forest if the field grazing was too good. This herd's area was to undergo an unplanned expansion in late 2009, when attempts to maintain a fence boundary across the river Liza failed following two successive years of significant flood events.

Abandoning this boundary allowed the herd to wander freely up to 550ha of the Middle Valley (Figure 19.2)

Figure 19.2 Second Galloway herd grazing the valley bottom
Source: Gareth Browning and John Gorst

In 2009, the tenant farmer managing the Silvercove herd suggested reducing his sheep flock at the eastern end of the valley under Great Gable and introducing a third herd into an area of 240ha known as Blacksail. This was partly driven by and made possible by the expansion of the Silvercove herd through breeding. The latter herd was now at the capacity for the site and the farmer suggested splitting this herd, taking some animals to the eastern end of the valley. So by 2012, we had free roaming cattle grazing nearly 1,000ha of Ennerdale.

Wild cattle – wilder people

As with the concept of wildness itself, the introduction of wild-roaming cattle challenged the cultural and traditional values we have of farm animals being regularly tended and managed to a defined end point: the production of meat. Their introduction required a change of philosophy, a standing back and waiting rather than being in control. There are three principal communities that were affected by the introduction of cattle: the farming community, visitors to the valley and the Wild Ennerdale partners as land managers.

To capture the farming community's experience, three of the valleys farmers were interviewed. Two of those interviewed now look after herds of extensive grazing cattle and while the third does not, they have been involved in sharing cattle grazing with one of the other two. When the introduction of extensive grazed cattle was first discussed with the farming community there were a number of common responses from existing farmers in the valley. The interviews explored these concerns and also how farmers felt now. These are summarised below.

Welfare was the area that solicited the strongest concerns including 'how would they get enough to eat?', 'would they maintain condition in winter?', 'would they roam much or just stand at the gate waiting to be fed?' and 'how would they cope?'. Thinking about the situation in 2011, those interviewed expressed pleasure and some surprise with how well the animals coped making comments such as 'they've done everything asked of them', 'no bother at all', 'they make a lot of decisions' and 'Galloway cattle know what they are doing next'.

There was concern that the size of the area that they could roam would make them difficult to manage. This can be seen in comments such as 'how far would they move?' and 'how would you find them?'. The answer to the first question has been that they move quite a lot and often some distance in a day. Finding the cattle can occasionally be difficult but they have learnt to recognise the sound of their farmer's vehicle and one farmer uses a loud yodel-like call to which they respond with their own bellow and rapidly make their way to where the farmer is.

The valley's farmers clearly saw that this was a change to their previous way of management. Comments such as: 'hadn't done anything like this before', 'frightened of it failing' and 'having cattle out all the time was a new concept', sum up a wider range of reactions. By comparison, farmers expressed different feelings once the cattle were established, one saying positively 'The cattle at Silvercove have changed the way we farm cattle at home'. Other comments included, 'the experiment's worked' and 'Dad was quite surprised it worked but pleased'.

In terms of members of the public, there have been many who have supported and welcomed the new beasts. The choice of Galloways with their thick curly hair has won them regular comparison with bears. There is even a YouTube video published in 2008 that documents the finding of the 'rarely sighted bears in the Lake District'. For many visitors the animals do often go unseen as the area they roam is quite extensive. A search of comments left by visitors in the Ennerdale YHA (Youth Hostel Association) diary revealed nothing mentioned about the cattle. Maybe they weren't seen or else the visitors didn't realise they were unusual.

When it was first suggested that the cattle be introduced there was concern that dog owners would be chased and people regularly frightened by cows with calves. This was especially because of the high number of visitors, estimated at around 60,000, and the many kilometres of rights of way that criss-cross the valley. Five years on from their introduction, there have been no formal complaints from members of the public. The worst criticism we have had informally is that the cows 'pooh on the footpaths'. One of the Wild Ennerdale partners did meet a couple on mountain bikes with a pair of 'pet wolves' on leads who had turned back from riding up the valley because the cattle were encamped on the forest road. However, they were not complaining but accepted that their wolves and the cattle would be wary of each other. Encouraged to follow the Wild Ennerdale partner's vehicle along the road, the bikers and their wolves made it through the herd that moved aside for the vehicle, and while interested in the wolves, were not agitated or threatening.

The introduction of cattle to the valley was instigated by the Wild Ennerdale partners and therefore it is safe to assume that the partnership had already decided to allow natural processes more freedom. However, the experience of their introduction has at times challenged the partnership. Initially the partners were keen to see Highland cattle introduced as they had been farmed in the valley in the 1930s and were the iconic 'wild' cattle breed. For a time we were quite focussed on this iconic species. However, concerns over handling horned cattle and the risk they posed to visitors, along with the farmers experience with Galloways, suggested they would be a more suitable breed. It was when we returned to our 'Vision and Guiding Principals' that we realised that we were being drawn by the iconic nature of Highland cattle when in fact introducing Galloways would deliver the same benefits and importantly would reward the enthusiasm of the farmer who was to take on the Silvercove tenancy.

One incident that reminded us all of our vision and principals was our attempt to maintain a stock-fenced boundary across the River Liza. This aimed at keeping the Middle-Valley herd from roaming across the eastern valley, where native broadleaf trees were being planted. While we fully expected to have to maintain this boundary we did not anticipate the complete change in our thinking within barely six months of the introduction of the Middle-Valley herd. In late October 2008, the river Liza experienced a significant flood event, one that made national newspaper headlines when the Original Mountain Marathon (OMM), held to the east of the valley, was cancelled for the first time in its long history. The river smashed through the fence across its path burying it under significant amounts of woody debris and gravel. The Wild Ennerdale partners spent much of the following year discussing whether to reinstate the boundary only to witness another headline grabbing flood event in October 2009, this time focussed on Cockermouth and Workington. The River Liza moved 20m or more in places across the valley bottom again, bringing more debris into the river system. It was after this event that we realised we should celebrate the power of natural forces and accept that the boundary was not sustainable. The Middle-Valley herd's area of roaming increased 100 per cent and has stayed the same since.

Wild cattle – wilder animals

The question we often get asked is 'Are they wild?', referring to the cattle themselves. While legally the animals are still domestic stock, a number of episodes have given us an insight into the development of a perhaps a more self-willed animal.

Our first herd of cattle arrived in the valley in 2006, and we had an inkling that one of them, the oldest matriarch, was already pregnant. She was, and gave birth unaided in the late spring. That she was unaided was not out of the desire of the farmer, but it was clear she didn't want help. Like many a wild animal she took herself away from the herd and gave birth in an area of scrub and bracken. We were all surprised and the farmer was very concerned when one day she went missing from the herd and could not be found. However, this has now become the norm and like it or not, we have all had to get used to absent mothers at calving time. Most cows are away for just a few days but the longest absence has been more than a week. Normally the cow rejoins the herd after a few days, returning regularly to suckle the calf that is left hidden in scrub, as if being sheltered from some predator.

Another story shows how the herd has become a close family unit just like with many wild animals. One of our Middle-Valley herd injured its foot to the point where it couldn't walk during its first year onsite. The Wild Ennerdale partners and the farm tenant discussed what should be done and as the cow seemed in little pain we decided to see if it would recover. For a while the cow did not move far at all, preferring to graze a very small area immediately around itself. During this period the rest of the herd exhibited very protective behaviour. They would graze away from the injured animal during the day but would always rejoin and stay with it during the night. As the injured animal improved and started to walk again members of the herd were often seen helping it by pushing it up steeper slopes. Unfortunately something happened after this point and the animal's health deteriorated again so we decided to remove the animal from the herd.

Lastly, a more recent story sheds another angle on the strong bonds that develop between animals in the herd. In late summer 2010, two herds managed by one farmer were both put to the same bull. The bull was allowed to roam with each herd for a couple of weeks. At the time the farmer and Wild Ennerdale partners noticed that when the bull was with the Blacksail herd roaming under Great Gable the bull was never seen interacting with the cows but instead always seemed to be alone. In spring 2011, when the cows were pregnancy tested none of the Blacksail herd was pregnant, yet four out of five of the suitable cows in the other herd were. Discussing this incident since, the farmer has surmised that the Blacksail herd included two young bull calves that while not sexually active, may have been blocking the new 'interloper' bull from interacting with the females in the herd. The other herd in which the bull was successful was made up entirely of females.

While the cattle have exhibited what might be described as 'self-willed' behaviour they have also exhibited tame domestic characteristics. Both herds have learnt to recognise the sound of their farmer's vehicle, often responding to its arrival with loud calling and sometimes appearing from a long distance away. Recently, one of the Wild Ennerdale partners changed vehicles from a small car to a larger four-wheeled drive type and this now attracts similar attention where before the car did not.

One of the farmers users a very loud call, almost but not quite like a yodel, to call the animals when they cannot be seen or found visually. The sound of this call seems to carry long distance as the cattle can be heard only just replying with their own loud call perhaps up to a kilometre away and eventually arrive sometimes five minutes or more after they have been called.

While the animals receive only minimal tending, the character of each of the three herds is different. They are for the most part-tame and far from being as wild, as say a wild deer. The Silvercove herd are very sociable and friendly, always keen to find out who the latest visiting group are. They are noisy too, bellowing out their recognition for their farmer's arrival. The Middle-Valley herd are a more quietly reserved group and are inclined to retreat if you pay them too much attention. The Blacksail herd have yet to establish a particular identity. Being only recently made up of cattle from the Silvercove herd, they have carried with them the sociable interested character and can often be found standing around walkers staying at the Blacksail YHA.

Wild cattle – wilder treescapes

The process of landscape change across Ennerdale is generally slow with tangible, touchable results only becoming visible at the landscape scale after a minimum of ten years. While extensive grazing of the first site at Silvercove is only in its fifth year, we are just starting to see the results on the ground. The impact of the Middle-Valley and Blacksail herds on the treescapes of the valley has not yet been discussed since they have not been active in their areas long enough to show significant results.

Before extensive grazing cattle were introduced into Ennerdale, the Wild Ennerdale partners decided to establish some baseline monitoring of the Silvercove site so that future managers would see the impact of cattle on the landscape. This focussed on four principal methods: photography, exclosures, vegetation quadrats and satellite tracking. The use of data from a satellite tracking collar fitted to one of the herd provides an estimation for the whole herd's activities as it moves around the landscape. Baseline vegetation quadrat surveys were completed before the cattle were introduced but have yet to be repeated and so are not reported on in this chapter. The use of photography is established and is be used to illustrate the main changes discussed, as are the satellite tracking data. The use of exclosures is maybe less well-known and is described below.

Exclosures are simply small fenced areas that aim to keeping grazing animals out rather than in. Typically the exclosures are no more than 0.015ha in size, being constructed from one 50m-long roll of stock net. They take less than a day to erect and can be easily moved if required. They have been extremely valuable in illustrating the impact of grazing on the developing habitat as they are very easy to visit and provide very tangible ongoing and live feedback. Simply walking around the site, comparing photographs and looking at the habitats inside and outside the exclosures, it is clear that cattle grazing is having a positive impact in three key areas. The impacts are in diversifying structure and species and in opportunities for change.

Figure 19.3 Exclosure showing impact of grazing
Source: Gareth Browning and John Gorst

Inside the enclosures located on the areas where conifers have been clear-felled the habitat is fast developing towards a closed-canopy woodland dominated by native broadleaves with a couple of non-native pine (*Pinus* spp.), larch (*Larix* spp.) and spruce (*Picea* spp.). Photographs of the exclosures from just before the introduction of cattle show how quickly woodland regeneration has established and dominated. During the first two to three years of grazing, the habitat inside the exclosures looked the more desirable with a mosaic of heathland, scrub and native tree species. However, in the last two to three years, this structure has changed significantly as trees have gained succession over shrub and ground vegetation and are shading the latter out.

Outside the exclosures the habitat is much more diverse, both in terms of species and structure (Figure 19.3). The cattle naturally spend more time in the lower valley bottom where forage is better and this has led to the lower lying areas being significantly grazed. Even so the vegetation is diverse and there are clumps of woodland regeneration and areas of scrub heathland. The upward growth of all species, apart from a low stocking of conifer regeneration, is being significantly kept in check by grazing. As you walk towards and onto the sloping ground, the distinction between open grasses, heathland, grazed scrub and established woodland is blurred and the future development uncertain. The cattle clearly have preferred and regularly used pathways up and across the slope along which they graze. These areas have regenerated with native woodland species notably birch (*Betula* spp.) but the cattle are keeping most regeneration under 1m high. The grazed access corridors seem to connect to more defined glades where grazing and perhaps soil type are limiting woodland regeneration significantly. These areas currently look destined to stay open, whereas the scrub corridors could develop as woodland if the cattle choose a different way across the site and stop grazing along the existing routes.

When the Silvercove herd were first introduced, data from the tracking collar showed that during the first few months they did not roam far at all. This seemed to be a response to ample grazing in the valley bottom brought about by the site not having been grazed before. As time passed, the herd have explored further and wider in their search for grazing and perhaps in response to weather and temperature. The tracking-collar data show, for example, how the herd react to temperature and how during cold periods they tend to stay more in defined areas and don't roam far. This is presumably to conserve energy. Such behaviour will have an impact on the treescape as where the herd spends more time, the opportunity for woodland to regenerate is much lower. Outside of the grazed routes and open glades there are areas of more dense woodland but very few that exhibit the same canopy closure as is seen within the exclosures. Aside from woodland regeneration, the habitats outside the exclosures are developing a much more diverse mix of shrub and heathland species, wetland and wet meadow. In addition to generating a diverse species mix and structure, it is clear that the natural process of opportunistic change is still ongoing. The herd's patchwork ground disturbance caused by social 'pushing and shoving', 'pathway grazing' and 'hill climbing' all create random and emerging opportunities for seedling germination and vegetation development. Areas of dense bracken under mature pine trees have been broken up by the cattle creating pathways; and within these are found patches of disturbed ground containing young seedlings of birch, whereas under the bracken there is no tree seedling regeneration just grass.

As this is not intended to be a scientific account we do not offer a species list to demonstrate the emerging diversity and raw detail of the changes. However, the sheer variety of texture, colour and mosaic of vegetation intermixed with trees and scrub is a joy to walk through and fascinating to watch develop.

Wild cattle – future natural landscapes

So we now have three roaming herds of cattle covering around 1,000ha of the valley. While they are still considered domestic in the eyes of the law, they roam over large expanses of landscape (Figure 19.4), with limited tending, choice in their daily foraging and the freedom to enjoy all that the Cumbrian weather throws at them. We would argue that they are some of the wildest animals in England.

The people involved in their introduction have become wilder as they have let go of past concerns and traditions. They now have the confidence to stand back and allow the cattle to decide their own routine or seek them out to gain joy from seeing their calves, their teddy bear faces, or to learn how they survive the winters.

As a partnership we continue to aspire to remove boundaries and give our cattle as large an area to roam as is possible. The experiment has only just begun; five years of grazing is just showing little glimpses of tangible change and benefit. The treescape of Ennerdale has benefited from these native black animals as they have lived up to the expectations we had for them to blur the boundary between forestry and faming, open and wooded landscape.

We do not know what the valley will look like in the future but one missing natural process is firmly back and helping to make Ennerdale a wilder valley for the benefit of people. Here nature will determine the detail and we, as stewards, can only marvel and be excited by what the future natural landscape may look and feel like.

Figure 19.4 Cattle crossing the river Liza
    Source: Gareth Browning and John Gorst

Bibliography and references

Browning, G. and Gorst, J. (2011) *Wild Cattle – Wilder Valley: Sharing Experiences from Introducing Extensive Cattle Grazing to a Lakeland Valley*, I.D. Rotherham and C. Handley (eds) *Animals, Man & Treescapes* Wildtrack Publishing, Sheffield, 91–108.

Vera, F.W.M. (2009) 'Large-scale nature development: The Oostvaardersplassen', *British Wildlife*, 20(5): 28–36.

Wild Ennerdale Partnership (2003) *Ennerdale Historic Landscape Survey*, Wild Ennerdale Partnership, Cumbria.

Wild Ennerdale Partnership (2006) *Wild Ennerdale Stewardship Plan*, Wild Ennerdale Partnership, Cumbria.

Winchester, A.J.L. (1987) *Landscape and Society in Medieval Cumbria*, John Donald Publishers, Edinburgh.

20 Treescapes: Trees, animals, landscape, people and 'treetime'

Luke Steer

Introduction

This chapter has been prepared, not from the perspective of an academic landscape historian or ecologist, but from the perspective of someone trained in forestry and arboriculture. Moreover the author was brought up on a Cumbrian dairy farm, lives in the Lake District and spends lot of time in the UK uplands. He has also visited the Alps, Karakorum in Pakistan, the Jammu and Kashmir province of India, the Atlas Mountains in Morocco and the Julian Alps in Slovenia. In this chapter I recount some observations and hypotheses.

We know that foresters go to great lengths to protect regenerating woodland from herbivorous mammals. How then was it possible for much of the countryside to be covered in woods, pasture woodland, wood-pasture and parkland, before people built walls, planted hedges or erected fences? Vera (2000 and 2002) puts forward an attractive hypothesis but it has not been accepted by all (Hodder *et al.*, 2005; Kirby, 2003 and 2004; Mitchell, 2005; Rackham, 2006).

The majority of the ancient semi-natural woodlands (ASNWs) in the Lake District are predominantly composed of sessile oak (*Quercus petraea* (Mattuschka) Lieblein) and located on bouldery valley sides (Figure 20.1). There are also a few oaks in the grazed intakes but the majority of the trees in these areas are ash (*Fraxinus excelsior* L.) and hawthorn (*Crataegus monogyna* Jacq.). Many of these areas appear to contain trees in a limited number of age classes – why is this?

In this chapter I review where trees are likely to regenerate in grazed upland landscapes and the events that allow them to do so. The chapter is divided into six main sections: a review of how site features affect where trees regenerate; the factors that can lead to tree regeneration events – treetime and tree and woodland cycles; a discussion about human influences on 'treescapes'; the current situation in the Lake District; some tentative recommendations for future tree planting; and suggested further research.

Figure 20.1 Brothers Water in the Lake District – a mosaic of cropped fertile land; steep, bouldery and infertile valley sides; and exposed leached hilltops
Source: Luke Steer

Site features

Yalden (1999) tells us that plants and animals quickly colonised the British landscape after the end of the last Ice Age and soils started to develop. It seems logical to expect that areas with better soil (deep, moist, fertile and aerated) supported lush vegetation, whereas those with poor soil were sparsely vegetated with plants that were often woody and/or unpalatable (Bardgett, 2005). Generally most species of grazing herbivores prefer to feed on lush nutritious vegetation and only visit other areas when passing through from one feeding place to another, or for shelter and defence. Different species of browsing herbivores feed on woody and herbaceous vegetation in varying proportions (Mason and Kerr, 2004; Mason *et al.*, 1999; Thompson, 2004; Vera, 2000). Wild animals also range over large areas. The impact that these animals had on the vegetation probably depended on their numbers; and their numbers would have largely depended on the winter carrying capacity of the landscape.

Work in Yellowstone National Park (for example: Beschta, 2005; Beschta and Ripple, 2010; Ripple and Beschta, 2006 and 2007; Ripple and Larson, 2000) indicates that when top predators – wolves in that instance – are reintroduced to an area, herbivorous mammal numbers are reduced. The remaining populations spend less time in any one area and this allows woody vegetation to develop. When British woodlands developed, wolves and other predators, including humans, still roamed the landscape.

Work by the Forestry Commission (Mason and Kerr, 2004; Mason *et al.*, 1999; Thompson, 2004) shows that tree seedlings are more likely to develop on infertile soils than fertile soils because of reduced competition from other vegetation. However, Fenton (2008) suggests that in upland Scotland, trees are more likely to grow on the fertile soils rather than infertile soils. Differences between these authors may not be related to soil fertility but to its organic matter content. Fenton (2008) suggests that mor soil types, high in organic material, inhibit tree regeneration, whereas the other authors suggest that low fertility mineral soils, such as those dominated by sand, enhance it. I therefore suggest that many of our ASNWs developed under the influence of one or a combination of the factors listed in Box 20.1.

Box 20.1 Factors, either singly or in combination, that can allow trees to regenerate in upland grazed landscapes

Low-fertility mineral soils (Figure 20.2)

Soil prone to drought (Figures 20.2, 20.4 and 20.6)

Waterlogged soil (alder and willow carr)

Steep slopes less easy for animals to traverse (Figures 20.1, 20.2, 20.3, 20.4 and 20.5)

Bouldery and craggy areas that provide a natural barrier to large herbivorous mammals (Figures 20.3 and 20.4)

I suggest that there would be a gradient from the areas favoured by grazing animals that, in turn, prevented tree and shrub regeneration, to those least favoured by grazing animals where trees and woodland would naturally regenerate (Figures 20.1 to 20.8). In addition to areas on this continuum are others that are unsuitable for both trees and grazing animals such as peat bogs. These will not be considered further in this chapter (Figure 20.2).

One or a number of the factors listed in Box 20.1, when animal pressure was sufficiently low, would create conditions suitable for tree regeneration. Humans then 'allowed' these trees and woodlands to remain because alternative uses for the land, principally agriculture, were limited due to the low productivity of the soil or difficulties in harvesting produce (Figures 20.4, 20.5 and 20.6).

The Triglav National Park in Slovenia could be thought of as an extreme version of the UK uplands with very steep-sided valleys. High up in these valleys there are some beautiful alpine meadows that were traditionally

Figure 20.2 Trees and hawthorn growing on a steep free draining area with thin mineral soil adjacent to a fertile area with deeper moist soil – the vegetation and the differential grazing patterns indicate the fertility and moisture status of the soil
Source: Luke Steer

used for grazing cattle and producing hay (Figures 20.5 and 20.8). There are no manmade boundaries around these alpine meadows; the edge trees have branches down to ground level and there is little of interest to cattle within the forest matrix.

After snow, I observed that the paths were covered with animal footprints but there were few footprints in the adjacent forest. Could these tracks be the same paths that were used by wild grazing animals when humans were still hunter-gatherers? People, when hunting, would follow them from alpine meadow to alpine meadow, similar to wolves following herds of large herbivores in North America or big cats following their prey in Africa. The alpine meadows are generally in bowls where, over time, a depth of fertile soil has developed, whereas, the soils in the forest are often thin or infertile and many are prone to either drought or water-logging.

I therefore suggest that the alpine meadows may have been open to a greater or lesser extent since the end of the last glaciation and before humans had a significant impact on the landscape and its ecology. When people started to farm the landscape and domesticate animals they would

Figure 20.3 Glen Brittle in Skye – trees are growing on the steep sides of the gorge where sheep graze less intensively than on the more easily traversed land adjacent to it; the soil on the moor is peaty mor whereas the soil in the gorge contains less organic matter and is more free draining
Source: Luke Steer

have continued to move them from alpine meadow to alpine meadow along the same paths once used for hunting prey.

In the British lowlands some of the oldest woodlands that contain ancient trees are on poor sandy soils. Examples are Sherwood and Staverton (Figure 20.6). Both Sherwood and Staverton are prone to drought but I expect that tree roots, particularly those of oak that dominate these woodlands, are able to grow deeper into the soil to obtain moisture than roots of competing vegetation.

Other ASNWs are on heavy clay soil that was hard to cultivate using traditional methods and Wistmans Wood in Devon is growing in a boulder-field that deters animals from staying in it for long periods.

Regeneration events

Vera (2000 and 2002) and Rackham (2006) suggest that before humans had a significant influence on the ecology of the British landscape it contained a variety of terrestrial habitats. These ranged from species-rich grassland grading into parkland with individual trees and copses, to woodland with glades.

Trees, animals, landscape, people, 'treetime' 287

Figure 20.4 Typical Lake District woodland with rocky, bouldery, thin and infertile soil prone to drought
Source: Luke Steer

Figure 20.5 Alpine meadow in Triglav National Park – here are no fences between the grazed areas and the trees; the trees are growing on steeper ground with less productive soils, often thin and prone to drought
Source Luke Steer

Figure 20.6 Sherwood Forest – the soil is very sandy, free draining and infertile
Source: Luke Steer

I imagine that the proportion of the land that was covered by grassland, wood-pasture and woods would have slowly but continually fluctuated over space and time (Figures 20.7 and 20.8). Over time, trees within closed-canopy woodland would have died and, initially, the canopies of their neighbours would join to fill the gaps, but eventually glades would develop and expand. In some instances these glades would have been kept open by grazing mammals and the vegetation in them would alter from shade-tolerant woodland vegetation to grazing-tolerant grassland vegetation. This would increase the winter carrying capacity of the landscape for large herbivorous mammals and their numbers would increase as a consequence. This process would have continued until what were once closed-canopy woodlands thinned to a greater or lesser extent to become pasture woodland or even wood-pasture.

Areas least suited to support grazing mammals might have continually remained as woodland until this day (Gulliver, this volume and Figure 20.4) but at their peripheries, over time, they probably expanded and contracted across the landscape (Figures 20.7 and 20.8). The extent of these fluctuations would depend on tree age, health and events that affected tree cover (such as infestations of herbivorous insects, tree diseases, gales and poten-

Figure 20.7 A dynamic Lakeland treescape – scrub and woodland 'invading' pasture between Embleton and Cockermouth in the Lake District; reduced stocking numbers are allowing gorse (Ulex spp.) and hawthorn (Crataegus monogyna *Jacq.*) to regenerate and these are protecting tree seedlings from grazing animals
Source: Luke Steer

tially fire, particularly bracken and heather fires). There would also be tree regeneration events regulated by the population size of grazing animals in combination with the winter carrying capacity of the landscape.

At different periods along this gradation a collapse of the grazing animal population would allow a pulse of tree regeneration that, in certain areas, but more likely those with the features listed in Box 20.2, would again become increasingly treed or even closed-canopy woodland. The more severe the population collapse of large herbivorous mammals, the more extensive the tree regeneration.

In Triglav National Park, Slovenia, numbers of grazing animals have reduced as the viability of agricultural businesses becomes increasingly stressed. Some meadows are being colonised by thorny shrubs (*Berberis vulgaris* L.) and trees (Figure 20.8).

Fenton (2008) suggests that in the Scottish Highlands the winter carrying capacity of the vegetation probably allows sufficient large herbivorous mammals, deer and farm animals to live there and prevent landscape-scale woodland regeneration. This view is not shared by all (Bennett, 2009; Peterken, 2009). However, it appears to be generally accepted that numbers of large grazing herbivores can prevent or significantly reduce tree

Figure 20.8 A dynamic Slovenian treescape in Triglav National Park – cattle numbers have reduced during the recent past and *Berberis vulgaris* L. and trees are invading the slightly steeper upper area, whereas the remaining cattle are concentrating their feeding in the lower areas
Source: Luke Steer

regeneration. This indicates that something is required to reduce animal numbers sufficiently, certainly below numbers currently experienced in upland Britain, to allow tree and shrub regeneration. Box 20.2 lists factors that could impact on the population size of large herbivorous mammals to a greater or lesser extent, either singly or in combination.

Box 20.2 Factors that can affect the population size of large herbivores for a given site, either singly or in combination

Predation

Winter mortality

Animal diseases

Animal parasites

Pests and diseases of food plants

Reduction in human population, potentially due to disease or agricultural viability

---

No doubt numbers of large herbivorous mammals have always fluctuated in both the short and long term. I suspect that generally they would have tended to increase in proportion to improving habitat as wooded and treed areas thinned. Eventually, potentially a number of centuries after the previous mass tree regeneration event, animal numbers could increase to levels where they would become vulnerable to a massive reduction in their population caused by one or a combination of the factors listed in Box 20.2. At such a time there would be an aging population of trees and some of the areas where trees once grew would be wood-pasture or even open grassland.

Rabbits (*Oryctolagus cuniculus* L.) prevented tree regeneration at Silwood Park for decades if not centuries prior to the introduction of the virus that causes myxomatosis (*Myxoma* spp.) in 1953. Myxomatosis caused catastrophic mortality of the rabbit population and by the late 1960s a new cohort of oaks had established (Dobson and Crawley, 1994).

I suggest that during periods with increasing animal numbers their core feeding areas would expand into areas with sub-optimal feed vegetation and these habitats would 'degrade'. The canopy of aging woodland would thin over time as trees died and high animal numbers prevented regeneration. The opening of woodland canopies would allow increased levels of light to the woodland floor and consequently the vegetation would alter from being dominated by woodland plants to light-demanding grazing-tolerant vegetation such as grasses and sedges. This alteration of vegetation would favour large herbivorous mammals and increase the area's winter carrying capacity for them and, in turn, they would prevent tree regeneration.

During warm summer weather, large herbivorous mammals often prefer to spend a disproportionate amount of time sheltering close to trees or in groups of trees growing in accessible areas with nutritious vegetation. This results in nutrient importation to those areas. Nettles (*Urtica dioica* L.), an indicator of soils with high fertility, are often found growing in and around groups of parkland trees for this reason. Enhanced soil fertility can have a detrimental effect on fungi that develop mycorrhizal relationships with tree roots and, in turn, the health of their host tree. This leads to a further decline of the tree canopy and an improvement of the quality and quantity of the vegetation for large herbivorous mammals.

Large numbers of animals often cause soil and vegetation disturbance and, if then followed by a reduction in their numbers, this would create opportunities for tree regeneration in areas less favoured by the remaining animals. Population maximums of large herbivorous mammals, followed by a reduction in their numbers sufficient to allow a pulse of tree regeneration, would only have to occur once or twice within the lifetime of a tree to allow sufficient trees to regenerate and maintain the presence of that species in the landscape; for oak this might be once or twice every 200–600 years or potentially longer.

At high population densities, large herbivorous mammals would come into contact with each other with increasing frequency and consequently increase the risk of spreading pests, parasites and diseases around the herd.

Also, high numbers of animals may increase the time and frequency they spend in areas where they had defecated and urinated and this also increases the risk of spreading pests, parasites and diseases. Parasite populations, such as ticks and intestinal nematodes, could then increase and often these are vectors for diseases, such as Lyme disease, which can amplify the detrimental effects of the parasite on host vitality and population numbers.

After a reduction in the numbers of large herbivorous mammals the remaining animals would concentrate in areas with the most suitable and nutritious vegetation rather than areas with poor soils and low-quality food plants. Consequently trees would regenerate in areas with the features listed in Box 20.1 due to reduced browsing and grazing pressure, and a mosaic of closed-canopy woodland, groups of trees and individuals would again develop in those areas. The mosaic or trees may mirror areas with infertile, thin, drought prone soil, boulders or steep gradients (Figures 20.1 to 20.8). The number and density of trees would depend on the complexity of site conditions, such as those listed in Box 20.1, in combination with the numbers and species of the remaining herbivorous mammals. Within closed-canopy woodland, the quality and quantity of feed vegetation of large herbivorous mammals would decrease and, at a landscape scale, the animal winter carrying capacity would also decrease. Where closed-canopy woodland formed it would remain as long as the gaps created by tree death and failure could be closed by neighbouring trees but, eventually, this may not be possible. Glades would again form and the vegetation within them would alter, so improving the quality and quantity of feed vegetation for large herbivorous mammals, and these would again prevent further tree regeneration until the next reduction in their numbers. However, some woodlands would occupy areas where animals would never or hardly ever favour, even when their numbers were extremely high. These would remain as woodlands (Gulliver, this volume) (Figures 20.1, 20.4 and 20.5).

The resulting treescape would contain a limited number of cohorts of trees from separate regeneration events: a woodland regeneration cycle partly regulated by topography and partly by large herbivorous mammals. This cycle is summarised as:

- After a period of high animal numbers they decrease, potentially due one or a combination of the factors listed in Box 20.2
- In a period of low animal numbers there is an episodic pulse of thorn and tree regeneration.
- Tree canopies expand and cast shade on ground vegetation, which consequently changes from being dominated by grazing-tolerant grasses to shade-tolerant woodland vegetation. This is less productive and of a lower feed quality than grazing-tolerant, light-demanding grasses, and thus reduces the stock winter carrying capacity of the landscape.

- Many decades pass with a relatively stable treescape. As trees die the crowns of their neighbours expand to occupy their vacated canopy space.
- Woodland canopies eventually thin as trees die and the gaps become too large for neighbouring trees to fill. Grazing animals utilise these gaps and the vegetation changes from being dominated by shade-tolerant woodland plants, to grazing-tolerant vegetation. These gaps are maintained by grazing animals and expand as other trees die. This process increases the winter carrying capacity of the landscape for large herbivorous mammals.
- There are minor thorny shrub and tree regeneration events if animal numbers decrease by small amounts, potentially for the reasons listed in Box 20.2, but generally the landscape becomes increasingly open due to thinning woodland canopies and tree loss, and this leads to a generally upward trend in grazing animal numbers.
- Eventually there is another reduction in animal numbers, due to one or a combination of the reasons listed in Box 20.2, and the remaining population concentrates in the areas most suited for grazing (good fertile soil and nutritious vegetation). Trees again regenerate in areas least suited for the grazing animals and the cycle continues.

With this model small frequent reductions in the populations of grazing animals would create a diverse tree age-class structure, whereas a long period without a significant crash in animal numbers would lead to a simplified tree age-class structure favouring long-lived light-demanding trees such as oak.

Disturbance caused by high numbers of herbivorous mammals that maintained and enlarged open areas and woodland glades would, when their populations collapsed, create conditions suitable for the regeneration of thorny shrubs and light-demanding tree species such as oak.

If numbers of grazing animals are insufficient to maintain glades in areas of dense woodland, such as steep river valleys, shade-tolerant, grazing-intolerant tree species could become established and the species make-up alter to become dominated by these shade-tolerant trees. This is happening today in enclosed Lakeland oak woodlands in which non-native beech trees were established in the nineteenth century to 'enhance' the landscape. Also it is occurring with naturally regenerated limes in river valleys and woodland gorges in the Coniston Lake catchment area (Pigott, 1993) or steep valleys in Helmsley Red Deer Park.

Human influence

Before humans had a significant influence on the landscape, treescapes would have had a few thousand years to develop and it is likely that some of the trees were huge. Felling large trees, even with a chainsaw, is hard work: imagine trying to fell and convert large trees with a stone axe or non-mechanised hand tools? It is more likely that early humans concentrated their efforts on small trees and the branches of larger ones. Therefore, if we believe that Neolithic people cleared large areas of trees and woodland, I suggest that it would probably have taken centuries, if not thousands of years. However, Vera (2000 and 2002) and Rackham (2006) suggest that the landscape was a mosaic of woodland, wood-pasture and grassland.

We know that large herbivorous animals can maintain open or semi-open landscapes, such as the Serengeti and, according to Fenton (2008), upland Scotland. As I mentioned earlier, foresters go to great lengths to exclude large herbivorous mammals from young and regenerating woodland because they can kill or damage high numbers of trees and render them useless as a timber crop.

For human populations to increase they had to farm the landscape efficiently to provide sufficient food. In the uplands, apart from the valley bottoms, pasture is the most efficient land use for food production. Fenton (2008) suggests that historically domestic stock might have replaced indigenous herbivores to a greater or lesser extent and the overall level of grazing may not have altered significantly, at least since about AD1700; however, others dispute this (Bennett, 2009; Peterken, 2009). Whichever point of view is correct, small numbers of large herbivorous animals can prevent tree and shrub regeneration in large areas of the countryside.

Records indicate that, at least up to AD1700, farmers in the Pennines, Lake District and Scottish borders kept more cattle and fewer sheep than today (Winchester, 2000). Cattle are less nimble than sheep and cause more soil disturbance. These factors probably increased the proportion of the uplands that was unsuitable for grazing and thus created opportunities for trees, copses and woodlands. The 'shadow woodlands' discussed by Rotherham (this volume) may date from this period. I suggest that the greatest challenges that early farmers had to overcome were: the limited winter carrying capacity of the uplands; and catastrophic collapses in animal numbers caused by the factors listed in Box 20.2.

Naturally wild animals probably ranged over large territories between their winter and summer grounds. Initially people mimicked this with the shieling system. Later they collected winter fodder: hay and 'tree-hay' (leaves and branches of leaves) (Quelch, pers. comm.) and 'stabled' the cattle over winter to reduce their energy use and the amount of food required to keep them alive – this was a great innovation. More recently human inputs, such as lime and fertilisers, have increased the productivity of some areas of land and the practice of importing fodder and feed from other locations has also increased the number of animals that upland farms can support.

Enclosures may have also increased the number of animals that could be maintained (Figure 20.10). Extensively grazed animals move from place to place in response to the quality and quantity of food plants and the amount of dung and urine in an area. Once enclosures were erected the animals could not move on and were forced to graze more intensively than if they were grazing extensively. Consequently the grass is probably grazed closer to the ground than in extensively grazed situations, thus further reducing opportunities for tree and shrub regeneration.

At the time of enclosure, many woodlands became 'fossilised' within the landscape and their areas have not expanded since, although some have contracted or been lost as trees died, the woodland canopies thinned and grazing animals were encouraged to feed in them (Rotherham this volume) (Figures 20.9 and 20.10).

Figure 20.9 A woodland boundary wall with some isolated trees beyond that may indicate the former extent of the woodland/wood-pasture prior to its enclosure as it naturally expanded and contacted – this is the boundary of the woodland shown on Figure 20.4
Source: Luke Steer

More recently large, relatively inexpensive farm buildings and veterinary science have allowed high numbers of animals to be kept in close proximately to each other with a reduced risk of winter mortality, parasites and disease. I suggest that this allows unnaturally high numbers of grazing animals to be maintained without a collapse in the size of their populations and this prevents pulses of tree regeneration.

The current situation in the Lake District

It strikes me that, due to a reduction in cattle numbers in the Lake District during the nineteenth and twentieth centuries, and an increase in sheep numbers, tree and woodland cover is now at an all-time low. Many of the existing trees date from previous times when the land was farmed less intensively, sometimes before the last enclosure walls were erected (Figure 20.10).

Typically, within Lakeland intakes, most of the trees are ashes and hawthorns but there are some oaks with trunk diameters of 80–120cm at 1.3m. Using the formula suggested by White (1998) these would be about 175 years old; was the last pulse of oak regeneration in these areas around 1837? I also have a section from a lapsed ash pollard that was cut in 2005; it is 63cm in diameter and has 172 annual rings – it was therefore was last pollarded in 1833.

One hundred and seventy five years ago was prior to the erection of some of the enclosure walls in the Lake District. However, the ring-garth walls, sometimes referred to as the head dyke, and some of the intake walls date from the sixteenth and seventeenth centuries (Winchester, 2000). Observations suggest that there has been little oak regeneration within the intakes and fields during the last 150–200 years. The oak tree depicted in Figure 20.10 is growing at Ecclerigg between Windermere and Ambleside. It has a trunk diameter a little over 1m and appears to predate the field boundary wall, which appears to be an 'Enclosure Act' wall.

There are hawthorns in many of the Lakeland intakes that could indicate a period of reduced animal numbers. These are often said to date from the Great Depression of the 1930s but could also date from periods with severe winters that killed significant numbers of animals. In areas where these hawthorns are growing there are few large-growing trees except, in some areas, a small number of ashes. These are often in bouldery areas and on rocky knolls. I suggest that this indicates that generally the succession from hawthorn scrub to a landscape containing mature trees was arrested by an unnaturally quick increase in animal numbers. This might have been due to intensified agriculture during and after World War II.

There are many meadows within the Lake District that contain groups of trees growing on rocky knolls; mainly oak but some ash and hawthorns. I consider it unlikely that these would have established if the fields were grazed unless they were protected by fences but, as I have not come across this in the literature, I consider it unlikely. More likely is that, at the time the trees colonised these knolls, the fields were hay meadows or arable and the vegetation on the knolls was not harvested. The current practice of cutting grass early for silage and then grazing the aftermath now prevents tree regeneration in these areas. During the 2005 gales, some of the trees on these rocky knolls blew over. If this trend continues trees on these rocky knolls will eventually disappear from our landscape.

Figure 20.10 Oak tree, Ecclerigg in the Lake District – the tree is incorporated into the field boundary wall that was probably constructed as the result of a parliamentary Enclosure Act; is the tree older than the wall?
Source: Luke Steer

In a number of valley-bottom fields bounded with dry-stone walls there are often trees around their perimeters, usually pollarded ashes. Again these may have established when the fields were hay meadows or arable and not used for stock grazing. Alternatively they may have been planted to be cropped as pollards.

There are areas of land in the Higher Level Stewardship agri-environment scheme where the thorny shrubs, gorse (*Ulex* spp. L.), hawthorn and blackthorn (*Prunus spinosa* L.), and trees, are regenerating in relatively large grazed enclosures. In some of these it appears that the slope angle is regulating whether the area is heavily grazed to a short turf or grazed infrequently enough to allow the thorny shrubs to become established and protect tree seedlings sufficiently to allow them to grow large enough not to be damaged by farm stock (Figure 20.7).

In many grazed areas within the Lake District and other parts of upland Britain, tree regeneration is currently non-existent and the existing trees are ageing. Observation suggests that many of the trees in Lakeland fields and intakes may date from before the land was enclosed. Some of these blew over in the gales in January 2005. The number of trees within meadows and traditional wood-pasture is therefore reducing. Financial incentives for tree planting in pasture appear to favour woodland creation that alters the established ecology and landscape character. I suggest that, in some of the areas where woodlands have been established during the last 40 years, there may never have been closed-canopy woodland within historic times – indeed, some of these woodlands have been established on prime agricultural land close to farms and hamlets in hay meadows that historically may have been arable.

It appears to me that there are huge differences in the objectives of the government grant-giving agencies and that grants available for tree planting in pasture areas are inappropriate to maintain the traditional landscape character and historic ecology. I consider that the current objectives of the grant-aided schemes will be detrimental to upland landscapes and their wildlife. Finally, the ancient semi-natural woodlands are invariably within enclosures and are now unable to expand and contract as they once did. Some are grazed by sheep and unable to regenerate even if there are large gaps in their canopies. However, many of the UK's enclosed upland woodlands are in sympathetic management and maintained with government grant aid. I consider that the future existence of these woodlands is not under threat.

Conclusions

The observations discussed in this chapter lead me to conclude that prior to the time when humans started to have a significant impact on the upland landscape it was a mosaic of meadows, wood-pasture, pasture woodland, copses and woodland. I also consider that the proportion of canopy cover fluctuated over time but areas least suitable for large grazing mammals may have continually remained as woodland, whereas areas most suitable for them remained open pasture.

Recommendations

On the basis of these observations, I make a series of recommendations:

1 Improve grant aid for planting individual trees and shrubs, or groups of trees and shrubs, in upland pastures and intakes rather than establish woodland. Small-scale exclosures would allow natural regeneration but, currently, there would be a risk that this practice may reduce grant aid available to farmers as it is based on area payments.
2 Carry out a site assessment prior to establishing trees in upland pastures and intakes and only establish them in infertile or rocky and naturally protected areas.
3 Do not establish woodland on land favoured by grazing animals. Instead, in these areas, establish trees individually or in groups with thorns.

4   Do not plant woodland on land that could be harvested for hay or silage. In these areas plant trees on rocky knolls and around the peripheries of the fields.
5   Do not plant too many trees! Leave places to plant trees in 50 or 100 years' time to mimic natural pulses of regeneration.

Suggested research

Further research could be carried out to ascertain the validity of the suggestions made in this chapter. Certain features, such as aspect, slope angle, soil type, depth, fertility, aeration and moisture retention, local climate and bouldieriness, could be assessed in a number of ASNWs and intakes that contain trees. This information could be inputted into a geographic information system (GIS) to produce a computer model of where woodlands could be expected under different levels of grazing. The GIS model could then be used to assess how the area of these woodlands may alter under different grazing pressures. It may also be possible to compare maps produced by the GIS model of potential woodland areas with historic maps and archaeological records and features.

Once a robust model has been developed it could be used to guide where trees, wood-pasture and woodland establishment should be carried out and at what densities. This information could be used to prioritise where financial incentives for tree planting, in the form of grant aid, should be allocated.

As discussed above I suspect that in enclosed upland pasture landscapes many of the trees may predate the erection of the walls and hedges. I also suspect that their numbers are declining and at an all-time low. I suggest that a landscape tree survey should be carried out of trees in fields and intakes to assess how robust or vulnerable they are. This information could then be used to allocate financial incentives for tree establishment, at appropriate densities, to ensure that trees remain landscape features for the next 200 years.

References

Bardgett, R. (2005) *The Biology of Soil: A Community and Ecosystem Approach*, Oxford University Press, Oxford.
Bennett, K.D. (2009) 'Woodland decline in Scotland', *Plant Ecology and Diversity*, 2(1): 91–3.
Beschta, R.L. (2005) 'Reduced cottonwood recruitment following extirpation of wolves in Yellowstone's northern range', *Ecology*, 86(2): 391–403.
Beschta, R.L. and Ripple, W.J. (2010) 'Recovering riparian plant communities with wolves in northern Yellowstone, USA', *Restoration Ecology*, 18(3): 380–9.
Dobson, A. and Crawley, M. (1994) 'Pathogens and the structures of plant communities', *Trends in Ecology and Evolution*, 9(10): 393–7.
Fenton, J.H.C. (2008) 'A postulated natural origin for the open landscape of upland Scotland', *Plant Ecology and Diversity*, 1(1): 115–27.

Hodder, K.H., Bullock, J.M., Buckland, P.C. and Kirby, K. (2005) *Large Herbivores in the Wildwood and in Modern Naturalistic Grazing Systems*, English Nature Research Report 648, Peterborough, UK.

Kirby, K.J. (2003) *What Might a British Forest-landscape Driven by Large herbivores Look Like?*, English Nature Research Report 530, Peterborough, UK.

Kirby, K.J. (2004) 'A model of a natural wooded landscape in Britain as influenced by large herbivore activity', *Forestry*, 77(5): 405–20.

Mason, B. and Kerr, G. (2004) *Transforming Even-aged Conifer Stands to Continuous Cover Management*, Forestry Commission Information Note 40, Edinburgh.

Mason, B., Kerr, G. and Simpson, J. (1999) *What is Continuous Cover Forestry?*, Forestry Commission Information Note 29, Edinburgh.

Mitchell, F.J.G. (2005) 'How open were European primeval forests? Hypothesis testing using palaeocological data', *Journal of Ecology*, 93(1): 168–77.

Peterken, G. (2009) 'Response to "A postulated natural origin for the open landscape of upland Scotland"', *Plant Ecology and Diversity*, 2(1): 89–90.

Pigott, C.D. (1993) 'The history and ecology of ancient woodlands', in P. Beswick and I.D. Rotherham (eds) 'Ancient woodlands – their archaeology and ecology – a coincidence of interest', *Landscape Archaeology and Ecology*, 1: 1–11.

Rackham, O. (2006) *Woodlands*, Collins New Naturalist 100, London.

Ripple, W.J. and Beschta, R.L. (2006) 'Linking wolves to willows via risk-sensitive foraging by ungulates in the northern Yellowstone ecosystem', *Forest Ecology and Management*, 230: 96–106.

Ripple W.J. and Beschta, R.L. (2007) 'Restoring Yellowstone's aspen with wolves', *Biological Conservation*, 138: 514–19.

Ripple W.J. and Larson, E.J. (2000) 'Historic aspen recruitment, elk, and wolves in northern Yellowstone National Park, USA', *Biological Conservation*, 95: 361–70.

Thompson, R. (2004) *Predicting Site Suitability for Natural Colonisation: Upland Birchwoods and Native Pinewoods in Northern Scotland*, Forestry Commission Information Note 54, Edinburgh.

Vera, F.W.M. (2000) *Grazing Ecology and Forest History*, CABI, Wallingford, UK.

Vera, F.W.M. (2002) 'The dynamic European forest', *Arboricultual Journal*, 26: 179–211.

White, J. (1998) *Estimating the Age of Large and Veteran Trees in Britain*, Forestry Commission Information Note 12, Edinburgh.

Winchester, A.J.L. (2000) *The Harvest of the Hills – Rural Life in Northern England and the Scottish Borders, 1400–1700*, Edinburgh University Press, Edinburgh.

Yalden, D. (1999) *The history of British Mammals*, T. & A.D. Poyser Ltd, London.

21 Creation of open woodlands through pasture

Genesis, relevance as biotopes, value in the landscape and in nature conservation in south-west Germany

Mattias Rupp

Introduction

In the three-year project (2008–2011) Open Woodlands Through Pasture, such woodlands have been searched for and examined to find out why this old land-use activity is kept alive and what effects it has on biodiversity. The spatial focus of the study is the federal state Baden-Wuerttemberg in south-west Germany.

Due to historical overuse of woodlands, wood-pastures (WP) are forbidden in Germany, but some can still be found by insiders. It has been possible to demonstrate that historical and modern WPs have little in common with each other. Nowadays they serve mainly as functional elements to support the farm business, the wellbeing of livestock and, when applied within the framework of a management concept, to increase biodiversity.

WPs can trigger reserved or even antagonistic reactions against livestock owners. This form of agriculture can lead to unnecessary clashes with the law. In light of these issues, it was decided that anonymity would be maintained and that study locations would only be referred to at the regional scale.

Theoretical background

The area of interest is the federal state of Baden-Wuerttemberg (Ba-Wue) in south-west Germany. This was determined in part by the project's financing through the Stiftung Naturschutzfonds Ba-Wue (Foundation for Nature Conservation). Due to a highly variable topography and a rich agricultural history, it was expected that remnants of open woodlands remained there and that new ones could be initiated.

Historical context of wood-pasture

In Middle Europe, various factors can create open woodlands: extreme edaphic conditions, mechanic and/or biological disturbances, advanced age of climax vegetation and agricultural influences. In Germany, agriculturally created open woodlands have been commonly occurring biotopes over recent centuries (Figure 21.1). Due to multi-functional uses and intensified land use, forests have been continuously over-cultivated and large areas went into decline or devastation. Consequently, sustainable forest management and laws such as Badische Forstgesetz (Forestry Law of Baden, 1833) were implemented. Since then, keeping livestock in forests has been forbidden and those doing so can be prosecuted (Liss, 1988; Sproßmann, 2009; Vera, 2000).

In the course of the Industrial Revolution, open land could also be used for keeping animals and to obtain fodder for winter. Hence, as it was no longer necessary to herd livestock in woodland and WP became a marginalised agricultural system. Correspondingly, multi-functional forest use also changed. Wood consuming crafts (charcoal, glassblowing, coal mining) were downsized or even terminated in south-west Germany (www.badische-seiten.de) and forests were rearranged, mainly into timber forests. To keep the hunting business running for the upper-class and aristocrats, great efforts were undertaken to drive the rural population and their economic activities out of the forests (www.adelegg.de). These circumstances led to the loss of semi-open biotopes with mosaic-like vegetation patterns.

Following World War II, many WP remnants were abandoned and succession took over or the biotopes were reforested as monocultures. A corresponding loss of biodiversity can be assumed (Liss, 1988; Sproßmann, 2009; Vera, 2000).

Recent situation

Relicts of, or recently established, WPs in Ba-Wue can be found on commons, remote areas of low mountain ranges, on unproductive soils or in some nature conservation areas. Nowadays, farmers, foresters and conservationists interact and create projects with the purpose of keeping animals in a species-appropriate environment and increasing local diversity in structures and biota. In this modern approach careful handling on the pastures, planned management and long periods of development are taken into consideration.

Those who run WPs lament the significant lack of ecological information resulting from the fact that the WP tradition was stopped around 180 years ago. They note that in order to create open woodlands, 'the wheel needs to be reinvented'.

Research questions and data acquisition

The guiding questions for this research were: 1) where can WPs be found and what are their characteristics?; 2) do WPs have effects on the diversity of local plant species and the variety of spatial structures?; and 3) are there further ecological (side-)effects?

Figure 21.1 Historical wood-pasture in Germany with different influences on vegetation and landscape: Deceleration of natural regeneration, leaf and litter harvest, active herding
Source: WebMuseum, Paris: Les très riches heures du Duc de Berry, www.ibiblio.org/wm/rh/4.html

Spatial research

In Baden-Wuerttemberg only very few locations with current WPs are known and literature barely exists. Nature conservationists and foresters in marginal areas were the most knowledgeable people to ask. Given that old breeds of farm animals are used in extensive agriculture, contacting breeder-clubs was an additional strategy for accessing information. Given concerns about misunderstanding and a possible clash with the law, all stakeholders wanted to get to know the researcher personally before sharing information. A very important aspect was therefore building confidence and visiting interviewees at their place of work.

Based on the proceeding work, the researcher's name was already familiar to prospective interview partners and the confidence built in the earlier phase of research eased communication. When the researcher rang the same people a second time and asked the same questions, also mentioning who he had met in the meantime, they were willing to provide information about more WPs. Once that had been agreed upon, the farmers shared further information or gave references to hidden WPs.

Guided interviews

To obtain more detailed information from the practitioner's perspective, guided interviews were conducted. The stakeholders interviewed were farmers (12) and their partners in government administrations (6). The questions were structured according to these guiding themes:

- biotope tradition, land use history;
- land-use management;
- vegetation, ecological effects;
- WPs in the supra-regional context;
- exchange of information, networking;
- farming and its future;
- possible conflicts, constructive criticism.

The interviews with practitioners mostly took place on the sites, as the farmers wanted to show the important features of their work in situ. As confidence building was important, interviews were spread over two or even three meetings, to allow researcher and interviewees to get to know each other better. The contents were written down, as interviewees had requested that the interviews not be recorded.

All administrators were interviewed in their offices and the interview followed the prearranged chronology. Time for confidence building was rarely necessary due to their familiarity with giving interviews. They did not fear misunderstanding to the same degree as did the farmers.

Plant-ecological fieldwork

The examined WPs required the following features:

- forest character with trees older than 50 years;
- pasture tradition of at least eight years to allow for the identification of floristic responses;

- sites designated as pasture for at least the next ten years (opportunity for monitoring);
- size of WP greater than 1ha;
- landscape work with machines only at the outset of projects and later just for safety reasons.

The sites showed individual compositions of location factors, pasture traditions and management concepts. Therefore a comparison of sites was impossible. The following conclusions are linked to management aims and selected localities. In all cases it was necessary to consider pre-existing pollution and to exclude some parts of the site.

On each site and in each stratum, criteria of homogeneity for landscape, soil and vegetation structures were defined. A land-use classification characterised the strata as: dense forest (dF) – abandoned, ongoing forestry or protected forests; light pasture woodland (lPW); and formerly pasture woodland (fPW).

Environmental conditions change abruptly at the boundary between forest and WP. In such cases, frequency analysis as a punctual site-specific method is suitable to compare the floristic diversity between the two biotopes (Tremp, 2005). In each stratum, six plots (1m² divided into 25 squares) were laid out at random. For each square a list of species was compiled, which covered *spermatophyta* (seed plants), *pteridophyta* (ferns), *bryophyta* (mosses), *lycopsida* (clubmosses) and *lichenes* (lichens) (Figure 21.2).

Figure 21.2 Frequency method with 6 x 25 species lists compiled per pasture woodland and per forest (left); and structural method with 3 x 4 vegetation layers and 3 x 100 steps compiled per pasture woodland and per forest (right)
Source: Oelke, 2010

Effects of livestock behaviour are evident in micro-topography, soil and vegetation composition as well as the morphology of bigger plant individuals. To describe the effects in each stratum, vertical and horizontal structures were recorded in plots of 2,500m² (lateral length 50m) on a mesoscale level. Taking into account the size of the biotope, between one and three such plots could be laid within. A modified version of the step-point method was applied to get even more detailed information (Evans and Love, 1957, Strauss and Neal, 1983), that is, designed to record structures on microscale level. This happened along two transects crossing each plot, each transect being 50m long and partitioned in 1m-wide steps. This allowed 100 records per plot. The structures were divided into the following groups:

- Direct effects – peeling, cropping on woody species, rubbing on soil and trees, depositions (excrement, fur), trampling.
- Indirect effects – plant species with mechanical and chemical defence strategies, fructiferous species, habitat changes and decaying of woody species, reaction of fauna (breeding borrows in old/dead trees, presence of burrowing animals and anthills).
- Anthropogenic effects – tracks and paths, deposits, cultural remnants (boundary stones, buildings, fences etc.).

Analysis and results

Spatial distribution and features

One hundred WP were found in Baden-Wuerttemberg. It is estimated that about another 50 can be found in marginalised regions. The presently known WPs display common features (combinations possible):

- All are situated in agriculturally and silviculturally unfavourable areas:
  - higher altitudes of low mountain ranges (500–1,300m above sea level)
  - steep slopes (inclines up to 35°)
  - dry/sandy/fast drying or wet/boggy soils
  - quarries, stone pits, riverbanks
- Traditional WP areas, where no other land uses occur
- Nature reserves
- As yet no rivalry with other land uses

WPs tend to cluster in regions with idealistic stakeholders who initiate projects in their area of influence. They attract further project partners. Recently initiated pasture projects are mainly conducted for the purposes of nature conservation, if possible combining the protection of plant species and endangered old breeds. Supporting landscape elements with recreational character can be another driving force.

The size of the wooded parts in WPs is usually between 1 and 7.5ha. They are connected with grassland to allow the animals to roam. None of the WPs shows degradation as the possible stocking capacity is intentionally not reached.

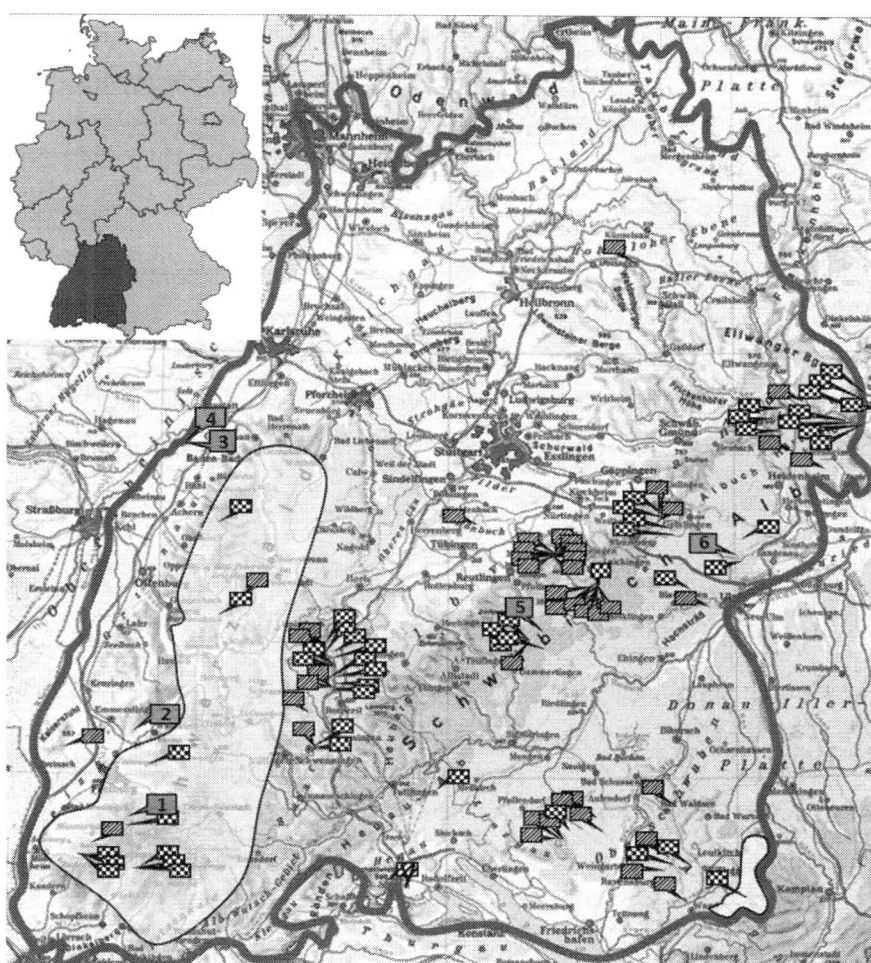

Figure 21.3 Topographical map of Baden-Wuerttemberg with WPs found in the project and schematic map of Germany, with Baden-Wuerttemberg highlighted

Source: www.dierke.de, modified, and www.gmk-net.de

Note: ▨ site known, visited (44); ▪ site examined closely (6); ▩ site visited, information gathered (50); ♫ region with traditional WPs. Many further sites expected.

Guided interviews

The following section presents a qualitative analysis of interviews with practitioners (12) and administrators (6). As the stakeholder groups interact, the findings from their statements are combined in this chapter.

Biotope tradition and land-use history

It is a common opinion among the stakeholders that modern WP should be orientated towards local specificities and be implemented as part of ongoing land-use management history. To avoid making WPs foreign bodies in the landscape, they use local breeds where possible and produce traditional products.

Land-use management

'Pure' WP in an historical sense can no longer be conducted as the forest sites are too small to feed livestock and the aims as such have become totally different. Consequently, WP always includes grassland and planned management is applied. Table 21.1 shows the main usage of modern WPs.

The expenditure of human labour to keep a WP running is high. Daily controls of herd and fences and keeping the herd accustomed to the farmer can be very demanding. In case of accident, it can be difficult to reach or provide help in remote or mountainous areas. Furthermore, keeping wild animals separated from domesticated ones occasionally requires expensive measures.

Some projects use disturbance from farm animals to create biotopes with increased structural diversity, which can in turn increase biodiversity. The challenges are to accept slow progress due to natural processes and to secure long-term finance.

Table 21.1 Structure, usage and effects of wood-pastures

Structures	Usage	Effects/benefits
Tree stands, thickets	Shelter from heavy rain, hail, heat	No buildings needed, better air circulation than in stables, therefore less illnesses
	Safe space during births and periods of social stress	Healthy calves and fawns, stress relief in the herd
Coniferous trees	Essential oils	Stress relief in the herd: reducing the burden of biting insects
Species rich in secondary phytochemicals	Bark and leaves as dietary supplement	Healthy animals, major reduction of endoparasites and illnesses High-quality meat
Topography, vegetation structure	Training	Good muscle development: – High-quality meat – uncomplicated births
		Build-up of solid social herd structures, better long-term handling, reduction of on-site accidents

Creation of open woodlands through pasture 309

Figure 21.4 Cow hiding from biting insects; continuous use of shrubs leads to their thinning
Source: Rupp, 2010.

Vegetation and ecological effects

If woods function only as shelters, hardly any effort goes into the development of biodiversity. When protection goals exist there are efforts to build up mosaic-like biotopes. The farmer's primary motivation is the desire to create sites – rich in flowers and fructiferous shrubs – that remind him of images from his adolescence. Management strategies are appraised autodidactically and are developed through learning about disturbance ecology and biodiversity in the farmer's leisure time. Administrators usually want to enhance biodiversity by reactivating the diaspore bank where it exists. Side-effects are biotope structures that attract many species weak in competition. Observed changes are the increased germination of blossom-bearing herbs, the presence of more insect species and consequently birds as well as associated fructiferous shrubs.

WPs in the supra-regional context

WPs are usually not connected to each other and appear like islands in the agricultural or silvicultural landscape. If WPs develop successfully, only very few sites can be enlarged due to a lack of manpower, rights of ownership and the legal situation. That is why open WPs do not yet play an important role in nature protection. The stakeholders are aware of the importance of this land use for biodiversity but as there is limited knowledge about such biotopes and they cannot assess the value of their activities in comparison to other conservation projects.

Exchange of information and networking

As the judicial situation is restrictive, the current pastures are rarely made known to outsiders. Usually the stakeholders are not connected to each other, meaning that awareness of their work is reserved to a certain circle of insiders. Should the legal framework become more supportive, they would seek a broader network of contacts and partners. They complain about the lack of information and demand scientific support to obtain information about how to improve modern WPs to support biodiversity.

Farms and their future

WP does not provide primary financial input into farm business. It does support the business as outlined above; however, these benefits only pay off over the long term. Broader political decisions, market conditions and other farm products determine the farm's future. In south-west Germany, remote areas face rural depopulation and farms are given up without successors. If local authorities integrate ecological pasture projects into a financially supported network of projects, farmers may be incentivised to continue the business. Furthermore, if our society is willing to pay for ecological services, farms could persist.

Possible conflicts and constructive criticism

Conflict fields are shown in Table 21.2. Interview partners also communicated ideas about how to solve these conflicts.

Plant-ecological fieldwork

Frequency analysis

Where surface water was present (streams, watering places), vegetation was nearly gone and the soil weakened due to trampling. Such sites were not covered by the frequency analysis. The stakeholders follow the emerging succession with interest and expect benefits for amphibians as soon as the animals move on, but they also ensure that these sites maintained their small size.

The floristic comparison between grazed and non-grazed forests is shown in Table 21.3. The stratum with maximal species number is high-

Table 21.2 Conflicts, stakeholders, constructive criticism and proposals for solutions

Stakeholders	Conflicts	Constructive criticism/proposals for solutions
Foresters	Damage of trees (mechanically, through micro-organisms)	Trees get damaged, but in WPs lucrative forestry will not be built up anyway. Damaged trees will provide habitat structures and support biodiversity. WPs could become part of forestry's commitment to sustain biodiversity at sites where reasonable
	Fences keep out game and complicate forestry	Modern fence systems allow wild animals and forestry vehicles to pass. Fence taken down following pasture period
Hunters	Simultaneous use of the site: – hunting impossible – game scared off	Effective time management necessary Statements cannot be verified by farmers and administrators, scientific research should be conducted
	Illnesses transferred between game and livestock	Veterinary laws have to be considered; farmers should build necessary installations to avoid contact during risky times. If necessary, pasture should be stopped
Conservationists	Endangerment of protected species (orchids, ground breeders, etc.)	Pasture management needs to be developed. A certain number of plants will be destroyed, others will be gained. In the long run pasture creates supportive conditions
	Change of present state	Change is inherent in any system. Conservation should include processes, not only states
Passersby, residents	Restricted mobility, dogs to be kept on leash	In nature reserves and on agricultural production sites, people are asked to stay on tracks and to keep dogs on a leash. This is common law and not specifically applied on WPs. The feeling of being restricted is subjective. One idea for addressing this issue is to offer information and guided trips over the site following the pasture period
	Old dumping grounds found	Local problem to be dealt with by local authorities
	Landscape change considered as ugly	Vegetation needs time to build up suitable societies. Effective communication in advance and on the site can contribute towards understanding of the changes

Table 21.2 Continued

Stakeholders	Conflicts	Constructive criticism/proposals for solutions
Livestock owners	Difficult legal situation, complicated applications for landscape conservation funding	Adapting forestry laws to local/regional circumstances would be beneficial. Alternatively, exceptional rules for biodiversity projects could be made in regions where forestry and extensive pasture are not competitors. Close cooperation between practitioners and administrators is helpful over long timespans
	Sabotage, theft of livestock	In many cases fences and solar panels get destroyed by vandals. Now and then animals get tormented or stolen. The more remote a site is, the more likely this may occur. There are no ideas on how to stop these problems. Financial compensation after a case might be considered
	Litter thrown into the fields	As this problem is caused by a lack of awareness in society, solutions are mainly limited to legal restrictions. Only fencing off parts of the pasture that are close to parking grounds or meeting points can help
	Unleashed dogs and dog excrement	Dogs can harm the herd and kill animals. Prosecution and financial compensation are needed. High-voltage fences with low wires can keep dogs away. Now and then cows defend the herd and can severely harm the dog

lighted in grey. The floristic similarities are described by the Sørensen-coefficient.

In five sites the grazed woodlands contain most species, up to more than double the number found in non-pasture forests. Probable reasons for this are better light supply in the herb layer, reactivation of diaspore banks and the input of diaspores by mobile farm animals and birds. Three types of species combine: woodland species overlap with grassland species, and additionally, hedge species of half-shade biotopes emerge.

Site number three, where the dense wood contains more species, is an exceptional case. The pasture's herb layer is densely covered with grasses (e.g. *Agrostis capillaris, Brachipodium pinnatum, Deschampsia flexuosa*). These species form a dense felt of roots and leaves and they dominate other species. Where livestock manages to break this felt, sandy meagre immature soil is exposed and only a few specialised plants, such as *Erophila verna*, can

Table 21.3 Floristic comparison between wood-pastures and adjacent forests

Site no.	Stratum	Species no.	Difference to max. (number/%)		Sørensen-coefficient	
1	dF	25	34	57/63	(dF – WP)	S = 0.57
	lPW	32	27	45/76	(WP – fWP)	S = 0.38
	fPW	59			(dF – fWP)	S = 0.38
2	dF	35	43	33/96	(dF – WP)	S = 0.32
	lPW	78				
3	dF	53			(dF – WP)	S = 0.64
	lPW	47	5	9/43		
4	dF	33	8	19/51	(dF – WP)	S = 0.46
	lPW	41				
5	dF	39	52	57/14	(dF – WP)	S = 0.35
	lPW	91				
6	dF	52	35	40/23	(dF – WP(J))	S = 0.56
	lPW(J)	80	7	8/05	(dF – WP(P))	S = 0.45
	lPW(P)	87			(WP(P) – WP(J))	S = 0.64

Note: dF = dense forest, no grazing; LPW = light pasture-wood, grazing; fPW = formerly pasture wood, no grazing; S = Sørensen-coefficient with 0 = max. difference, 1 = max. similarity; J = *Juniperus communis* wood; P = *Pinus sylvestris* wood.

grow. Adjacent woods with mature soil and slight disturbance through forestry support more species.

In summary, mechanical opening of the canopy and associated pasture and can change biodiversity. Usually the stakeholders act too gently as strong disturbance can be applied in the first years to allow species weak in competition to move in. The grazing pressure can then be reduced gradually.

Structural analysis

In summary, in grazed woodlands more structural inventory could be found, as in adjacent forests. This observation pertains to both direct and indirect effects.

DIRECT EFFECTS FROM LIVESTOCK

Loss of woody species could not be detected. All sites show species-rich undergrowth. In the WPs most of the woody species were gnawed but still quite vital. Goats bite nearly every woody species; some species are particularly popular and become totally gnawed, for example *Euonymus europaeus* and *Fraxinus excelsior*. Less popular are *Acer* species. Cattle mainly eat young and sprouting woody species but do not bite as efficiently as goats. Cattle influence older trees by rubbing and leaning on them; they can even thin thickets by forcing themselves through. They do this predominantly in stands of coniferous trees to get the essential oils on their fur to protect themselves against biting insects. Sheep browse mainly sprouting woody species and gnaw young twigs but leave older parts. Due to their natural behaviour, sheep tend to prefer open land. If they are to have an impact on the forest it is necessary to paddock them there.

Pasture size determines the way the animals take on the species and how intensively they gnaw. If alot of space is provided, total consumption of individual trees or shrubs can seldom be detected. When the animals are accustomed to being herded mainly on grassland, in the first year they do not use the given spectrum of fodder plants in a WP. Over the years they eat plants that at the beginning were untouched. It was observed that cows start browsing *Prunus spinosa* shrubs in a similar manner to goats and even graze stands of thistles. It is estimated that a lack of minerals makes goats gnaw bark of *Juniperus communis* and massive *Pinus sylvestris* of 50cm or more in diameter, but in these cases only a small number of individuals are totally consumed. In spring, young leaves of *Betula* and *Frangula alnus* are preferred, mainly as a means to fight endoparasites.

The animals create paths and stick to them over the years. Consequently the pasture becomes rich in patterns of differing usage intensity. Goats and cows create rubbing spots on the soil that benefit some fauna. Wild animal (for example foxes) droppings of variously advanced decomposition can be found, which indicates regular use of these spots to warm up or even to sleep.

Over the years, defecation encourages the growth of nitrophilous plants at spots of regular defecation. These plants grow densely and screen smaller animals from view. They also constitute energy-rich fodder for both domesticated and wild animals. Domestic animal excrement differs from that of wild animals and offers substrata for dung flora and fauna.

INDIRECT EFFECTS

On WPs old trees are left to die. All stages of decomposition can be found on one individual, mainly from pioneer species such as *Betula* and *Salix*. The tree offers a realm for xylobiontic organisms, with leaking sap used by many insects. Lying deadwood can become nurse logs and offer shelter for invertebrates. For instance, antlions (*Myrmeleontidae*) were found on two pastures.

The mosaic-like pattern, decaying trees and increased invertebrate fauna attracts birds. They work as vectors and bring diaspores of fructiferous woody species. Species such as *Lonicera xylosteum, Sambucus nigra, Rosa* or *Ilex aquifolium* can accumulate along paths, around exposed and decaying trees where droppings are preferably left. In the WPs more fructiferous species could be found than in adjacent plots.

An increased presence of burrowing animals and anthills was detected. We estimate that the combination of more light and therefore warmth, nutrients from droppings, enriched soil and a wider plants species spectrum supports the accumulation of soil fauna.

The mosaic-like pattern is also built from plant species with mechanical and chemical defence strategies. *Prunus* and *Crataegus* species form thickets that can also contain *Ilex aquifolium*, *Rosa* and *Rubus* species. Plants with chemical defence strategies gather in more open sections and along paths. On site one, the toxic *Gentiana lutea* increased its population enormously in the grazed area. In the other pastures for example, *Thymus* species and *Oreganum vulgare* emerged at more dry spots.

ANTHROPOGENIC EFFECTS

To initiate thinner, more open woodland character, trees are cut, often left or piled up and tracks are made. Around gates, cars compact soil, fodder is provided and the soil turns eutrophic. The animals can reveal contamination, deposits and cultural remnants, such as overgrown boundary stones, old dumping grounds or military items, e. g. metal bars. In all pastures old barbed wire was found. Where these items could not be removed, and where they posed dangers, they were fenced out.

Conclusions

Within a three-year project, 100 WPs were found in Baden-Wuerttemberg. The overview demonstrates that economically unattractive areas and nature reserves are the most suitable sites for establishing this type of extensive agriculture. Wet soils and areas with operating forestry seem inappropriate. Sites with endangered species require special handling. WPs are conducted using modern approaches and are not at all comparable with historically known pasture systems. They offer structural elements that benefit animals and therefore the farm business.

Among the sites, six were examined closely. The field studies showed that they can gradually increase the structural inventory and local biodiversity. To widen the floristic spectrum in the pastures, high grazing pressure and strong mechanical influences on the root layer should be undertaken over a number of years and then gradually reduced. Pausing the grazing for some years can be considered in pasture management at some locations.

Complications arise from the existing legal framework and lack of knowledge, extending from the fact that WP has been forbidden in the region for around 180 years. Stakeholders ask for scientific support and legal adaptation. They also desire networking through different media and mentoring from a neutral actor e.g. a university. Other problems occur in the confrontation between different stakeholder groups, however, effective communication and bottom-up strategies can address most of these in advance.

WPs are not keeping farms running but they do enrich businesses. If the wish to maintain and develop WPs arises, the economic survival of farms must be guaranteed through broader political decisions and a favourable market situation for agricultural goods in general. As properly managed

WPs can increase biodiversity, they could be integrated within sustainability and biodiversity concepts in forestry, agriculture and landscape management.

Bibliography and references

Bunzel-Drüke, M., Böhm, C., Finkh, P., Kämmer, G., Luick, R., Reisinger, E., Riecken, U., Riedl, J., Scharf, M. and Zimball, O. (2008) *Wilde Weiden*, Praxisleitfaden für Ganzjahresbeweidung in Naturschutz und Landschaftsentwicklung, Bonn.

Conradi, M. and Plachter, H. (2001) *Analyse ökologischer Prozesse in Weidelandschaften und ihre naturschutzfachliche Beurteilung mit Hilfe skalendifferenzierter Strukturanalysen*, in Gerken, B. and Görner, M. (eds) (2001) *Neue Modelle zu Maßnahmen der Landschaftsentwicklung mit großen Pflanzenfressern*, Praktische Erfahrungen bei der Umsetzung, Natur- und Kulturlandschaft 4, Höxter, Jena, pp132–46.

Evans, R. and Love, M. (1957) 'The step-point method of sampling. A practical tool in range research', *Journal of Range Management*, 10: 208–12.

Grossmann, H. (1927) *Die Waldweide in der Schweiz. Dissertation. Betreut von Prof. Dr. H. Knuchel und Prof. Dr H. Badoux.* Zürich, Eidgenössische Technische Hochschule.

Kapfer, A. (1995) 'Der Einfluss der Beweidung auf die Vegetation aus der Sicht des Naturschutzes, in Wieder beweiden? Möglichkeiten und Grenzen der Beweidung als Maßnahme des Naturschutzes und der Landschaftspflege', *Beiträge der Akademie für Natur- und Umweltschutz Baden-Württemberg*, Band 18, pp27–36.

Liss, B.-M. (1988) 'Versuche zur Waldweide – der Einfluss von Weidevieh und Wild auf Verjüngung, Bodenvegetation und Boden im Bergmischwald der ostbayerischen Alpen, München, (Schriftenreihe der Forstwissenschaftlichen Fakultät der Universität München und der Bayer', *Forstlichen Versuchs- und Forschungsanstalt*, 87.

Michels, C. and Spencer J. (2003) 'Waldweide im New Forest, in LÖBF-Mittleilungen', *Landesanstalt für Ökologie, Bodenordnung und Forsten Nordrhein-Westfalen, Recklinghausen*, 04/03, pp53–8.

Sonnenburg, H., Gerken, B., Wagner, H.-G. and Ebersbach, H. (2003) *Hutewaldprojekt Solling*, Ein Baustein für eine Ära des Naturschutzes, Höxter.

Sproßmann, H. (2009) 'Extensive Waldweide in Thüringen-Waldfrevel oder ein innovatives Landnutzungsmodell?', *Forst und Holz*, 64(2): 32–7.

Strauss, D. and Neal, D.L. (1983) 'Biases in the step-point method on bunchgrass ranges', *Journal of Range Management*, 36(5): 623–6.

Tremp, H. (2005) *Aufnahme und Analyse vegetationsökologischer Daten*, 1, Auflage, Stuttgart.

Vera, F.W.M. (2000) *Grazing Ecology and Forest History*, CABI, Wallingford.

22 Woodland grazing with cattle

Results from 25 years of grazing in acidophilus pedunculate oak (*Quercus robur*) woodland

Rita M. Buttenschøn and Jon Buttenschøn

Introduction

In Denmark, up to 10 per cent of the forest-preserved land may be used for woodland grazing (Danish Forest Act, 2004). Woodland grazing is now considered an important tool in the creation and maintenance of diverse and stable woodlands. Prior to this, from the turn of the nineteenth century, woodland grazing was forbidden in Denmark. For about 200 years this resulted in a division of the landscape into agricultural and forestry land with distinct boundaries between land use.

In this study we investigate the effect of cattle grazing on structure and vegetation composition in the oak-dominated woodland of Skovbjerg (Wood Hill) situated in Mols Bjerge National Park. Skovbjerg is one of the very few remnants of ancient woodland in Mols Bjerge. It is surrounded by young scrub and open grassland. Skovbjerg contains a series of woodland development, represented by remnants of ancient woodland, and typical succession processes of abandonment of arable land into grassland and heathland followed by gradual encroachment of scrub and woodland. Whereas succession in former periods appears to have been cyclical, carried by varying impacts from grazing and timbering, it has over the last century become increasingly less influenced by agricultural use (according to the study of maps and aerial photographs; Buttenschøn and Buttenschøn, unpublished a and b). The woodland type is old acidophilus oak woodland with pedunculate oak (*Quercus robur*) (henceforth, oak) (Code 9192 in the Habitats Directive: Council Directive 92/43/EC), and is characterised by a low diversity of vascular plants but is rich in epiphytic lichens.

Over the last 25 years, 15ha of Skovbjerg has been grazed yearly during autumn (October to January) at a moderate grazing pressure (1–1.3 heifer or cow mean weight 500kg/ha) by crossbred beef cattle (Figure 22.1). Autumn grazing was chosen to allow flowering of the field-layer species, so enhancing the seed-rain and seed dispersal by the livestock. Based on dung-pat densities, the cattle grazing pressure is estimated to vary by a factor of 40 within the fencing. Initially the area was composed of woodland, scrubland and open grassland, each occupying approximately a third of the fencing. Presently the scrubland is developing into proper woodland, while the open grassland is characterised by fragments of scrubland and solitary trees. Roe deer is the only wild ungulate present. The roe deer population is estimated to be approximately 20 deer km^{-2} in Mols Bjerge. Pellet group counts indicate an even distribution of roe deer inside and outside the cattle-grazed area. Estimated from the roe deer density in the larger area, the deer grazing pressure is one tenth of the cattle grazing pressure. The present study was made on woodland developing on arable and grazed land over the last 90 to 100 years, its development being accelerated by the cessation of grazing 45 to 50 years ago.

Figure 22.1 Forest cattle grazing in grassland with fragments of scrubland and solitary oak
Source: R.M. Buttenschøn

In the present study we compare grazed and ungrazed vegetation over a gradient of gradual cessation of agriculture with subsequent woodland encroachment, and with ancient oak woodland. The vegetation analyses were made in permanent transects across the vegetation types from ancient woodland to open grassland. The field-layer analysis was undertaken in 1 x 1m plots. The tree and bush analysis was in 10 x 10m plots. In the woody species analyses we tally the individuals of the different species in reference to four size groups: 1 = current year's seedlings, 2 = saplings <0.5m, 3 = saplings 0.5–2m, and 4 = trees >2m. Concurrent to the tally, we estimate the impact of browsing on the individuals on a four-step scale: 0 = no browsing, 0.5 = slight browsing, 1 = medium browsing, 2 = heavy browsing (Buttenschøn and Buttenschøn, 1985).

Wood and underwood

Our study demonstrates distinct differences between the cattle-grazed and ungrazed woodland. By abundance, the sources of browse ranked as oak, aspen (*Populus tremula*) and beech (*Fagus sylvatica*). By over all size groups, most woody species were browsed more heavily in the grazed area, oak being a notable exception (Table 22.1). Average browsing pressure >1.5 was critical to long-term survival of populations of saplings in most woody species. Saplings of species with protective thorns or spikes withstood a higher browsing pressure over long periods. However, the differences in browsing impact between the grazed and ungrazed areas must be envisaged in terms of the between-area difference in grazing pressure, which is tenfold higher in the grazed area. Relative to grazing pressure, the browsing impact in the ungrazed area is larger. This is consistent with the grazing strategy of the two grazer species, one, roe deer, being extreme browsers, the other, cattle, being generalist grazers (Hofmann, 1989). The browser is very selective during forage uptake and concentrates uptake on matter with high cell content and high digestion rate (Illius and Gordon, 1993), ranking browse, forbs to grasses. The forage targeting of the browser, accordingly, is primarily focused on browse, secondarily on field-layer vegetation in growth phase. The generalist grazer's targeting is more focused on field-layer vegetation (van Dyne *et al.*, 1980) and need to engage increasingly less foraging selectivity to achieve an adequate diet with increasing body weight (Illius and Gordon, 1993). This is consistent with the browsing by cattle as an integral part of foraging governed by choice of field-layer sub-swards (Buttenschøn, 2007; Buttenschøn and Buttenschøn, 1982). Choice of woody species for browsing is secondary to sward selection.

Table 22.1 Average browsing pressure on important woody species in Skovbjerg

	Grazed			Ungrazed		
	Saplings <0.5m	Saplings 0.5–2m	Trees >2m	Saplings <0.5m	Saplings 0.5–2m	Trees >2m
Cytisus scoparius	1.53[a]	1.52[a]	0.5	0.67[c]	1.17	
Fagus sylvatica	0.61[ab]	1.15[a]	1.08	0.29[c]	0.74	
Malus sylvestris	0.41[b]	1.25		0.38		
Populus tremula	1.08[a]	1.10	0.01	0.50[c]	0.92	0.00
Prunus spp.	0.57[b]	1.56	1.11	0.55		
Quercus robur	0.74[b]	1.56	0.15	0.68[c]	1.60	0.05
Rosa spp.	0.41[b]	1.00[a]	0.07	0.50[c]	0.00	
Sorbus aucuparia	1.13[a]	1.50	0.25	0.62		0.25

Note: 0 = no browse, 0.5 = slight browse, 1 = medium browse, 2 = heavy browse; [a] significant difference between grazed and ungrazed, [b] significant difference between size group within grazed, [c] significant difference between size group within ungrazed. Probability: p <0.05.

Miller et al. (2007) have shown that browsing impact on saplings is higher on individuals with high nutrient status of the sapling, high palatability of the surrounding vegetation and low abundance of surrounding vegetation. The high to low nutrient status of the saplings in Miller et al.'s experiment was achieved by the nutrient supply during potted growth, but would in natural settings reflect the micro-site growth conditions of the sapling. This is largely expressed through soil nutrients and light availability and browsing prehistory, over which long-term heavy browsing may imbalance the photosynthetic to structural biomass ratio. Similar gradients of browsing related to accessibility are seen in the centripetal decline of impact in dense sapling stands (Buttenschøn and Buttenschøn, 2003).

In the grazed area, the browsing horizon on larger trees lies between 1.5 and 2m, whereas there is no distinct browsing horizon in the ungrazed woodland. This level was established during the initial years of cattle grazing. The openness under the browsing horizon allows more light to penetrate to the grazed woodland floor and a denser field-layer to establish. It also provides growth conditions for a larger spectrum of species. Whereas we generally find a persistent or fluctuating bank of woody species saplings in the ungrazed area, there is sufficient light for growth of saplings in large parts of the grazed area even if the higher browsing impact slows down growth and reduces individual success.

There are differences in the browsing pressure on the small and large saplings in most species. This is to a large extent related to limited growth the first few years after germination. Common broom (*Cytosus scoparius*) with nitrogen fixation, and aspen, with root suckers, produce appreciable accessible browse from the first year on. Species that are protected by thorns and spikes are generally browsed less than unprotected species.

A preliminary serial light measurement in August 2011 in the 72 field-layer analysis plots of the grazed and ungrazed woodland shows that the light regime in the grazed area is much more variable than in the ungrazed. The measurements were made using an ACCUPAR LP-80 (Decagon Devices Inc.), which makes six individual measurements along the 80cm-long sensor. Whereas on average 15.8 per cent of the light reaches the ground in the grazed area, only 9.5 per cent reaches the ground in the ungrazed. The measured average leaf area indices, however, are 3.69 and 3.55, and show that the overall leaf cover does not vary significantly between the two areas. The greater average light penetration supports the suggestion that the intra-area variation in the grazed area is larger with higher occurrence of light and shade conditions. The higher light penetration may in part be ascribed to the browsing horizon, the similar leaf-area index values to higher underwood density and, locally, dense carried canopies of honeysuckle (*Lonicera periclymenum*) in the grazed area. Over the long term, the light indication (Ellenberg et al., 1991) of the field-layer vegetation decreases at a rate of 0.025 and 0.039 per year in the grazed and ungrazed areas. Initially, at the beginning of the grazing, the value was 6.5 (extrapolation of trendlines). In

2011, the values were 5.80 and 5.51. The intra-area spatial variation in light influx and the difference in long-term linear decrease in light are important for vegetation development at ground level.

For germination, woody species have different light requirements, varying from almost full light to only a few percent of ambient levels (Table 22.2). The light penetration in both areas allows shade and semi-shade species to germinate widespread in both areas, whereas in the area suitable for semi-light and particularly light-demanding species, germination is reduced notably from the grazed to the ungrazed area. Ellenberg *et al.* (1991) state that the light-indicator value primarily covers germination. Growth and particularly survival in a persistent sapling bank is possible for many woody species at lower values than indicated by Ellenberg *et al.* (1991). The comparative low recruitment rate seen in most woody species in the ungrazed area can only partly be explained by low light regimes at ground level.

Table 22.2 Light requirements for germination of woody species at Skovbjerg assembled from Ellenberg *et al.* (1991) – many of the species germinate some weeks before the sprouting oak-aspen-canopy reaches full cover

Light group (light-indicator value)	Light requirement (%)	Species
Shadow (3–4)	>2	*Acer pseudoplatanus*
		Fagus sylvatica
		Prunus avium
		Ulmus glabra
Semi-shadow (5–6)	>10	*Crataegus monogyna*
		Lonicera periclymenum
		Populus tremula
		Prunus cerasifera
		Prunus serotina
		Prunus spinosa
		Rosa spp.
		Sorbus aucuparia
Semi-light (7)	>20	*Quercus robur*
Light (8)	>50	*Cytisus scoparius*
		Juniperus communis
		Malus sylvestris

The germination of many woody species is promoted by grazing through seeds being disseminated by the cattle (Figure 22.2). In autumn, ripe seeds are available to be eaten and trodden into the ground by the cattle and the autumn grazing leaves the seedlings undisturbed during the first season of growth. Furthermore, germination on fibre-rich cattle dung-pats, which may persist and be repulsive for up to a year and a half, greatly enhances seedling survival. While 88 per cent of woody species seedlings on cattle dung survived the first two years, only 26 per cent survived outside the pats (Buttenschøn et al., 2008). Buttenschøn and Buttenschøn (1998) found that the survival expectancy of crab apple (*Malus sylvestris*) seedlings germinating on cattle dung-pats rises with decreasing grazing pressure. The grazing pressure at Skovbjerg is sufficiently low to allow an appreciable survival of crab apple and eventually a number of individuals will grow to a height beyond the reach of the cattle. Similar developments are seen in connection with hawthorn species (*Crataegus laevigata, C. monogyna*), roses (*Rosa canina, R. dumalis* and *R. rubiginosa*), cherry plum (*Prunus cerasifera*) and sloe (*Prunus spinosa*), which also are disseminated through dung-pats in grazed areas (Buttenschøn and Buttenschøn, 1985 and 2003). The survival expectancy is even higher with these species.

Over the study period, the extinction rate of some species is higher than the recruitment rate (Figure 22.3). This reflects tendencies of actual extinction and thinning of canopy-forming individuals as a result of growth or fluctuations in underwood populations of banks of persistent saplings.

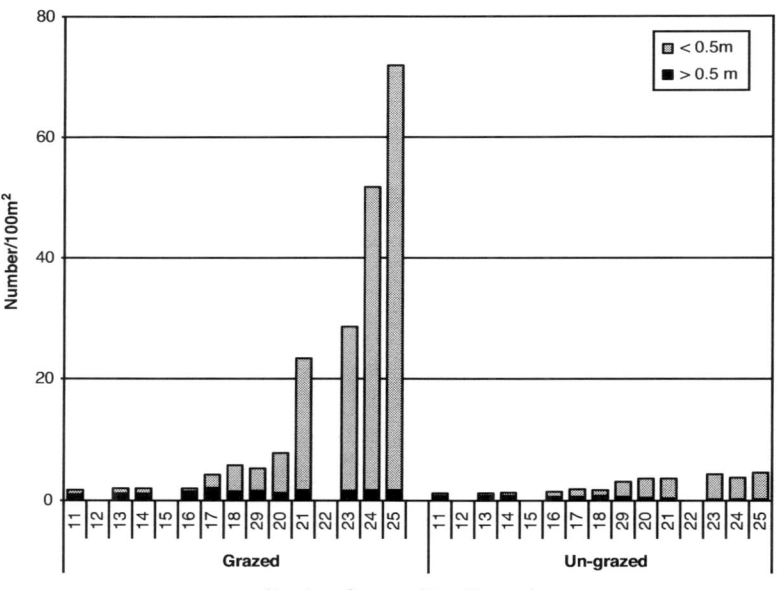

Figure 22.2 Density of animal spread woody species (*Malus sylvestris, Prunus avium, P. cerasifera, P. serotina, P. spinosa, Rosa* spp. and *Sorbus aucuparia*) in grazed and ungrazed oak woodland
Source: Buttenschøn and Buttenschøn

The latter is seen in the oak and aspen populations. There is no appreciable net recruitment in most species in the ungrazed woodland. Beech and rowan (*Sorbus aucuparia*), the more shade-tolerant species, do however, increase in number. In contrast, some species have an appreciable net recruitment in the grazed woodland, partly due the creation of seedbeds and spreading of seeds by the cattle. This is also due to the higher influx of light to the woodland understoreys. The population of apple, rose and oak survive significantly better (p <0.05) under grazing than without. The grazed populations of the former two species increase, whereas the grazed oak population is stable. Sapling growth beyond 0.5m marks the size where saplings successfully may be protected from browsing by forming a dense brush of twigs; in some species this is assisted by spikes and thorns and they are less susceptible to damage by trampling. On reaching 2m, the saplings are above the browsing reach if the trunk can withstand bending under browsing. Within the study period, some saplings of beech, apple, *Prunus* and rose grew beyond 0.5m in height in the grazed area; a few saplings of beech and apple even beyond 2m. In the ungrazed areas some saplings of beech grew beyond 0.5m and 2m in height.

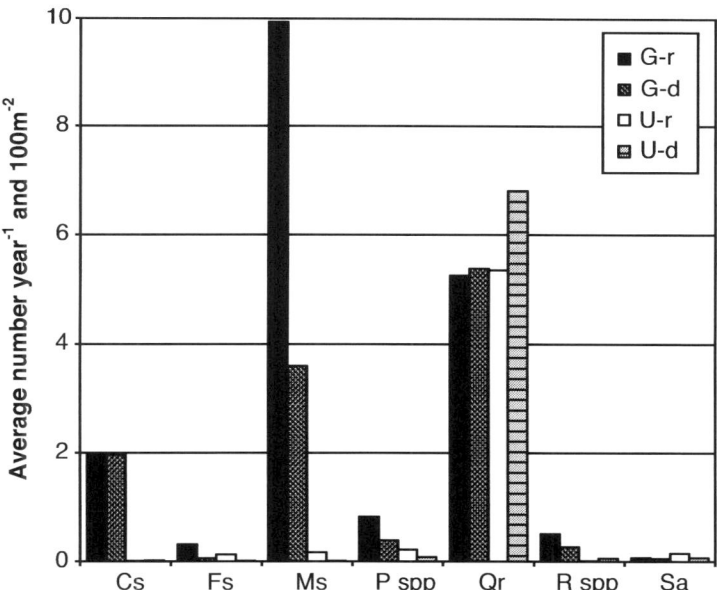

Figure 22.3 Average annual recruitment (-r) and death (-d) rate in grazed (G) and ungrazed (U) oak woodland

Note: The species are, left to right, *Cytisus scoparius, Fagus sylvatica, Malus sylvestris, Prunus* spp. (*avium, cerasifera, serotina, spinosa*), *Quercus robur, Rosa* spp. (*canina, dumalis, rubiginosa*) and *Sorbus aucuparia*. The rate is individuals per $100m^2$.

Source: Buttenschøn and Buttenschøn

Field-layer

The grazed field-layer is influenced by the higher influx of light to the surface, the grazing-induced structural diversity of the sward (grazing per se, trampling and patchy nutrient reallocation by excreta) and by the cattle's dissemination of seed. Cattle dung-pats cover about 1 per cent of the total area, which receive between 10 and 20 per cent of the nutrients for recycling. The pats contribute with some 5,000 to 10,000 viable seedlings per hectare per year, representing at least 10 per cent of the Danish higher plant flora (Buttenschøn et al., 2008; Buttenschøn and Buttenschøn, unpublished).

The species number in the grazed woodland is higher than in the ungrazed (Tables 22.3 and 22.4). This applies to woodland character species, semi-shadow tolerant species, other herbaceous species and woody species. The woodland field-layer in the grazed area has become denser during the study, with increasing cover in all the above mentioned species groups, whereas the field-layer is stable in the ungrazed area, but shifting towards higher contribution of woodland character species.

Table 22.3 Numerical presence of species in grazed and ungrazed oak woodland

Species category	Only grazed	Both	Only ungrazed	Sum
Woodland species	2	9	1	12
Semi-shadow species	6	15	2	23
Other species	17	15	5	37
Woody species	4	10	2	16
Sum	29	49	10	88

While the typical woodland species have a more or less parallel colonisation and expansion under both management regimes, the habitat zone of semi-shade or woodland boundary species expands widely under the autumn grazing. This, for instance, is the case with crested cow-wheat (*Melampyrum cristatum*) (listed as vulnerable in the *Danish Red Data List*; Wind and Pihl, 2004). This has expanded massively in the fragmented woodland alongside the old woodland and now is 'present' to 'abundant' along the length of the woodland–grassland interface (Figure 22.4).

Differences in colonisation rate under the two managements appear to be associated with the management dependency of the species in question (*sensu* Ellenberg et al., 1991). This dependency may reflect a number of different species' traits such as tolerance to grazing, necessity of trampled seedbeds, seed dispersal and germination strategies and shade tolerance. Husbandry grazing provides or creates many of these conditions.

The species density increases linearly in young developing woodland, be it grazed or ungrazed ($p < 0.001$ for both areas) (Figure 22.5). The initial

Table 22.4 Significant increase and decrease (a <0.001, b <0.01 and c <0.5, based on linear regression over time) of the species in young oak woodland based on correlation between time since 1987 and average cover index

Increase	Grazed	Ungrazed	Decrease	Grazed	Ungrazed
Agrostis capillaris*	a	a	Calluna vulgaris	a	c
Carex pilulifera*	a		Carex pilulifera*		c
Dactylis glomerata*	a		Cytisus scoparius*	c	
Festuca rubra*	a	a	Deschampsia flexuosa*	a	a
Galeopsis bifida	b		Galium verum*	a	a
Galium aparine	b	b	Hieracium pilosella	a	
Holcus mollis		b	Hieracium vulgatum	b	
Lactuca muralis	np	b	Hypochoeris radicata	a	np
Lonicera periclymenum		a	Juniperus communis	a	
Luzula pilosa*	a	b	Luzula campestris*	b	b
Maianthemum bifolia	a	a	Melampyrum pratense*		c
Malus sylvestris*	a		Pimpinella saxifraga	b	
Melampyrum cristatum*	b		Poa pratensis*		c
Oxalis acetosella	b	a	Polypodium vulgare	a	a
Poa trivialis*	a	np	Rumex acetosella	b	
Populus tremula		c			
Prunus cerasifera*	a				
Quercus robur	c				
Rubus idaeus		b			
Stellaria graminea*	a				
Stellaria holostea*	a	a			
Torilis japonica	b	np			
Trientalis europaea	b	c			
Urtica dioeca*	a				
Veronica chamaedrys*	a				
Veronica officinalis*	b				
Viola riviniana*	a				

Note: Only species that occur in more than 5 per cent of the plots in at least one year were tested. np is not present. * species found germinating on dung-pats but also having other dispersal strategies.

rate of increase and the density, however, are significantly higher in grazed woodland. From the data for the ancient woodland at Skovbjerg we suggest that density levels reach a maximum and then decline as the group 'other species' declines with falling light influx.

The qualitative composition of the grazed as well as ungrazed vegetation changes over time (Table 22.5, frequency section). Accordingly, there are ongoing changes in the field-layer in general representing change from open

Figure 22.4 Crested cow-wheat in grazed vegetation at Skovbjerg
Source: R.M. Buttenschøn

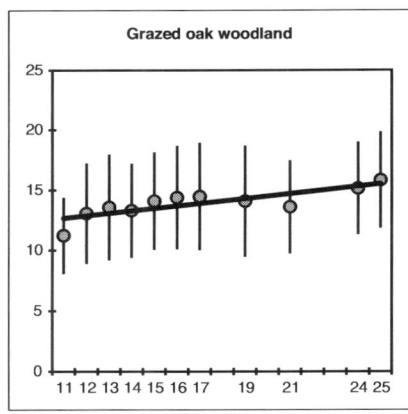 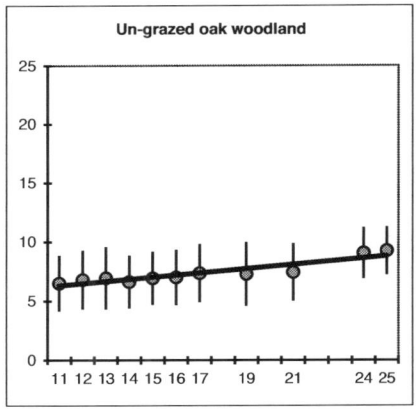

Figure 22.5 Development in species density (average species number per m^2) in grazed and ungrazed oak woodland 11 to 25 years after grazing was initiated
Source: Buttenschøn and Buttenschøn

Table 22.5 Identity tests (t-test and F-test) comparing arrays of grazed and ungrazed field-layer vegetation in 1997 and 2011

	Grazed 1997 to grazed 2011	Ungrazed 1997 to ungrazed 2011	Grazed 1997 to ungrazed 1997	Grazed 2011 to ungrazed 2011
Frequency				
t-test	0.0023	0.0038	0.000000016	0.000023
F-test	0.003	0.000000078	0.000058	0.12
Cover index				
t-test	0.87	0.012	0.00000071	0.0023
F-test	0.64	0.000005	0.000075	0.87

Note: The tests are made on groups of frequency and cover index data, representing qualitative (expressing similarity in distribution) and quantitative (expressing similarity in abundance or dominance) composition. The cover is based on the index values 1 to 6, the six steps in the Hult-Sernander cover index scale. The values are p levels. Vegetation groups with different management regimes that have $p < 0.05$ in both tests indicate that they are different vegetation sub-types. Vegetation groups with the same management that have $p < 0.05$ in both tests indicate significant development of composition and structural variation over time.

grassland (60–70 years ago) to young woodland (Table 22.4). The composition of the grazed and ungrazed vegetation is very different after the first decade of grazing (1997 tests), a difference that appears to decline by 2011. The variation in qualitative composition of the grazed and ungrazed vegetation in 2011 is not significantly different. This suggests that some early phase, small-scale vegetation elements are disappearing from the vegetation with aging of the woodland and many species in both areas are becoming more evenly distributed. This is consistent with the underlying development from open land to woodland field-layer seen under both management regimes.

Final thoughts on the impacts of grazing and browsing

The quantitative composition and the intra-matrix variation of the grazed area do not change significantly over time (Table 22.5, cover index section). This suggests that the vegetation processes in the early stages of the grazed woodland rather is characterised by replacement of light species with more shadow-tolerant species, whereas features such as cover dominance remain the same. In the grazed vegetation, dominance is often suppressed and a situation of co-dominance of a number of species prevails, but to some extent with replacement of composite species. This agrees with the stress-tolerant-ruderal-competitor growth strategy of many of the grazing tolerant species (Grime, 2001). Contrary to these observations, there are significant changes in the quantitative composition in the ungrazed vegetation from 1997 to 2011.

This represents adjustment processes to the declining light influx to the field-layer under which woodland character species and shade-tolerant species gain importance (Table 22.4). In 1997, there was a significant difference between the quantitative composition of grazed and ungrazed vegetation, supporting the assertion that the first decade of grazing not only changes the qualitative but also the quantitative composition of grazed vegetation. By 2011, this difference is only significant for the quantitative composition. The pattern of dominance is becoming more similar but the species dominating differ between the grazed and ungrazed vegetation.

Thus, on introduction of grazing in developing woodland, we see two main changes in vegetation: 1) change introduced by the disturbance of grazing, and 2) change associated with declining light reaching the field-layer. The post-introduction disturbance effect is pronounced in the early stages, while decreasing light appears to be more important later in development. As mentioned above, the influence of decreasing light is more pronounced in the ungrazed area. The diversifying effect of the disturbance of grazing, however, is apparent in the long term.

Strandberg *et al.* (2005) looked at oak woodland succession in relation to management and conclude that unmanaged oak scrub over time succeeds into shady oak-beech woodland on raw humus. Managed (selectively cut, coppiced or grazed) scrub succeeds to more homogeneous oak woodland on mull. There is significantly higher light penetration, species richness and pH in the managed sites. In the grazed woodland the organic layer is significantly thinner. The same trend of development is seen in Mols Bjerge, where Skovbjerg lies. The old, undisturbed woodland fragments in the area are, via oak-beech woodland, succeeding into shady beech woodland. The underwood is sparse and the thin field-layer indicates acidic soil. The difference in soil character between the managed and unmanaged woodland is not only important for nutrient cycling but also for the vegetation the soil may support.

Thus, from the 25-year study we conclude that grazing of woodland with cattle will eventually result in the build-up of a diverse understorey of scrub enriched by animal-disseminated species. Without grazing the understorey building is sparse and consists mostly of a persistent bank of small saplings. The field-layer is also more diverse under grazing and the characteristic woodland species, the semi-shade tolerant and less shade-tolerant species gain in abundance and cover. The cattle are important for development as they disperse seeds in their fur or via dunging, tread seeds into the ground and reallocate nutrients in a patchy pattern, which diversifies growth conditions. Finally, the establishment of the browsing horizon on the larger trees allows more light to the surface.

References

Buttenschøn, J. and Buttenschøn, R.M. (1982) 'Grazing experiments with cattle and sheep on nutrient poor acidic grassland and heath: II grazing impact', *Natura Jutlandica*, 21: 19–27.

Buttenschøn, J. and Buttenschøn, R.M. (1985) 'Grazing experiments with cattle and sheep on nutrient poor acidic grassland and heath: IV establishment of woody species', *Natura Jutlandica*, 21: 117–40.

Buttenschøn, J. and Buttenschøn, R.M. (2003) 'Langtidseffekten af husdyrgræsning, II. Skovudvikling under husdyrgræsning', in Austad, I., Hamre, L.N. and Ådland, E. (eds) *Gjengroing av kulturmark*, Bergen Museum, og HSF, Bergen, pp61–72.

Buttenschøn, J. and Buttenschøn, R.M. (2011) 'Woodland grazing with cattle – the effects of twenty-five years grazing', in Rotherham, I.D. and Handley, C. (eds) *Animals, Man and Treescapes*, Wildtrack Publishing, Sheffield, pp122–7.

Buttenschøn, R.M. (2007) *Græsning og høslæt i naturplejen*, Miljøministeriet, Skov og Naturstyrelsen, www.skovognatur.dk; www.sl.life.ku.dk

Buttenschøn, R.M. and Buttenschøn, J. (unpublished a) 'Study on land use at Skovbjerg based on Geographic Survey maps from 1860/70- and 1900-series and aerial photographs from 1945, 1954, 1960, 1972 and 1985', Skovbjerg dyrkningshistorie.doc, unpublished.

Buttenschøn, R.M. and Buttenschøn, J. (unpublished b) 'Data from cattle dung-pat study on four grassland and one grazed woodland 1987/88 (Skovbjerg) and 1999/2000', draft paper, unpublished.

Buttenschøn, R.M. and Buttenschøn, J. (1998) 'Population dynamics of *Malus sylvestris* stands in grazed and un-grazed, semi-natural grasslands and fragmented woodlands in Mols Bjerge, Denmark', *Annales Botanici Fennici*, 35: 233–46.

Buttenschøn, R.M., Odgaard, B, Buttenschøn, J. and Hansen, J.B. (2008) 'Fra hedeplantage til lysåben græsningsskov', *Skoven*, 03: 124–8.

Ellenberg, H., Weber, H. E., Düll, R., Wirth, V., Werner, W. and Paulissen D. (1991) '*Zeigerwerte von Pflanzen in Mitteleuropa*', *Scripta Geobotanica*, 18: 1–258.

Grime, J.P. (2001) *Plant Strategies, Vegetation Processes and Ecosystem Properties*, 2nd edition, John Wiley and Sons Ltd., Chichester.

Hofmann, R.R. (1989) 'Evolutionary steps of ecophysiological adaptation and diversification of ruminants – a comparative view of their digestive-system', *Oecologia*, 78: 443–57.

Illius, A.W. and Gordon, I.J. (1993) 'Diet selection in mammalian herbivores: Constraints and tactics', in Hughes, R.N. (ed.) *Diet Selection – An Interdisciplinary Approach to Foraging Behaviour*, Wiley-Blackwell, Chichester, pp157–81.

Miller, A.M., McArthur, C. and Smethurst, P.J. (2007) 'Effects of within patch characteristics on the vulnerability of a plant to herbivory', *Oikos*, 116: 41–52.

Strandberg, B., Kristiansen, S.M. and Tybirk, K. (2005) 'Dynamic oak-scrub to forest succession: Effects of management on under-story vegetation, humus forms and soils', *Forest Ecology and Management*, 211: 318–28.

van Dyne, G.M., Brockington, N.R., Szocs, J. and Ribic, C.A. (1980) 'Large herbivore subsystems', in Breymeyer, A.I. and van Dyne, G.M. (eds) *Grasslands, Systems Analysis and Man*, International Biological Programme, Volume 19, Cambridge University Press, Cambridge, pp270–560.

Wind, P. and Pihl, S. (eds) (2004) *The Danish Red List*, The National Environmental Research Institute, Aarhus University, http://dmu.dk (updated April 2010).

23 Ancient trees, grazing landscapes and the conservation of deadwood and wood decay invertebrates

Keith Alexander

Introduction

Deadwood and wood decay (saproxylic) invertebrates are widely acknowledged as one of the two most threatened ecological groupings of invertebrates across Europe – the other being wetland invertebrates (Koomen and van Helsdingen, 1996). This has recently been recognised by the European Union with their commissioning of the *IUCN European Red List of Saproxylic Beetles* (Nieto and Alexander, 2010) – the first ever *IUCN Red List* focused primarily on an ecological rather than taxonomic grouping of species. Despite this recognition, there is notably little widespread understanding among invertebrate specialists of the fundamental role of grazing in maintaining the habitat structures required by many of these invertebrates.

The significance of open-grown conditions

A high proportion of saproxylic invertebrates are dependent on the host trees having sufficient space to develop their full potential in terms of crown development and natural death and decay of the various woody tissues produced (Alexander, 2008). Under natural conditions, that space tends to be provided by patterns of grazing and browsing by wild herbivores – and influenced to some extent by their predators and parasites (Vera, 2000), while in cultural landscapes that role is appropriated by people. Managed large herbivores may be part of the cultural landscape but human land use may provide suitable structural conditions without grazing, albeit incidentally. All too often, however, the cultural landscape does not favour the development of the tree's full potential and the associated invertebrates disappear – become extinct.

Two of the key factors in the tree's development for wood decay invertebrates are heartwood decay and lateral branches that eventually die and decay while still attached to the tree (Alexander, 2007). Without grazing to maintain open conditions around the tree – or other management in the case of some cultural landscapes – the tree will die young, over-shaded by the expanding crowns of younger neighbours, as it goes into natural retrenchment (Green, 2007) and before heartwood decay has progressed fully, and lateral branch development will be very limited.

This is especially acute for light-demanding tree species such as oak, but also significant for shade-tolerant trees such as beech or hornbeam. Although shade-tolerant trees are able to regenerate under closed-canopy conditions and are better able to compete with neighbours for light, they may not produce lateral branch growth, and are still vulnerable to overshading once natural retrenchment begins or where crowns are reduced through storm damage. Ancient beeches and hornbeams tend to be found either in naturally grazed forest or cultural wood-pasture situations.

Open conditions are also important for those saproxylic insects that have a requirement for refuelling at blossom (Alexander, 1999; Kirby, 1992). Nectar provides a source of carbohydrates that fuels flight activity while pollen is protein-rich and good for egg production. Not all saproxylic insects however visit blossom, and blossom may not be essential for all of those that do. The key blossom-providing plants tend to be flowering shrubs such as hawthorn and elder, or tall herbs with large white flowers such as hogweed. These are not available under closed-canopy conditions – even where present, hawthorn does not flower in shade.

It has also been demonstrated that other deadwood invertebrate species are also dependent on light to fulfil their lifecycles – many longhorn beetles that develop in fallen branches require those branches to fall into well-lit situations (Vodka *et al.*, 2009).

Shade is important for many other species, however, and a mosaic of light conditions is clearly required in order to conserve the whole fauna. Certain saproxylic Diptera are believed to be associated particularly with shade, especially among the Limoniidae (crane-flies) and Mycetophilidae (fungus gnats), as are certain beetles, for example among the Eucnemidae (false click beetles) and Leiodidae (slime mould beetles). The optimal balance between light and shade is however not known.

Awakening appreciation of the significance of grazing

The study of deadwood invertebrates has long revealed the relationship with open treescapes maintained by grazing animals. While this fact is inescapable, it has rarely been enunciated by the field entomologists concerned. The earliest British localities found to be rich in exciting rarities were Windsor Great Park and Forest, Sherwood Forest, Epping Forest and the New Forest – all sites with a long and more or less continuous history of livestock grazing, at least until very recently. The pattern repeats it across other European countries, for example Fontainebleau Forest in France, the Hasbruch in Germany and Białowieża Forest in Poland – all former wood-pasture systems (Vera, 2000).

Many of the key British sites were included in *A Nature Conservation Review* (Ratcliffe, 1997) and it was only then that the government nature conservation agency at the time, the Nature Conservancy Council (NCC), initiated a project to collate and analyse the records for the key deadwood beetle species. This project was named 'The fauna of the mature timber habitat' and the research was carried out for the NCC by the Institute of Terrestrial Ecology (ITE) during the late 1970s. The study was to examine whether there were more sites that were of value for conserving the invertebrate fauna associated with old trees, especially in pasture woodlands. This study was intended as a counterpart to surveys being made, independently, of the epiphytes in such areas (Harding and Rose, 1986).

The narrow entrenched entomological view and progression in its erosion

It was the publication of Harding and Rose (1986) that stimulated the first widespread appreciation of grazing as an important conservation management tool for deadwood invertebrates. However, Speight (1989), in his very influential 'Saproxylic invertebrates and their conservation', written for the Council of Europe, considered pasture woodlands as a degraded form of forest, where 'only very old trees remain, surrounded by grazed grassland in which there has been no regeneration of tree species for centuries'. He did not comprehend the key role of sustainable grazing in generating the open wooded landscapes with open-grown trees that were known to be exceptionally species-rich in saproxylic invertebrates. He was not alone either. As recently as 1996, it was beginning to be recognised that while open-grown trees are important for saproxylic invertebrates, grazing was still seen more from the point of view of the problems for tree health arising from root damage and bark-stripping damage to tree bases (Alexander *et al.*, 1996). Grazing by large herbivores was not seen as part of the process that produced naturally open-grown trees and the natural forest conditions required by saproxylic invertebrates. This important aspect is a very recent feature of conservation thinking and one that has nowhere near permeated the whole of the conservation movement. The publication of Vera (2000) was particularly instrumental in this major change in thinking but there is still considerable resistance.

The recent *European Strategy for the Conservation of Invertebrates* (Haslett, 2007) is very typical of the failure to recognise the need for integrated land management and separates out the impacts of agriculture and forestry. It does not mention grazing in the forestry section, but does comment: 'Maintaining the dynamic, open-mosaic nature of forests is essential for invertebrate conservation, particularly saproxylic species, and must be a central aim of woodland conservation policy and practice (Alexander, 2005)' but does not consider how this might be achieved. Most conservation management manuals still focus entirely on mechanical clearance as a means of maintaining open-forest conditions.

An examination of the proceedings from the many recent European conferences on saproxylic invertebrates reveals notably few mentions of the importance of grazing (Barclay and Telnov, 2005; Bowen, 2002; Buse et al., 2009; Mason et al., 2003; Vignon and Asmodé, 2008). This partly reflects the fact that most research is carried out in forestry areas, where grazing has long since been excluded, and where the issues tend to be protecting deadwood habitats from forestry operations. There is clearly a major issue of polarisation, of circular thinking and lack of integrated land management.

It is very clear that ecological understanding among invertebrate specialists has been, and still largely is, entrenched and failing to grasp the significant role of grazing in creating and maintaining the forest structures required by deadwood invertebrates – entomologists are interested in insects rather than land management and vegetation dynamics. However, this problem is not confined to entomologists but is prevalent across the natural sciences – the breadth of understanding shown in Vera (2000) is largely unique although there are encouraging signs elsewhere, for example Jeschke (2006). The problem is especially apparent in the approach to nature conservation enshrined in the Habitats Directive and Natura 2000 (see below).

The *IUCN European Red List* does however highlight the real landscape management issues (Nieto and Alexander, 2010). The conclusions include first, the main long-term threats identified as habitat loss in relation to the decline of veteran trees throughout the landscape, as well as lack of land management targeted at promotion of recruitment of new generations of trees; and second, the crucial need to raise awareness among conservation professionals and resources managers about the needs of saproxylic organisms, as they depend on the dynamics of tree aging and wood decay processes, which in turn have implications for land management; non-intervention or minimum intervention in former wood-pasture can prevent the renewal of old trees and be very damaging, and livestock grazing can be essential to maintain adequate habitats.

Role of grazing landscapes – a case study of hermit beetle

Increasingly saproxylic beetle researchers are appreciating the role of grazing in maintaining habitat structure. The addition of hermit beetle (*Osmoderma eremita*) to the Annexes of the Habitats Directive has made a major contribution in stimulating scientific studies of this rare and threatened beetle and its habitat. This research has been led by Swedish specialists, as the old oak stands across the cultural landscapes of southern Sweden retain Europe's northernmost substantial populations of this beetle. The beetle develops deep within large accumulations of wood mould within hollow, standing, living tree trunks, and requires concentrations of suitable trees within flight distance of each other. Ranius and Jansson (2000) have demonstrated that abandonment of grazing and the woody re-growth that followed was harmful to many saproxylic beetle species, not just hermit beetle, resulting in lower abundances and hence greater risk of local extinction. Their recommendation was to cut the secondary growth back and to restore grazing management. This is one of the earliest scientific papers on saproxylic beetles to result in a recommendation for restoration of grazing specifically for saproxylic beetle conservation, although Ranius and Nilsson (1997) had earlier commented that cessation of management of pasture woodland might be unfavourable to the species, but probably the indirect influence of forestation is more harmful. Interestingly, the hermit beetle also lives in active wood-pasture systems at the southern edge of its range in Mediterranean Italy (Chiari, 2011), demonstrating that the Swedish experience does not just reflect the colder climate there and a resulting need for greater sun penetration to the tree trunks.

There are few other studies where grazing and its impacts on vegetation structure have been seriously researched with regard to saproxylic invertebrates. Wilde (2005) reported on preliminary results from a study of the impact of increased shade in Epping Forest. The key European 'near-threatened' beetle *Ampedus cardinalis* was only found in the open situations and not in shade – like hermit beetle it develops in the products of heartwood decay. Grazing restoration work is already being progressed here.

Conservation – strict forest reserves and the Natura 2000 Network

Conservation management for saproxylic invertebrates across Europe tends to concentrate on the symptoms rather than the causes – volumes of deadwood, availability of veteran trees, sunny glades and rides, etc., rather than addressing the manipulation of natural processes at landscape scale. Europe is now almost entirely a cultural landscape – as opposed to a 'natural' one – and the actual meaning of this expression 'natural processes' is as vague and ill-defined as 'natural' itself. And yet this is all too commonly the language of scientific conservation.

In the late 1990s, vegetation ecologists developed ideas about the need for 'minimum intervention reserves' in order to study 'natural processes' and these ideas resulted in two reports being published by English Nature (Mountford, 2000; Peterken, 2000). These reserves were seen as 'assigned indefinitely to a management policy designed to minimize the impacts of people; their composition, structure, patterns and wildlife would henceforward be almost wholly determined by natural factors and processes'. None of this was adequately explained and discussed, particularly what these 'natural processes' would be and how they differed from processes in the cultural landscape. The list of proposed reserves included many of the former wood-pasture sites that were known to be of greatest interest for saproxylic invertebrates – those sites that required active management in order to conserve their saproxylic interests were proposed for abandonment in support of hypothetical vegetation science. Fortunately this was never fully adopted in Britain.

However, the ideas found influential support, and inadequate resistance, among European agencies. Many of the best saproxylic invertebrate sites, all former wood-pastures, have been declared Strict Forest Reserves (SFR) and are now under non-intervention management. Despite the original intention for monitoring, no baseline data were gathered on the current status of saproxylic invertebrates, and of course no follow-up survey has been carried out in order to adequately assess the changes that have resulted. It is one of the most urgent conservation issues in saproxylic invertebrate conservation that SFRs should be subject to grazing by large herbivores or at least confined to sites that do not have significant current interest for saproxylics.

SFRs are in reality actually part of the cultural landscape since they primarily reflect the prevalent hypothesis among vegetation ecologists that large herbivores are not important in natural processes in natural forests. This is merely a hypothesis and one that has been severely undermined by Vera (2000). These reserves have little to do with 're-wilding' as much of the wild processes are no longer operational within the cultural landscape of Europe. This is especially so with regard to the former influences of wild herbivores and their predators and parasites. Abandonment of land management by people will not restore 'natural processes' as too many of the key factors have changed irreversibly. The results of non-intervention in the modern cultural landscape may be of scientific interest but this should not be at the cost of biodiversity conservation.

These SFRs have been incorporated into the Natura 2000 Network, which itself is dominated by vegetation ecology. Wood-pasture is not recognised as a habitat type of European importance and so the best examples are very under-represented in the Network – sites tend to be selected to a considerable degree on the basis of their ground flora characteristics rather than their saproxylic assemblages (Bergmeier et al., 2010; Nieto and Alexander, 2010). The Natura 2000 Network is unfortunately largely based on the all too-widespread botanical arrogance that it is only necessary to conserve vegetation and 'natural processes' in order to conserve biodiversity as a whole. The evidence in support of this assumption is however noticeably lacking.

Conclusions: Landscape conservation aspects

Nature conservation today has refocused on landscape-scale conservation and developing greater linkages between sites to counteract the fragmentation and isolation in the past. This is entirely sensible but needs to retain an understanding of more small-scale issues. A fully integrated landscape-conservation management approach may be very desirable but equally may not be achievable. It is important that site quality and site condition are maintained at the local level while seeking the larger-scale objectives.

Unlike many habitat types, open-grown trees are achievable across landscapes, making linkages particularly achievable (theoretically). The presence of open-grown trees also does not preclude other grazed habitat types such as grasslands or heathlands. Time factors are still crucial of course as the heartwood decay habitat will take hundreds of years to develop within newly established trees of species such as oak and beech. This does make conservation of all existing veteran trees, throughout the landscape, even more important; in the past these have largely been dismissed as unimportant as they are too sparsely distributed to support rich assemblages of saproxylic invertebrates. If they are now recognised as part of future linkages then their true value can be acknowledged. There remains a scarcity of scientific evidence however that the saproxylic beetle assemblages are sufficiently mobile to be able to penetrate landscapes thin in habitat. Studies have been carried out on a small number of key rarities, but the European assemblage comprises many thousands of species. If open-grown veteran trees really are to become more widespread in the future, then conservation needs to maintain today's pockets of rich fauna in good condition for the next few hundred years at least; landscape-scale conservation of saproxylic invertebrates is only achievable if resources are maintained – and preferably increased – at site conservation level. And there are still the issues outlined in the preceding section to deal with.

The conservation of saproxylic invertebrates is a highly complex subject, with many fundamental issues in need of clarification and widespread acceptance. There is still a long way to go before grazing, in particular, is widely accepted as a key management tool.

References

Alexander, K.N.A. (1999) 'The invertebrates of Britain's wood pastures', *British Wildlife*, 11: 108–17.

Alexander, K.N.A. (2005) 'Wood decay, insects, palaeoecology and woodland conservation policy and practice – breaking the halter', *Antenna*, 29: 171–8.

Alexander, K.N.A., Green, E.E. and Key, R. (1996) 'The management of overmature tree populations for nature conservation – the basic guidelines', in H.J. Read (ed.) *Pollard and Veteran Tree Management* II, Corporation of London, London, pp122–35.

Alexander, K.N.A. (2007) 'Old growth: Ageing and decaying processes in trees', in I.D. Rotherham (ed.) 'The history, ecology and archaeology of medieval parks and parklands', *Landscape Archaeology and Ecology*, 6: 8–12.

Alexander, K.N.A. (2008) 'Tree biology and saproxylic Coleoptera: Issues of definitions and conservation language', in V. Vignon and J.-F. Asmodé (eds) *Proceedings of the 4th Symposium and Workshop on the Conservation of Saproxylic Beetles*, held in Vivoin, Sarthe Department, France, 27–9 June, *Revue d'Écologie*, Supplément 10, pp9–13.

Barclay, M.V.L. and Telnov, D. (eds) (2005) 'Proceedings of the 3rd Symposium and Workshop on the Conservation of Saproxylic Beetles, Riga/Latvia, 7–11 July, 2004', *Latvijas entomologs*, Supplementum VI.

Bergmeier, E., Petermann, J. and Schröder, E. (2010) 'Geobotanical survey of wood-pasture habitats in Europe: Diversity, threats and conservation', *Biodiversity Conservation*, 19: 2995–3014.

Bowen C.P. (ed.) (2002) *Proceedings of the Second Pan-European Conference on Saproxylic Beetles*, People's Trust for Endangered Species, London.

Buse, J., Alexander, K.N.A., Ranius, T. and Assman, T. (eds) (2009) *Saproxylic Beetles – Their Role and Diversity in European Woodland and Tree Habitats. Proceedings of the 5th Symposium and Workshop on the Conservation of Saproxylic Beetles*, Pensoft, Sofia, Bulgaria.

Chiari, S. (2011) 'Ecology of the hermit beetle (*Osmoderma eremita*) in Mediterranean woodlands', unpublished PhD thesis, Roma Tre University, Rome, Italy.

Green, T. (2007) 'Stating the obvious: The biodiversity of an open-grown tree – from acorn to ancient', in I.D. Rotherham (ed.) 'The history, ecology and archaeology of medieval parks and parklands', *Landscape Archaeology and Ecology*, 6: 48–52.

Harding, P.T. and Rose, F. (1986) *Pasture-woodlands in Lowland Britain: A Review of their Importance for Wildlife Conservation*, Abbots Ripton, Institute of Terrestrial Ecology (Natural Environment Research Council), UK.

Haslett, J.R. (2007) 'European Strategy for the conservation of invertebrates', *Nature and Environment*, No. 145, Council of Europe Publishing, Strasbourg.

Kirby, P. (1992) *Habitat Management for Invertebrates: A Practical Handbook*, Royal Society for the Protection of Birds, Sandy, UK.

Koomen, P. and van Helsdingen, P.J. (1996) 'Listing of biotopes in Europe according to their significance for invertebrates', *Nature and Environment Series*, No. 77, Council of Europe, Strasbourg.

Jeschke, L. (2006) 'The oaks in natural and cultural landscape and the management of oak habitats in Germany' in *The Oak – History, Ecology, Management and Planning. Proceedings from a conference in Linköping, Sweden, 9–11 May 2006*, report 5617, Naturvårdsverket, Sweden, pp38–9.

Mason, F., Nardi, G. and Tisato, M. (eds) (2003) 'Proceedings of the International Symposium "Dead wood: a key to biodiversity"', Mantova, May 29–31 2003', *Sherwood*, 95, Suppl. 2.

Mountford, E.P. (2000) 'A provisional minimum intervention woodland reserve series for England with proposals for baseline recording and long-term monitoring therein', *English Nature Research Reports*, No. 385, English Nature, Sheffield.

Nieto, A. and Alexander, K.N.A. (2010) *European Red List of Saproxylic Beetles*, Publications Office of the European Union, Luxembourg.

Peterken, G.F. (2000) 'Natural reserves in English woodlands', *English Nature Research Report*, No. 384, English Nature, Sheffield.

Ranius, T. and Nilsson, S.G. (1997) 'Habitat of *Osmoderma eremita* Scop. (Coleoptera: Scarabidae), a beetle living in hollow trees', *Journal of Insect Conservation*, 1: 193–204.

Ranius, T. and Jansson, N. (2000) 'The influence of forest regrowth, original canopy cover and tree size on saproxylic beetles associated with old oaks', *Biological Conservation*, 95: 85–94.

Ratcliffe, D.A. (1997) *A Nature Conservation Review*, Cambridge University Press, Cambridge.

Speight, M.C.D. (1989) 'Saproxylic invertebrates and their conservation', *Nature and Environment Series*, No. 42, Council of Europe, Strasbourg.

Vera, F.W.M. (2000) *Grazing Ecology and Forest History*, CABI Publishing, Wallingford.

Vignon, V. and Asmodé, J.-F. (eds) (2008) 'Proceedings of the 4th Symposium and Workshop on the Conservation of Saproxylic Beetles, held in Vivoin, Sarthe Department, France 27–29 June 2006', *Revue D'Écologie*, Supplément, 10: 9–13.

Vodka, S., Konvicka, M., and Cizek, L. (2009) 'Habitat preferences of oak-feeding xylophagous beetles in a temperate woodland: Implications for forest history and management', *Journal of Insect Conservation*, 13: 553–62.

Wilde, I. (2005) 'The shadier side of Epping Forest: Saproxylic Coleoptera Research' in M.V.L. Barclay and D. Telnov (eds) 'Proceedings of the 3rd Symposium and Workshop on the Conservation of Saproxylic Beetles, Riga/Latvia, 7–11 July, 2004', *Latvijas entomologs*, Supplementum VI: 124–25.

24 The future potential of wood-pastures

Iris Glimmerveen

Introduction

Wood-pastures are intricate, usually ancient, landscapes; intimate mixes of both trees and pasture. Quelch (2010) defines them as: 'Ancient wood pastures in the uplands are areas of grazed pasture, heath or open hill with a scattering of open-grown veteran trees, some of which may have been pollarded in the past'. Like Vera (2007), he also described wood-pasture as: 'a "savannah" kind of landscape, being intermediate in character between woodland and (hill) grazing and depending on both' (P. Quelch, pers. comm.). As wood-pastures are so variable, there are probably other ways of describing them and although there might be some difference in emphasis, they still relate to the same broad habitat.

The author had the opportunity to be closely involved with these unique landscapes over about eleven years. This was through the discovery and subsequent management of the Geltsdale wood-pasture in Cumbria (Glimmerveen and Clark, 2008) and later with Thornthwaite Hall, Cumbria (Glimmerveen, 2009). It also involved visiting several additional upland wood-pastures such as Troutbeck, Borrowdale, Langdale in Cumbria, Elan Valley (Wales) and Glen Finglas (Scotland), and others throughout Europe. I have come to realise that they are special, but complex landscapes. This is because wood-pastures are dynamic features requiring space for all their phases and for these phases to progress from one to another. These include 1) dense thorny scrub to protect young trees from grazing (Figure 24.1); 2) mature trees providing seeds for the next generation; 3) ancient trees providing standing and lying deadwood (Figure 24.2); and 4) open spaces for scrub and trees to regenerate into. Wood-pastures are complicated further as on the whole they contain four elements: 1) grazing, 2) trees, 3) archaeology and 4) wildlife. Depending on how intensely the wood-pasture has been worked over time, each of these elements and their interactions shape the character of the wood-pasture. It generally takes a long time before such interactions become visible or recognisable in the landscape and their clarity further depends on the level of management that has taken place in the past or even

340 *Iris Glimmerveen*

Figure 24.1 Sloe scrub protecting tree saplings and young trees in Geltsdale's wood-pasture
Source: Iris Glimmerveen

Figure 24.2 Hollow ancient rowan, bent over touching lying dead birch, Geltsdale
Source: Iris Glimmerveen

currently. However, these interactions not only make for a beautiful landscape, but also create unique opportunities to make use of it.

The development of an ancient wood-pasture can be interrupted or its existence threatened by a wide range of factors, usually involving neglect, a change of land use, limiting its space, intensification of management, or others. This is likely to become worse with the onset of climate change (Table 24.1).

Table 24.1 Threats to wood-pastures with their associated effects

Wood-pasture threats	Effect on wood-pasture
Loss of expertise of managing mixed system	Change from mixed to single system
Perceived as not producing economic gain	Change of land use – intensification
Lack of political will	No capital to conserve/retain wood-pastures
Lack of space	Its system cannot function dynamically
Lack of (tree) regeneration	Wood-pastures dying out of old age
Lack of awareness of landscape/habitat and its importance	Neglect and conversion
Sea level rise	Less available land – conversion/urbanisation
Microclimate temperature and moisture changes	Unsuitable habitat for some or all biota

The UK government aims to expand the number of wood-pasture sites with 150 by 2015 and 300 by 2020. This is compared to a 1998 baseline (webarchive.nationalarchives.gov.uk) and although a small proportion of this might be achieved by reclassifying existing woodland or pasture sites as wood-pastures, the majority will have to be created 'from scratch'. Achieving this target through wood-pasture creation is much more likely if there was 'something in it' for the landowner. It is therefore important to find ways to explore how a wood-pasture system could be used to fulfil its maximum potential in a sustainable manner. In this way the conservation value of the land is maintained (or even improved) and the landowner benefits from the management efforts. The wood-pasture can then serve us for years to come. This chapter explores the intricate and multi-functional nature of wood-pastures and how people can use them sustainably.

The potential effects of and use for grazing

Many large grazers, such as cattle, horses, deer and sheep not only eat grass, but also eat leaves and bark to supplement their diet (Figure 24.3). Young tree leaves are particularly palatable and seedlings and saplings are repeatedly browsed. If there are many mouths to be fed, this continues until the trees have grown tall enough to put their leaves out of the grazer/browser's reach, the trees are protected by a fringe and mantle of thorny bushes, or by manmade barriers. Without any protection some trees will die and this explains why low tree-density is a characteristic of wood-pastures.

The effect of trees growing up in the presence of continuous grazing can further lead to them developing a 'basal skirt', as every year suckers produced by the tree are grazed off by either stock or wild animals. With this process, a layer of wood is laid down causing the tree to produce 'swelling' at the bottom of its trunk. Repeatedly eating leaves from lower branches will further create a browse line, which combined with grass eating, will keep the woodland relatively open, at least at the low level.

This interest of stock in tree material was why people pollarded their trees. This enabled them to protect both the tree and to feed the stock its dried leaves or 'leaf hay' at times of low grass production and in a controlled manner. The fact that the practice of pollarding has persisted in Cumbria until as recently as the 1970s (C. Brown, pers. comm.), shows that both leaves and branch materials (particularly of the more palatable species such as ash) were considered valuable resources in terms of both minerals and nutrients for stock. Grazing pressure is thus instrumental in shaping wood-pasture trees.

Figure 24.3 Swaledale nibbling ash bark in Langdale, Cumbria
Source: Iris Glimmerveen

Large grazers further benefit themselves from the habitat with trees they help to create (Table 24.2) as it allows them to: eat a varied diet, grazing on a wide variety of herbs, grasses and woody materials; shelter from wind, rain and sun; and maintain good condition and therefore increase the chance of producing healthy offspring.

Table 24.2 Effects of shelter on livestock

Livestock system	Response to shelter
Lamb mortality	10–50 per cent reduction in mortality
Milk production	10–20 per cent increase in yield when cows graze in shelter
Suckler cows	8 per cent decrease in feed requirements to maintain liveweight with shelter
Suckler calves	20 per cent decrease in feed requirements to maintain liveweight with shelter
Beef cattle	10–30 per cent decrease to maintain liveweight with shelter
Newly shorn ewes	12 per cent mortality in one flock shorn one to five days before a storm and turned out in exposed conditions
Pasture productivity	10 per cent increase in productivity

Source: Data from Hislop et al., 1999

Grazing with livestock not only results in a harvest of a wide range of meat and/or milk products (Table 24.3), but could potentially also bring a specific type of habitat in favourable/good condition with high biodiversity, and as a spin off also create ideal deer and game habitat. This so-called 'conservation grazing' thus effectively uses grazing animals as a tool to manage or direct the ecological development of certain habitats, but it relies on the herd manager to understand the functioning of the ecosystem he/she is working with and the impact the grazer will have on that habitat. For this kind of management to also be economically viable, the herd manager would further have to be aware of the potential meat and/or milk products the herd can produce. He or she further needs to know how to market them, as the animals generally produce less weight per carcass, but do carry a traditional breed, free range and/or organic premium from certain buyers (www.thescottishfarmer.co.uk).

Dairy farmers also may have a contract with certain buyers, which could pay them 28–30p/l for their milk. This is considered a good price for intensive dairy systems in the UK, because they produce milk with a generally low fat content (<5 per cent). More traditional dairy breeds (e.g. Jersey) will demand some premiums, as their milk has considerably higher butterfat content (S. Hendry, pers. comm.). Further costs for a range of livestock can be found at www.fwi.co.uk, which is regularly updated as these costs fluctuate tremendously.

Table 24.3 Grazing harvest from livestock

Meat produce	Price at market	Price in shop	Dairy produce	Price in shop
Lamb (liveweight)	£1.80–2.20/kg	£15.00/kg	Semi-skimmed cows' milk	£0.78/l
Lamb (deadweight)	£4.50–4.90/kg			
Venison	£2.50/kg	£20.00/kg	Vanilla ice cream	£0.40/l
Traditional-breed beef (deadweight)	£2.60–2.80/kg		Crème fraîche	£3.30/l
'Ordinary' beef (deadweight)	£3.00/kg	£8.25/kg	Yoghurt	£1.50/kg
Ewe	£93 each	£4.78/kg	Butter	£4.40/kg
Two-year old Scottish Highlander bulls	£2175 average		Halloumi cheese	£8.68/kg
Eight-year old oxen sold for logging (Romania)	€1700 each		Organic mature cheddar cheese	£8.84/kg

Source: Data from www.fwi.co.uk and www.tesco.com

At first sight Table 24.3 might suggest that the produce from conservation grazing is valued less than that of the more intensive systems. However, it does not include any of the habitat management benefits the cattle are achieving and is therefore not really a true comparison. To quantify these benefits is difficult, as they include features such as diverse vegetation structure, benefits to ground-nesting birds, to black grouse in particular, and dunging benefits for insects. The insects in turn feed a variety of bird chicks, again including black grouse. Cattle tracks are valuable for hens to move their chicks through. Other environmental services, such as improving soil health and water quality, are also extremely valuable and therefore should also be taken in account, but again they are difficult to quantify.

To get an approximation, however, Knott (pers. comm.) has worked out the income and cost figures for an 80ha Forestry Commission site in Scotland. This is grazed with 12 Scottish Highlanders for the specific aim of maintaining and enhancing the habitat and hence the population of the chequered skipper butterfly (*Carterocephalus palaemon*). He has compared these with figures from other sites that are mechanically managed, i.e. strimming, for similar purposes (Table 24.4).

Although the £17.25 per hectare is not a high income level, it still compares very favourably with the cost of £112.50 for the mechanical strimming. Moreover, it is likely that private enterprises, as opposed to the Forestry Commission, could benefit from the woodland option within the Scottish Rural Development Programme (RDP) grant. This would generate a further £104–111/ha for butterfly habitat improvement, or alternatively £87/ha for basic grazing, which are in addition to, for example, the beef cattle scheme, which gives £100/calf each year if born into the herd. Rates for England and Wales RDP may vary. When combined with the rural development grants, conservation grazing can thus be a profitable enterprise.

Table 24.4 Management cost comparison of conservation grazing versus strimming for an 80ha site

Income/cost calculations	Amount
Beef income from a herd of 12 = £1,100 x 12 =	£13,200
Replacement calves = £225 x 12 =	–£2,700
Supplementary concentrates, based on 12.5kg/day for 180 days = 2,250kg @ £240/tonne =	–£540
Labour to set out concentrates based on 2 hrs/day @ £12/hr x 2 for on-costs for 180 days =	–£8,640
Total conservation grazing profit	£1,320 (£17.25/ha)
Strimming cost based on contract of 2.4 man days/week for 6 months @ £150/day =	£9,000 (–£112.50/ha)

Source: Data from K. Knott, 2011

The potential use of trees and shrubs

Like any tree, wood-pasture trees have a wide range of functions, such as production of oxygen, locking up carbon dioxide, reduction of noise and dust etc. However, the open-grown shape of a wood-pasture tree further allows for the creation of substantially sized branches with bends, which are perfect for use in the wooden shipbuilding industry, roofing timbers of churches and other buildings, as well as the bespoke furniture trade (Table 24.5).

Table 24.5 Estimated timber value of an open-grown tree

1 butt = 33.33 hoppus ft (= 1.2m³)
(good straight planking quality @ 12' long and 20" quarter girth)
At £10/hoppus feet the butt is worth:
£330 for oak or
£150 for ash at roadside

Source: Data from D. Frost, 2011

Many other products can be created from wood-pasture timbers (Table 24.6), each of which has a specific end use and therefore takes precise skills and expertise to be produced. For example, for an oak swill basket the oak needs to be cut to size, steamed in a long tub over a fire for four hours and then split into strips and used to weave the baskets (www.oakswills.co.uk). Although these baskets are created in a traditional manner, they have retained their usefulness through time (Figure 24.4). Wood turners particularly favour 'burr' timber, which is created through repeated browsing of epicormic shoots, as it makes for characterful finished pieces (Figure 24.5) and can therefore be highly valued (www.jonathanleech.co.uk).

Figure 24.4 Owen Jones creating an oak swill basket at Talkin Treemendous, Cumbria
Source: Iris Glimmerveen

Figure 24.5 Burr elm bowl turned and finished by Danny Frost
Source: Iris Glimmerveen

Table 24.6 Value of some processed tree products

Product	Price	Unit
Firewood	£50.00	0.5 tonne bag
Charcoal	£3.80	3kg bag
Large elm burr bowl	£245.00	each
Oak swill basket	£52.00	22" each
Garden bench	±£250.00	4ft each
Oak beam fireplace	±£200.00	each
Roofing shakes/shingles	£80.00	m^2

Source: Data from I. Taylor, 2011 and J. Leech, 2011

Reinstating traditional practices, or importing them from other places, could well add to the product range of trees. For instance 'leaf hay' was produced from pollarded trees in Scandinavia, where during summer branches with leaves were cut, dried and stored. This was so that additional nutrients could be available when the grass sward was least productive, i.e. during drought or winter times (Quelch, 2009). However, Martin Clark (pers. comm.) recently found in south Romania up a wooded valley near Baile Herculanum, close to the Danube:

the most stunning wood pasture on a recognized route to a high pasture, so the cattle are taken through and graze on the way. The farmer/cowherd is still pollarding the trees – to get (amongst other things) poles for hayricks and firewood. Most interesting is the use of a hollow beech tree for cooking. He has cut a draft/smoke hole at around two metres height and the inside is gently smouldering – the fuel being the dead and rotten wood. The bole of the tree was warm to touch, but the forester told me later that day that this was common practice for cooking and smoking food and trees were not killed by it as long as the draft up the hollow trunk was controlled.

Pollarded trees further have the advantage that they can produce materials on a relatively short cycle while the main bole keeps growing; it is therefore inherently a very productive system of good straight and similar sized timber. In Cumbria there is still living memory of trees being pollarded in the 1970s (C. Brown, pers. comm.) and the National Trust now has an active pollarding programme. This is informed by initial tree surveys and is leading to prioritised tree management. Objectives for this work include conservation and associated biodiversity, landscape and cultural heritage reasons. Often the cut material is purposefully left in fields for much appreciated stock-browse, before usually being removed for firewood (J. Hooson, pers. comm.). Use for leaf hay or other value adding products has not been considered as yet.

Shrubs have traditionally been coppiced (i.e. cut at stump and then allowed to re-grow) by people for a wide variety of purposes. Some of these uses have continued to the present day, i.e. paling made from sweet chestnut and hazel spars for thatched roofs. Other uses are seeing a revival, i.e. dead hedging materials, to protect tree seedlings from grazing, and hazel hurdles, while others still are given a complete new lease of life, i.e. willow rods to weave living arbor seats or large-scale sculptures. A wide range of such coppice products can be found on www.coppice-products.co.uk.

Another shrub produce that has seen a revival in recent years is the use of some shrubs for their materials in food or as an active ingredient in herbal remedies. For instance hawthorn berries (Figure 24.6) are used to make syrups or gels (wildforager.survivalistssite.com), but can also combat heart disease, angina, diarrhoea, dysentery and kidney inflammation (www.herbalremediesinfo.com). Both websites give a wealth of other uses for our native shrubs and trees.

The benefits of wood-pastures to archaeology

Continuous and relatively light grazing further benefits archaeological features contained in a grazed area. This is because no ploughing is needed and therefore the features buried within or below the surface of a site are not disturbed, other than perhaps some rubbing from stock on protruding

Figure 24.6 Hawthorn berries, Geltsdale
Source: Iris Glimmerveen

items, which might dislodge some of the smaller pieces. It may therefore not come as a surprise that in the uplands wood-pastures with important archaeology have been found on hill slopes and/or tucked away in remote places, such as Geltsdale. This is as opposed to the more fertile valley bottoms where often there has been more disruption. Furthermore the archaeology is visible with grazing and therefore gives the opportunity for it to be recorded and protected.

Ancient trees contained in a wood-pasture site can further contribute to or inform the archaeological study of a site. This is because they are usually the longest-living component of the site and therefore their shape, habit, growth form and size, together with tree coring data (should it be available), can reflect past management. For instance, lapsed coppiced or pollarded trees are evidence of those management practices in the past, while repeated pollarding will further show up as a cyclical widening and narrowing of tree ring patterns (Mills *et al.*, 2009). Pollard tree location can also be indicative of past land use (Figure 24.7), as tall well-maintained cropping ash pollards are mostly on the lower slopes, i.e. within easy access to (past) farmers and livestock (Quelch, 2007).

This means that a wealth of past information can be contained within wood-pasture sites, which with care can be unlocked and combined with historical information, so that today's people can gain insights into the sustainable way past people made use of these places.

Figure 24.7 Pollarded ash in Langdale's medieval landscape setting
Source: Iris Glimmerveen

The benefits of wood-pastures to wildlife

The open-grown crown of a wood-pasture tree further creates a large volume of three-dimensional space, which can be occupied by a wide variety of wildlife to provide nest and roosting spaces for birds, bats and other animals alike. It further provides the all-important shade for (grazing) animals and space for them to shelter from adverse weather conditions (Table 24.2).

Table 24.7 Foodstuffs from trees and bushes for wildlife

Animal	Preferred food
Insects	Nectar, e.g. rose
Large tortoiseshell butterfly	Aspen, poplar, willow
White admiral butterfly	Bramble
Vapourer moth	Hawthorn, hazel, lime, oak
Mistletoe marble moth	Mistletoe
Song thrush	Yew, sloe, elder, guelder rose
Waxwing	Guelder rose
Nuthatch	Hazel nut
Bullfinch	Rowan berry
Black grouse	Birch buds and berries
Fieldfare and redwings	Berries (hawthorn, rowan, juniper)
Common dormice	Bramble
Badger	Juniper berries
Mice and voles	Berries

Shrubs and bushes also form an integral part of a wood-pasture; thorny bushes in particular are instrumental to the regeneration of broadleaved trees in the presence of grazing, as they can form a 'fringe and mantle' to a broadleaved sapling, thereby protecting it from browsing by large grazers (Vera and Buissink, 2004). These and other bushes or small trees also provide a good variety of foodstuffs for wildlife, such as berries and nuts (Table 24.7). They further attract a wide variety of insects, which form food for small songbirds and mammals alike; together they contribute to the wood-pasture's biodiversity.

Grazing by large ruminants (Figure 24.8) further benefits other wildlife in many different ways, as they:

- cut through dense herb layer with their hooves, thereby creating small spaces of bare soil in which seeds can drop and germinate;
- transport seeds in their coats;
- create open space or glades, providing sheltered sunny microhabitats, which is good for most flora and fauna, flowers, fungi and butterflies in particular;
- create diverse edge habitat, providing good habitat for a wide variety of food-bearing bush trees;
- allow trees to gain the opportunity to develop their full crown;
- ensure dappled shade on the tree stem, creating good microhabitat for lichens, mosses, bats and spiders;
- dung and thus provide a good additional food source for invertebrates, who in turn 'fuel' the wood-pasture's food web.

For these reasons, large grazers are used as 'tools' in conservation grazing to restore and maintain different types of habitat including wood-pastures in favourable condition. Animals used for this type of grazing should be adapted to local conditions and it therefore follows that local, native breeds should be preferred, whether cattle, sheep, deer or horses, as at least in theory they are most suitable. It is further important to graze with a diverse age- and sex-structure of a herd, as that allows the veteran cattle, for example, to lead the group well and know what to do in adverse weather conditions. Mixed sex herds allow bulls to become more solitary and range higher up the hill, while the females stay on the lower land (Quelch and Foster, 2005). These behavioural changes cause significant effects on grazed land, for example to set out its territory a bull creates scrapes that, whether wet or dry, add to the diversity of the available microhabitats. A 'Woodland Grazing Toolbox' has been devised by the Forestry Commission to help decide what type of grazing is best for a woodland/wood-pasture (www.forestry.gov.uk). In the process they are likely to generate good quality meat and/or milk, which if not completely organic is likely to be produced close to natural conditions.

The future potential of wood-pastures 351

Figure 24.8 Highland cattle herd in Anloo, Netherlands
   Source: Katherine Owen

However, this kind of management relies on the herd manager to fully understand:

- the functioning of the ecosystem his/her herd is to graze;
- the specific type of monitoring required for the desired key species and habitats;
- which type of stock is most suitable to the site/ecosystem;
- type of grazer available;
- health and welfare of the herd(s);
- length of time the herd should be deployed;
- marketability of stock produce (breeding animals, meat, milk);
- immediate positive and adverse impacts on the ecosystem.

These factors combined will determine whether the desired outcome for the habitat can be achieved and whether such management is sustainable in the long term (S. Hendry, pers. comm.).

The benefits of wood-pastures to people

Wood-pastures are thus valued by people for their trees, grazing, wildlife and archaeology. It is particularly the old wood-pasture remnants we see today that are extremely important, as persisting over a long time, they have been allowed to grow old and therefore give opportunities to study ecological and manmade processes over time and to relearn how people managed such sites in the past. They further give a sense of history, continuity of traditional ownership and a sense of place. Just for these reasons alone they are well worth conserving. This, along with the wealth of ancient trees, biodiversity and archaeology, they may contain.

However, if we are to sustain the benefits of wood-pastures in the future, then we should not only protect and rediscover what still persists today, but also enhance them. This is necessary so that they can develop to their full potential, both in terms of ecology and benefits to people. Should our existing wood-pastures turn out to be too precious because of their bio-cultural historic landscape values to be turned over to productive use, however extensively, then it seems that now is the time to create new ones. Already in 2003, the then manager of Royal Society for the Protection of Birds Geltsdale Nature Reserve and the author made use of two consecutive Forestry Commission grants (Joining and Increasing Grant Scheme for Ancient Woodlands (JIGSAW) and the Woodland Creation Grant) to plant Bruthwaite, a 160ha new native woodland. As Geltsdale contains a population of black grouse, the Forestry Commission allowed planting at relative low densities, which will enable the site to be turned over to wood-pasture once the trees are well established. A wide range of locally native tree-species mixes were used to maximise the benefits to wildlife. In addition, a relatively small section (2ha) was added with a view to produce hazel coppice. The trees are growing well and it won't be long before a cattle herd can be allowed in.

In Wales, the Woodland Trust's woodland creation team have also recognised that wood-pastures are beneficial to many situations. They are therefore piloting information days to encourage farmers to create new wood-pastures or to enhance existing ones (Figure 24.9). This pilot will take place over a year to gauge the response and outcome (K. Owen, pers.comm.).

With skill, this system will enable us to produce many food and timber items we need close to home. The approach reduces both food and timber miles, while also making the most productive use of the land. It is recognised that such land use will need to be economically viable too, which should be possible to achieve with even a single, albeit a niche-market, product. This is much more likely if an income is derived from any combination of produce available. Should we be able to do this in a sustainable and stable manner, then we may even enjoy the wood-pastures and their produce in the long term.

In addition to all the benefits mentioned above, people generally like wood-pastures. Their open spaces make it safe for them to be in, while the high biodiversity gives them a good chance to see wildlife at a safe distance. The varied components, woodlands, bushes, single trees and open spaces make for a beautiful landscape that is easy to enjoy by both visitors and locals alike.

Wood-pastures are thus ideal places for people to be in. This could be because within wood-pastures people can fully experience for themselves what it is like to be an integral part of the idyllic pastoral scene where an

The future potential of wood-pastures 353

Figure 24.9 Wood-pasture training event at Elan Valley, Wales
Source: Iris Glimmerveen

animal is peacefully grazing in an open space surrounded by bush fringed woodland and majestic trees. A little stream bubbles through the glade and you yourself can pick at will from the plants or trees to quench either hunger or thirst. Add to this the notion that at any moment you can come face-to-face with an amazing wild animal. It is perhaps no surprise that, subconsciously at least, people are reminded of arcadia, where you can live in harmony within the ideal landscape, where everything you require is within easy reach and where you can be drawn with peace of mind to either tree-enclosed spaces for inward looking contemplation, or open spaces for blue-sky thinking, or to simply rest safely underneath a tree. No wonder then that many artists have been inspired to use wood-pasture landscapes to portray their paradise and that a small wood-pasture, the Borkener Paradies in north-west Germany (Figure 24.10), is actually named after it.

Conclusion

Managing mixed land-use systems such as wood-pastures is clearly more difficult than managing single ones. I have already shown that considerable skills and expertise are required to ensure that for instance both trees and grazers can thrive and be productive in a wood-pasture site. However, when maximising the benefits of all of the wood-pasture elements, i.e. wildlife, people and archaeology in addition to timber and grazing, is considered, then specific skills for each discipline will have to be drawn in and

354 *Iris Glimmerveen*

Figure 24.10 Borkener Paradies, north-west Germany
Source: Iris Glimmerveen

combined. This is so that a workable strategy can be devised and followed. Of course, such a strategy cannot be achieved overnight and nor does it have to because on the whole existing wood-pastures are relatively robust and new ones need time to develop. Therefore there should be time for a manager (or management team) to learn the required skills and plan in a way so that product lines can be added gradually as and when both the site and the manager are ready to make use of them.

As the circumstances/objectives of a site and its owner will be different in each case, it will be impossible to have specific recommendations at the ready for any site manager to draw on. So like a manager, we will have to 'relearn' from the past and from each other how a wood-pasture can be managed and created, but I believe that with the collective expertise, knowledge and experience from professionals and practitioners alike we should be able to create for instance a 'curriculum' of wood-pasture management and creation courses, so that at least we can start to transfer this knowledge to those who need it. With the 'Living off the Landscape' project (www.woodlandinspirations.eu/projects/future-project.aspx) Woodland Inspirations and its European partners are aiming to put in place the structure for such courses so that we can pass on and disseminate our knowledge of worked mixed landscapes, including wood-pastures, widely, thereby improving the chance of existing and new wood-pastures to achieve their full potential and to serve us in perpetuity.

Acknowledgements

I would like to thank, C. Brown, retired from National Trust, M. Clark, Director, Grampus Heritage and Training Limited, Cumbria, D. Frost, furniture designer and maker, S. Hendry, grazing and livestock manager, Cowal and Trossachs Forest District, Forestry Commission Scotland, J. Hooson, wildlife and countryside adviser, National Trust (North West Region), Grasmere, Cumbria, K. Knott, environment and conservation manager, Lochaber, Forestry Commission Scotland, K. Owen, senior verifier, Ancient Tree Hunt, Woodland Trust, K. Owen, Seeds of inspiration Photographer, P. Quelch, Peter Quelch Woodland Services, and I. Taylor from Lakeland Coppice Products.

References

Anon. (1998) *Lowland Wood-pasture and Parkland Habitat Action Plan*, English Nature, Peterborough.

Glimmerveen, I. (2009) 'Thornthwaite Hall's wood pasture', unpublished report for Natural England, Penrith, Cumbria.

Glimmerveen, I. and Clark, M. (2008) *Geltsdale's Wood Pasture*, East Cumbria Countryside Project, Cumbria.

Hislop, M., Gardiner, B., Harriet, P. and Bailey, R. (1999) *The Value of Shelter Woods on Farms in Cumbria*, Forest Research, York.

Mills, C.M., Quelch, P. and Stewart, M. (2009) 'The evidence of tree forms, tree-rings and documented history around Bealach nam Bo, Loch Katrine', client report for FCS (Cowal and Trossachs District) (available from the authors).

Quelch, P. (2007) 'Seathwaite wood pasture. Bassenthwaite wood pasture project (4)', unpublished report, Grampus Heritage and Training Limited, Wigton, Cumbria.

Quelch, P. (2009) 'Woodland archaeology in Cumbria – stone gate stoups and drying leaf hay in Norway', unpublished poster presentation at Time Honoured Trees, Cumbria Wood and Forestry Festival, Cumbria.

Quelch, P. (2010) 'Upland wood pastures', *Landscape Archaeology and Ecology*, 8: 172–7.

Quelch, P. and Foster, S. (2005) 'Glen Garry woodland grazing project site visit', *Livestock in woods. Newsletter*, Spring: 7.

Vera, F.W.M. (2007) 'The wood-pasture theory and the deer park: The grove – the origin of the deer park', *Landscape Archaeology and Ecology*, 6: 107–12.

Vera, F. and Buissink, F. (2004) *Wildernis in Nederland. Het verhaal van bossen en beesten*, Tirion Uitgevers BV, Baarn.

25 A strategic view of the issues for wood-pasture and parkland conservation in England

Suzanne Perry

Introduction

Ancient environments, such as medieval parks and their veteran trees, are critically important for the safeguarding of deadwood invertebrates, epiphytic lichens, wood-rotting fungi, associated grassland and of course the veteran trees themselves. Effective conservation policy needs to be at the core of the delivery of future visions of grazed wooded or treed landscapes within their historic and cultural context. This chapter considers the approaches fostered by Natural England to conserve parkland and other wood-pastures in England.

Wood-pasture and parkland origins

Wood-pasture and parkland are lands that have been managed through a long-established tradition of grazing in a manner that allows the survival of multiple generations of trees. Veteran trees or shrubs are likely to be present, and these have usually been exploited in the past. The trees and shrubs can occur as scattered individuals, small groups or as an almost-complete canopy cover, frequently in a mosaic with open habitats such as grassland and heathland.

Whether wood-pasture existed as a distinct landscape type in the wildwood is strongly debated (see for example Hodder *et al.*, 2005; Vera, 2000); either way wood-pastures and parklands do contain some old growth features of the former natural landscape (Alexander *et al.*, 2002). However, wood-pasture and parkland sites as they occur today are cultural landscapes, shaped by centuries of human activity, often clearly derived from medieval parks, royal forests or common grazings (Rackham, 1980, 1998 and 2010). The open habitats tend to be highly managed alongside for example meadows, coppiced woodland and heaths.

But what is wood-pasture?

Most habitats within the UK Biodiversity Action Plan (BAP) are defined by their botanical composition, but because wood-pasture is a form of land use its composition and structure can vary widely (Figure 25.1) (http://jncc.defra.gov.uk/page-5706). Overall it is a mosaic habitat valued for its trees and the plants and animals that they support. Grazing animals are fundamental to the existence of this habitat. Key features include veteran trees, grazing animals, the presence of microhabitats including large-diameter hollowing trees and decaying wood, nectar sources for invertebrates, open habitat and continuity in terms of individual trees and their management. Debates as to what is or is not wood-pasture are not unimportant since they can affect eligibility for grant aid.

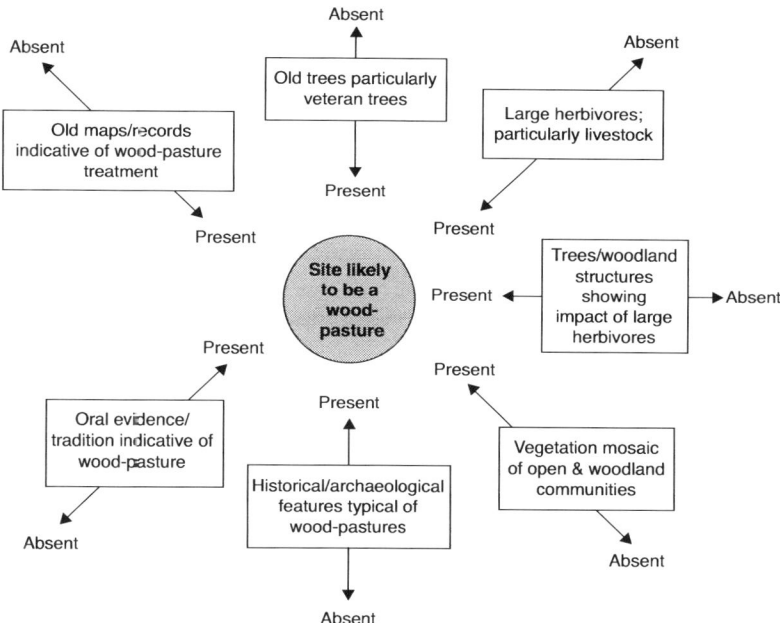

Figure 25.1 Diagram developed by Neil Sanderson to guide decisions about whether a site is wood-pasture – wood-pastures in good condition are likely to have most factors scoring towards the centre of the circle
Source: Suzanne Perry

Wood-pasture and parkland was identified as a priority for developing a Habitat Action Plan under the UK Biodiversity Action Plan (Defra, 1994 and 2007; UK Biodiversity Group, 1998). These summarised what was known about a habitat, set targets and identified actions required in a standard format according to:

- Distribution of the habitat in Britain and abroad
- Extent and abundance

- Conservation activity: protected sites, wider countryside
- Understanding the key sites for species and their needs
- Improved site management
- Future threats and challenges

Although the UK BAP has been superseded by country biodiversity strategies with new objectives (e.g. Biodiversity 2020 (Defra, 2011a)), these original action plans are still useful reference sources. Importantly, wood-pasture and parkland is also recognised in the legislative lists of habitats and species considered to be of principal importance for the conservation of biodiversity (in England and Wales these lists are respectively published under Sections 41 and 42 of the Natural Environment and Rural Communities Act 2006; in Scotland the lists are published Section 2 of the Nature Conservation (Scotland) Act 2004). Since the production of the first Habitat Action Plan (HAP) in 1998, interest in the cultural and social value of different habitats has developed, and recently in ecosystem services more generally (http://uknea.unep-wcmc.org/; UKNEA, 2011); hence in this account the contribution wood-pasture makes to these topics is considered.

Distribution and abundance in Britain and abroad

The original drafts of the HAP referred to 'lowland' wood-pasture and parkland and the habitat was assumed to occur mainly in England. Since the early 1990s, our understanding of the habitat has improved and it has been identified as occurring throughout the uplands as well, albeit in somewhat different forms (Quelch, 2001; Stiven and Holl, 2004) as well as in Northern Ireland (Alexander et al., 2007).

The habitat, including veteran trees, also occurs on the Continent (Bergmeier et al., 2010; Read unpublished: summary of Churchill Foundation sponsored trip to Spain: (http://frontpage.woodland-trust.org.uk/ancient-tree-forum/atfinternational/spain/spain.htm); Smith and Bunce, 2004). Considerable work has been undertaken on the distribution of veteran tree populations in southern Sweden (www.tradportalen.se) (Beckman, 2006; Claesson and Ek, 2008; Naturvårdsverket, 2004); large areas of wood-pasture have been found in the Pyrenees (Green, pers. comm.); scattered examples occur through Central Europe but the habitat again becomes common in the Balkans and eastern Mediterranean countries (Green, pers. comm.). Even apparently unpromising areas such as Flanders have been found to hold significant populations of veteran trees (Vandekerkhove et al., 2011). Some sites in Europe have only recently been picked up, for example wooded meadows in Estonia visited during the summer of 2011 by George Peterken and Keith Kirby (Kirby, pers. comm.).

Britain is of European importance for the large number of old broad-leaved trees still surviving (Read, 2000). Indeed, the Ancient Tree Forum strategic plan states that 'Britain may be home to 80 per cent of Northern

Europe's ancient trees', although the firm evidence to substantiate this statement has proved difficult to source.

How much is there?

Wood-pasture and parkland sites are often linked ecologically to the wider landscape, for example through the presence of veteran trees in hedgerows or scattered trees in fields. The variable nature of the habitat, combined with different opinions of what should be included in a site, have made it difficult to produce maps of where wood-pasture occurs, or figures for its extent. During the late 1970s, surveys of what were then called 'mature timber habitats' were carried out for the Nature Conservancy Council to identify important saproxylic invertebrate and epiphytic lichen sites (BLS, 1982; Harding and Rose, 1986). During the late 1990s and early 2000s, such survey work was updated and expanded to improve our understanding of sites and species, knowledge of the distribution of important areas, the value of particular sites and the ecology of individual species. Examples of such work include:

- *Ancient Tree survey of Staverton Park and the Thicks SSSI*, Staverton (Sibbett, 1999)
- *The Veteran Trees of Birklands and Billhaugh, Sherwood Forest, Nottinghamshire* (Clifton, 2000)
- *County Surveys of Parkland: The Staffordshire Experience* (Webb and Bowler, 2001)
- *Thames and Chilterns: Parkland and Wood Pastures with Veteran Trees* (Alexander and Lister, 2003)
- *A Provisional Inventory of Parklands and Wood-pasture in the East Midlands* (Harvey *et al.*, 2004)
- *A Provisional Inventory of the Wood-pasture and Parkland Habitat in England* (Lush, still in preparation for publication: Lush, M.J., Perry, S.C., Lush, C.E. (2012) England provisional wood-pasture and parkland inventory Report, Natural England (unpublished))

The distribution of veteran trees more generally, which may sometimes indicate former wood-pasture, is becoming clearer through the results of the Woodland Trusts Ancient Tree Hunt (Woodland Trust, 2012). In this project volunteers have recorded over 100,000 ancient and veteran trees throughout the countryside of the UK. Clusters of such trees could highlight the presence of to date unrecorded parkland and wood-pasture.

A full assessment of the coverage of wood-pasture and parkland sites is not possible because of the uncertainties over the extent and total distribution of the resource. However, there is often a wood-pasture element in existing woodland SSSIs, and it is unlikely that many large sites remain to be identified and notified (Natural England, 2008).

Conservation activity – protected sites

Large areas of wood-pasture and parkland, with their attendant veteran tree interest, are included within the protected site series. Under UK protected sites legislation, many 'woods' protected through notifications such as SSSIs in England Wales and Scotland, and Areas of Special Scientific Interest in Northern Ireland (ASSIs) (NCC, 1989) contain relic wood-pasture. In England it is estimated at least 17,617ha of the habitat are included within SSSI boundaries, together with a further 4,100ha in the New Forest Special Area of Conservation alone.

While veteran trees (Read, 2000) are now recognised as important in their own right, in the past conservation interest tended to focus on their role as hosts for rich communities of epiphytic lichens and deadwood invertebrates, including two species listed in Annex II to the EC Habitats Directive (violet click beetle, *Limoniscus violaceus*, and stag beetle, *Lucanus cervus*). The rot holes and decay in these trees also provide opportunities for hole-nesting birds and bats.

Table 25.1 List of sites put forward from the UK considered to contain broadleaved wood-pasture and parkland (a case can be made that many of the native pinewoods are also wood-pastures)

Habitat	Site	Country	Unitary authority	Area (ha)
9120*	Burnham Beeches	England	Buckinghamshire	382.76
	Ebernoe	England	West Sussex	234.93
	Epping	England	Essex	1,604.95
	The Mens	England	West Sussex	203.28
	New Forest	England	Hampshire/Wiltshire	29,262.36
9190**	Birklands and Bilhaugh	England	Nottinghamshire	271.84
	Staverton Park and the Thicks, Wantisden	England	Suffolk	81.45
	Windsor Forest and Great Park	England	Bracknell Forest; Surrey; Windsor and Maidenhead	1,687.26
91A0***	Exmoor and Quantock Oakwoods (part)	England	Devon; Somerset	1,895.17
	Elan Valley Woodlands	Wales	Powys	439.53
	Coille Mhor (part)	Scotland	Highland	311.23
	Kinlock and Kyleakin Hills (part)	Scotland	Highland	5,266.96
	Northern Pembrokeshire Woodlands (part)	Wales	Pembrokeshire	315.68

Note: *9120 Atlantic acidophilous beech forests with *Ilex*; **old acidophilous oak wood with *Quercus robur* on sandy plains; ***old sessile oak woods with *Ilex* and *Blechnum* in the British Isles.

Wood-pastures and parklands were not among the habitats originally listed as of European conservation concern under the EU Habitats Directive (92/43/EEC). Subsequently, '9070 Fennoscandian wooded pasture' was added as a priority habitat in the late 1990s, but this category is not applied to habitats in pre-existing EU countries. Nevertheless, large areas of wood-pasture and parkland were included within the list of UK SACs (Table 25.1) (Brown et al., 1997; McLeod et al., 2005) because they represented rich examples of 'old growth'-type woodland features; their wood-pasture nature is recognised in their management plans.

There may well be other wood-pasture sites that are of equivalent biodiversity value that are not yet within the protected site series because the original selection guidelines (NCC, 1989) did not allow for selection based on veteran tree population alone. Trials of evaluation methods for veteran tree sites from 2000 onward led to agreed selection criteria being approved in 2006 (Castle and Mileto, 2005). These guidelines are available from the Joint Nature Conservation Committee website at: http://jncc.defra.gov.uk/pdf/SSSIs_Chapter02.pdf. These criteria (Table 25.2) may also provide a basis for standardising comparisons of other sites.

Conservation activity – other policies, regulations and voluntary measures

The protected site series is unlikely to ever cover all wood-pasture and parkland sites and must be complemented by general land-use policies and regulations and by voluntary approaches. The main mechanisms are listed in Table 25.3.

Table 25.2 Agreed selection criteria for evaluation methodologies – recommended veteran tree site assessment protocol

Field measure	Possible thresholds		
	High value	Medium	Low value
Primary assessment criteria			
Number of veteran trees	>100	10–100	
Number of ancient trees	>15	<15	0
Number of trees >1.5m dbh	>15	5–15	<5
Secondary assessment criteria			
Extent of site	>50ha	11–50ha	10ha or less
Tree cohort continuity (assessed by tree size)	At least 1 cohort per 100 years similar spp and distribution to veterans	Future generations present but gaps in cohorts/new generations do not reflect spp/distribution of veterans	Large gaps in cohorts/ veteran trees only

Table 25.2 Continued

Field measure	Possible thresholds		
	High value	Medium	Low value
Visible deadwood (standing and fallen and including rot holes, hollow trunks etc)	Abundant	Present but evidence of removal	Little present
Ground vegetation	Unimproved grassland/ semi-natural woodland	Semi-improved or significantly disturbed	Arable, improved or suppressed (bare)
Veteran trees near by (sites and trees in the landscape)	Adjacent	Within 1km	>1km away
Diversity within veteran tree population (species, form, age, situation)	Diversity in at least three characteristics (species, age, form and situation)	Diversity in two characteristics or significant diversity in one characteristic	Little diversity
Associated species interest (e.g. lichens, saproxylic invertebrates)	Known to be high	Some interest known	
Documented habitat continuity – historical continuity	Documentary evidence of habitat continuity (several centuries)		
Potential	Interest likely to remain high or increase in short to medium term	Interest likely to remain moderate in short to medium term	Interest likely to remain low or decline in short to medium term

Other field measures

Density of veteran trees (over site)
Species composition of veterans
Scrub (incl. bramble and hawthorn)
Site management/threats
Water-bodies/wetland habitat
Shape
Surrounding land use
Local pollution load

Table 25.3 Listing of protection mechanisms in England

Legislation	Protection mechanism	Responsible body
Council Directive 92/43/EEC known as the Habitats Directive adopted in 1992, transposed into national law by the Conservation of Habitats and Species Regulations 2010 (as amended)	Special Area of Conservation/Special Protection Area	JNCC
Wildlife and Countryside Act 1981/Countryside and Rights of Way Act 2000/Natural Environment and Rural Communities Act 2006)	Site of Special Scientific Interest	Natural England
National Heritage Act 1983–2002	Registered Parks and Gardens	Historic Buildings and Monuments Commission also known as English Heritage
Wildlife and Countryside Act 1981 and Natural Environment and Rural Communities Act 2006	Section 41 flora and fauna protected	Natural England
Town and Country Planning Act 1990 (as amended)	Conservation Area Section 106 agreements Tree Preservation Order	Local planning authority
Town and Country Planning (General Development Procedure) Order 1995 (as amended)	Planning conditions	Local planning authority
Occupiers' Liability Acts of 1957 and 1984	Deals with lawful visitors and trespassers respectively	
Forestry Act 1967	Felling licence. Veteran trees particularly at risk because fellings for safety reasons are exempt	Forestry Commission
EC Directive 85/337/EEC for removal of trees/planting	Environmental Impact Assessment (Forestry) Regulations 1999	Forestry Commission
Ancient Monuments and Archaeological Areas Act 1979	Scheduled Ancient Monument	English Heritage
Planning Policy Statement number 9: Biodiversity and Geodiversity Conservation replaced in 2012 by National Planning Policy Framework		UK government advice to local planning authority

Table 25.3 Continued

Legislation	Protection mechanism	Responsible body
Inheritance Tax Act 1984	Inheritance tax exemption for 'land of outstanding scenic, historic and scientific interest'	Revenue and Customs with advice from Natural England
National Trust Act 1907 and	Inalienable land owned by National Trust	National Trust
Health and Safety at Work etc. Act 1974		Health and Safety Executive
Environment Act 1995 and the Hedgerows Regulations 1997	Protection of historically important hedgerows	Local authority
Town and Country Planning Act 1990 (to be amended by the Planning Act 2008)	Planning permission	Local planning authority
Voluntary approach through RDPE Environmental Stewardship	Agri-Environment Grant Aid Scheme	Natural England
Landscape designations and the European Landscape Convention Florence 2000		Council of Europe

Understanding of key sites for species and needs

Of the species associated with woodland habitats in England, many are associated with large, open-grown trees with spreading crowns and structural variability, which tend to be well-developed in wood-pasture areas (Alexander, 1999 and 2011; Webb *et al.*, 2010). The majority of these species consist of invertebrates, fungi and lower plants.

Most of the restricted and very restricted species are lichens and invertebrates having a strong association with veteran trees. Specific research on and surveys of individual species have been carried out in the UK (Alexander, 2002a; Coppins and Coppins, 2002; Hammond and Harding, 1991; Rayner, 1992; Smith, 2007 and 2011; Webb, 2002 and 2004) but we have also drawn on work undertaken overseas, most notably in Scandinavia (Jonsson *et al.*, 2005; Ranius and Jansson, 2000).

Improved site and tree management

Prior to the 1990s, relatively little was published on managing veteran trees, many of which had been regularly worked as pollards at one time but had been neglected for many years (Mitchell, 1989). Experience of veteran tree management at sites such as Burnham Beeches in Buckinghamshire (Read, 1996) and Epping Forest in Hertfordshire has been collated and published in the *Veteran Trees: A Guide to Good Management* handbook (Read, 2000), and promoted through publications such as the series of management practice guides published by the Woodland Trust in conjunction with the Ancient Tree Forum (www.ancient-tree-hunt.org.uk/ancienttrees/managing/guides) and for Environmental Stewardship guidance (Natural England, 2010). At the same time arboriculturists have been improving tree management practice encouraged by the ideas of Shigo (1999).

Procedures such as haloing of veteran trees to reduce the competition from younger growth and crown reduction are now common place in nature reserve management plans. (Haloing involves the removal of younger growth from the immediate surrounds of veteran trees). The success or otherwise of haloing of veteran oaks at Windsor Great Park has been followed up by Alexander *et al.* (2010), who found that the position of the tree and the speed of the haloing affects tree survival. However, papers such as this, reporting on the outcomes of management actions, are seldom published. Alongside management work such as this, the value of deadwood and how and where it should be conserved is better understood (Alexander, 2002b and 2005; Forestry Commission, 2002, Life in the deadwood. A guide to managing deadwood in Forestry Commission Forests; Kirby, 2001; Kirby and Drake, 1993; Kirby *et al.*, 1998).

The deliberate use of grazing animals for managing mosaics of semi-natural vegetation such as wood-pastures is recognised as valuable (Chatters and Sanderson, 1994) and was encouraged through the Grazing Animals Project (now Grazing Advice Partnership) (www.grazinganimalsproject.org.uk). For example site managers have reintroduced grazing to their wood-pasture sites, for example at Epping Forest in Essex and Savernake Forest in Wiltshire (Gibson, 1997). Elsewhere, wood-pasture has been designed into a future landscape for example at Knepp Castle in Sussex (www.knepp.co.uk/), where 3,500 acres (1,418ha) of land are being restored using an extensive grazing system to drive habitat change. Similarly at Ennerdale in Cumbria an extensive grazing system has been introduced to manage the landscape in which trees are an important feature.

Future threats and challenges

The threats identified in 1998 for this habitat largely remain relevant today, and are listed in Box 25.1. For example, the lack of younger generations of trees is producing a skewed age structure, leading to breaks in continuity of deadwood habitat and loss of specialised dependent species. Planting of replacement trees for those veterans we are losing is still an issue. Defra (2010) reports a serious decline in isolated hedgerow trees, and even their recommended replacement rate of 30,000 trees per annum will only retain the current estimated total of 1.6 million trees identified in the Countryside Survey 2010. At present rates only 2,000 trees are being planted, and the estimate has been made that by 2050 we will have lost all our hedgerow trees. Similar studies have not been undertaken for other non-woodland trees, but we could surmise that losses are greater than replacements, and that the population is in decline. It is still possible to see monoliths in a field surrounded by arable crops.

Box 25.1 Main threats identified in the UK Habitat Action Plan

- Lack of younger generations of trees is producing a skewed age structure, leading to breaks in continuity of deadwood habitat and loss of specialised dependent species;
- Neglect and loss of expertise of traditional tree management techniques such as pollarding leading to trees collapsing or being felled for safety reasons;
- Loss of veteran trees through disease such as Dutch elm disease, physiological stress such as drought and storm damage and competition for resources with surrounding younger trees;
- Removal of veteran trees and deadwood through perceptions of safety and tidiness where sites have high amenity use, forest hygiene, the supply of firewood or vandalism;
- Damage to trees and roots from soil compaction and erosion caused by trampling by livestock and people and car parking;
- Changes to groundwater levels leading to water stress and tree death, resulting from abstraction, drainage, neighbouring development, roads, prolonged drought and climate change;
- Changes in the countryside and its management that have lead to isolation and fragmentation of the remaining parklands and wood-pasture sites in the landscape. Many of the species dependent on old trees are unable to move between these sites due to their poor powers of dispersal and the increasing distances they need to travel, as well as the loss of pasture through conversion to arable and other land uses; inappropriate grazing levels: under-grazing leading to loss of habitat structure through bracken and scrub invasion; and over-grazing leading to bark browsing, soil compaction and loss of nectar plants.

Cessation of traditional management such as pollarding continues to influence the condition of the habitat, although our understanding of the requirements of veteran trees has increased and site managers may undertake operations such as haloing to release hulks enclosed by plantation trees, and re-pollarding veterans that may not have been worked for 90 plus years. Some research has been undertaken and written up for key sites across the country, but the lessons learned from such activities are often not recorded in a format that can be shared with the managers of veteran trees on other sites. The loss of grazing from sites has resulted in the growth of scrub around many veterans, which may also result in the loss of important lichen assemblages that rely on the light, open conditions provided by wood-pasture. Damage to trees and roots from soil compaction and erosion is also an issue.

Changes in the countryside and its management have led to isolation and fragmentation of the remaining sites continues to be an issue, although the process of addressing this is starting byimplementing the recommendations in Defra's Natural Environment White Paper published in 2011 (Defra, 2011b)

In recent years, the impact of new diseases has increased. Tree diseases such as acute oak decline are now present throughout the English countryside. We do not yet fully understand this disease or its potential impacts (Brasier, 2008; Kirby et al., 2010; Rackham, 2008). It is known to infect and in some cases kill oak trees, and is present on important sites in England. Oak processionary moth at Richmond Park is also having an impact on the health of the trees in their potential survival. Detail about these pests and diseases can be found at www.forestry.gov.uk/pestsanddiseases

Climate change may also pose a threat to the survival of this habitat by threatening the future of the veteran trees it supports (Hopkins et al., 2007; Smithers et al., 2008). Most veteran trees, by definition, have survived a range of extreme weather events, for example the Major Oak in Sherwood Forest, the Bowthorpe Oak in Northamptonshire, the many oak trees in Windsor Park. In extreme years there is a gradual attrition of old trees and there is no guarantee they will tolerate future changes. Mitchell et al. (2007) report that the risk of direct impact from climate change on wood-pasture and parkland is low but the strength of evidence is poor. The habitat is essentially determined by its management history, but climate change, particularly drought, is potentially a threat to the old veteran trees. The persistence of the habitat will depend on management decisions such as replanting and choice of species. Surveys undertaken at Moccas during the drought summers of 1989 and 1990, and following the high winds of January and February 1990, reveal that in total ten trees were lost from the park. Damage to a further 28 trees was recorded (Harding and Wall, 1999). This highlights the importance of developing new generations of trees and on many sites 'replacement' trees are being encouraged through planting or natural regeneration.

The levels of replacement needed depend on the degree of generation gap that is considered acceptable for long-term sustainability of populations of invertebrates and lichens living on or in the trees; and what levels of tree survival are assumed across time. Figure 25.2 illustrates the numbers of trees per hectare needed to ensure continuity of one veteran per hectare.

If, for example (Figure 25.2a), the desired gap between age classes is no more than 50 years and only three out of four trees (75 per cent) make the transition then the population would need to have 7.5 x 50-year old trees, 5.5 x 100-year old trees etc. to support one veteran per hectare. This gives a total of 27 trees per hectare across all age classes. If an age gap of 150 years, is acceptable, but only half the trees survive from one age class to the next, then 8 x 50-year old trees per hectare are needed to give the population structure shown in Figure 25.2b (15 trees per hectare across all age classes). These equate to the average death rates across all age classes of about one tree per decade per hectare (Figure 25.2a), or one tree per two decades per hectare (Figure 25.2b). These may therefore be conservative and so higher rates of replacement need to be considered.

Wood-pasture and parkland in a wider social and landscape context

Wood-pasture and parkland are poorly documented in the recent National Ecosystem Assessment (UKNEA, 2011) because of their mixed nature. However, they do have the potential to contribute to a range of different goods and services, in particular those that refer to cultural values. There may be scope for making analogies with modern agroforestry. Montagnini and Nair (2004) suggest that by growing trees and crops on the same piece of land, the value of that piece of land for carbon sequestration is increased when compared with each habitat alone.

The aesthetic appeal of wood-pasture has long been recognised. The Reverend Francis Kilvert visited Moccas Park in 1876 when he described the trees in the park as 'those grey, gnarled, low-browed, knock kneed, bowed, bent, huge, strange, long-armed, deformed, hunchbacked, misshapen oak men that stand waiting and watching century after century'. Moccas became the first parkland national nature reserve in England in 1978. Before this, ecologists did not regard parkland as a form of woodland, and parks were not considered worthy of conservation effort – Moccas was dismissed as woodland. In the 1960s, key entomologists and conservation agency staff demonstrated 'the under-estimation both of the degree to which the supposed "oak forest" (Moccas Park) was as much a product of cultural as of natural processes, and of the importance of parkland for nature conservation' (Harding and Wall, 1999). Today Moccas Park is recognised as one of the most important sites for saproxylic invertebrates in England.

Discussion

The habitat and its importance are now better recognised through both conservation and forestry sectors, and increasingly with the general public. In England, significant progress has been made in recent years in understanding how much of the habitat exists and where it is located, and with such awareness comes improved condition. The Corporation of London has done much to contribute to our understanding of the management of sites with veteran pollards (Read, 2000), and others have reported on the impact of haloing of veterans (Alexander *et al.*, 2010).

Agri-environment schemes in most countries of the UK recognise the value of this habitat and contain options to allow improved management to be financially supported. In England, for example, large parkland sites such as Grimsthorpe Park in Lincolnshire and Calke Abbey in Derbyshire

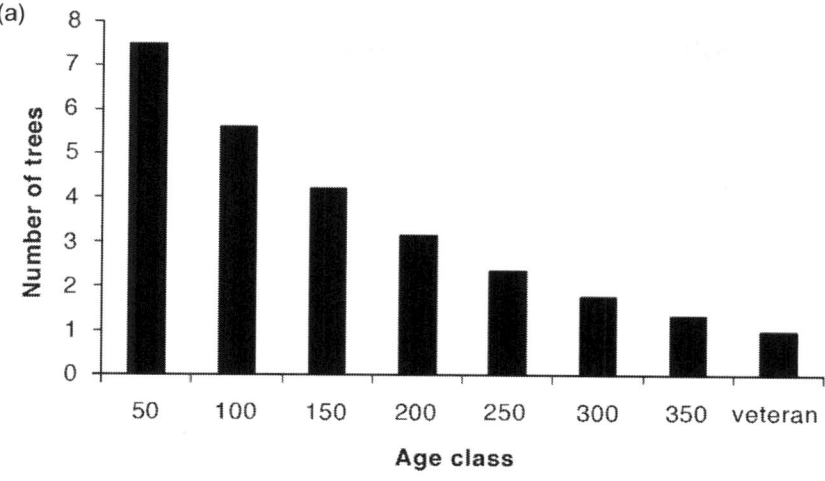
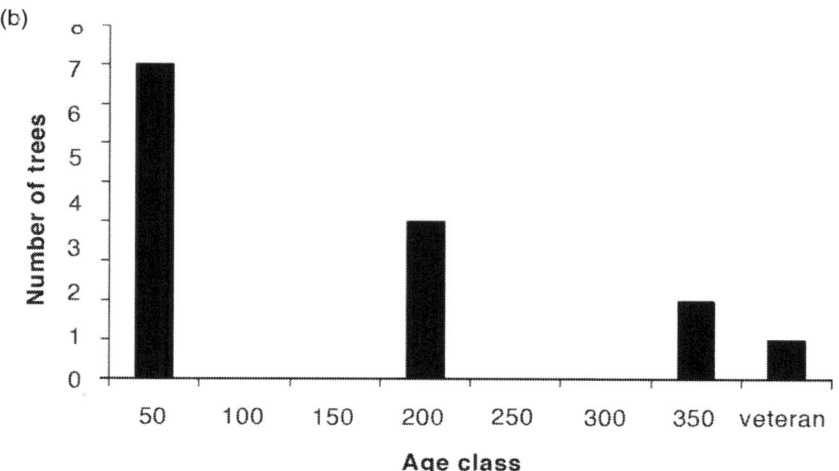

Figure 25.2 Number of trees per hectare in different age classes needed to sustain one veteran tree per hectare
Note: (a) Age class gap set at 50 years, with probability of 75 per cent survival between age classes. (b) Age class gap set at 150 years, with probability of 50 per cent survival between age classes.
Source: Suzanne Perry

are benefiting from the Higher Level Stewardship Scheme, and the New Forest is now the largest agri-environment agreement in Europe, by area. Funding has been committed for the next ten years to help safeguard the traditional grazing methods used in the New Forest. Table 25.4 indicates the number of agreements supporting wood-pasture and parkland since 2005.

Table 25.4 Higher Level Stewardship Scheme agreements supporting wood-pasture and parkland options between 2005 and 2012

Higher Level Stewardship options	Number of agreements containing selected options	Option area (ha)
HC12 Maintenance of wood-pasture and parkland	45	1,452
HC13 Restoration of wood-pasture and parkland	43	7,362
HC14 Creation of wood-pasture	13	233
Total	87	9,047

Source: Natural England Genrep accessed May 2012

Experience of managing sites and the trees within them is much improved with work by organisations such as the National Trust at Hatfield Forest and the City of London at Epping Forest and Burnham Beeches put into practice. Outreach work is constantly increasing understanding (Nunez *et al.*, 2012; Read, 2000). These studies are valuable in promoting the importance of management to improve site condition and more of these sorts of reports need to be written up and made available to other site managers.

At the same time as the value of the habitat is being recognised, the knowledge and skills required to maintain it are disappearing. For example, across Europe extensive pastoralism is in decline, which has resulted in the loss of open habitat, and this is particularly reflected in the UK. However, it is extremely difficult to quantify this when the habitat is often not recognised. This has improved since the advent of BAP and the Veteran Tree Initiative and subsequently the Ancient Tree Forum.

Looking to the future, three main themes can be identified where trees exist in the landscape outside woodlands but where they are managed to make a significant contribution:

1 Grazing to give diverse landscapes – wider use of grazing with trees as an element, for example at Ennerdale where trees are not necessarily veterans and management is not undertaken for the benefit of the trees, or Knepp Castle where a parkland landscape with parkland trees is being deliberately created;
2 Managing veteran trees and their replacements (which could be hedgerow or in-field trees) – cultural preservation of veteran trees is important but we have lost the traditional management techniques such as pollarding, which are part of that historic value;
3 New productive uses of agroforestry – where trees are part of a productive future landscape.

Veteran trees exist in many situations across the UK. Wood-pasture and parkland priority habitat sits at the centre of them all, but not all these scenarios contain veteran trees or the habitat as defined by the HAP, and not all will be important for their survival. Only by identifying the sites of value and putting effort into their long-term survival will the future of the habitat be assured.

Bibliography and references

Alexander, K.N.A. (1999) 'The invertebrates of Britain's wood pastures', *British Wildlife*, 11(2): 108–117.

Alexander, K.N.A., Smith, M., Stiven, R. and Sanderson, N. (2002) *Defining 'Old Growth' in the UK Context. A Report Prepared by the Ancient Tree Forum*, English Nature Research Report, No. 494, Peterborough.

Alexander, K.N.A. (2002a) 'The noble chafer *Gnorimus nobilis* in Gloucestershire – a report on the 2002 survey', unpublished report for the Peoples' Trust for Endangered Species, London.

Alexander, K.N.A. (2002b) *The Invertebrates of Living and Decaying Timber in Britain and Ireland: A Provisional Annotated Checklist*, English Nature Research Report, No. 467, English Nature, Peterborough.

Alexander, K.N.A. (2005) 'Wood decay, insects, palaeoecology, and woodland conservation policy and practice – breaking the halter', *Antennae*, 29: 171–8.

Alexander, K.N.A. (2011) *A Review of the National Importance and Current Condition of the Saproxylic Invertebrate Assemblages at Birklands and Bilhaugh Sites of Special Scientific Interest (SSSIs), Sherwood Forest, Nottinghamshire*, Natural England Commissioned Report, NECR072, Natural England, Sheffield.

Alexander, K.N.A. and Lister, J.A. (2003) *Thames and Chilterns: Parkland and Wood Pastures with Veteran Trees Phase I – A Provisional Inventory 2002/03*, English Nature Research Report No. 520, English Nature, Peterborough.

Alexander, K.N.A, Hope, J.C.E., Lucas, A., Smith, J.P. and Wright, MA. (2007) *Wood Pasture and Parkland Scoping Study*, Environment and Heritage Service Research and Development Series. No. 08/01, Northern Ireland Environment and Heritage Service, Belfast.

Alexander, K.N.A., Stickler, D. and Green, T. (2010) 'Is the practice of haloing successful in promoting extended life? A preliminary investigation of the response of veteran oak and beech trees to increased light levels in Windsor Forest', *Quarterly Journal of Forestry*, 104(4): 257–66.

Beckman, M. (2006) *Grova lövträd på Kinnekulle. Resultaten från trädkartering 2002-2004. Länsstyrelsen i Västra Götalands län*, (Large deciduous trees at Kinnekulle. The results of a mapping exercise 2002–2004), Rapport 2006: 94, Lansstyrelsen, Vastra Gotalands Lan.

Bergmeier, E., Petermann J. and Schroder, E. (2010) 'Geobotanical survey of woodpasture habitats in Europe: Diversity, threats and conservation', *Biodiversity Conservation*, 19: 2995–3014.

BLS (British Lichen Society) (1982) *Survey and Assessment of Epiphytic Lichen Habitats*, Commissioned Research Report, Nature Conservancy Council, Peterborough.

Brasier, C. (2008) 'The biosecurity threat to the UK and global environment from international trade in plants', *Plant Pathology*, 57(5): 792–808.

British Lichen Society (1993) *Revised assessment of epiphytic lichen habitats 1993*, Joint Nature Conservancy Council (JNCC), Report No. 170, Peterborough.

Brown, A.E., Burns, A.J., Hopkins, J.J. and Way S.F. (1997) *The Habitats Directive: Selection of Special Areas of Conservation in the UK*, Joint Nature Conservation Committee, Report No. 270, Peterborough.

Castle, G. and Mileto, R. (2005) *Development of a Veteran Tree Site Assessment Protocol*, English Nature Research Report No. 628, English Nature, Peterborough.

Chatters, C. and Sanderson, N.A. (1994) 'Grazing lowland pasture woodlands', *British Wildlife*, 6(2): 78–88.

Claesson, K. and Ek, T. (2008) *Skyddsvärda träd i Östergötland – 1997–2008*, (Trees of Nature Conservation Value in Östergötland – 1997–2008), Länsstyrelsen i Östragötaland. Rapport 2008, No. 13, Länsstyrelsen i Östragötaland, Linkoping.

Clifton, S.J. (2000) *The Veteran Trees of Birklands and Bilhaugh, Sherwood Forest, Nottinghamshire*, English Nature Research Report No. 361, English Nature, Peterborough.

Coppins, A.M. and Coppins, B.J. (2002) *Indices of Ecological Continuity for Woodland Epiphytic Lichen Habitats in the British Isles*, British Lichen Society, London.

Defra (Department for Environment, Food and Rural Affairs) (1994) *Biodiversity: The UK Action Plan*, Defra, London.

Defra (on behalf of the UK Biodiversity partnership) (2007) *Conserving Biodiversity – the UK Approach*, Defra, London.

Defra (2010) *Trends, Long Term Survival and Ecological Values of Hedgerow Trees: Development of Populations Models to Inform Strategy*, Defra, London.

Defra (2011a) *Biodiversity 2020: Strategies for England's Wildlife and Ecosystem Services*, Defra, London.

Defra (2011b) *The Natural Choice: Securing the Value of Nature*, Natural Environment White Paper, Defra, Defra, London.

Forestry Commission (2002) *Life in the Deadwood: A guide to Managing Deadwood in Forestry Commission Forests*, Forestry Commission, Edinburgh.

Gibson, C. (1997) *Reintroducing Stock Grazing to Savernake Forest: A Feasibility Study*, English Nature Research Report No. 224, English Nature, Peterborough.

Hammond, P.M. and Harding, P.T. (1991) 'Saproxylic invertebrate assemblages in British woodland. Their conservation significance and its evaluation', in H.J. Read (ed.) *Pollard and Veteran Tree Management*, Corporation of London, London, pp30–37.

Harding, P.T. (1978) *An Inventory of Areas of Conservation Value for the Invertebrate Fauna of the Mature Timber Habitat*, CST Report No. 160. Nature Conservancy Council, Banbury.

Harding, P.T. and Rose, F. (1986) *Pasture-Woodlands in Lowland Britain – A Review of their Importance for Wildlife Conservation*, Institute of Terrestrial Ecology, Monks Wood Experimental Station, Huntingdon.

Harding, P.T. and Wall, T. (eds) (1999) *Moccas: An English Deer Park: The History, Wildlife and Management of the First Parkland National Nature Reserve*, English Nature/Veteran Tree Initiative, English Nature, Peterborough.

Harvey, P., Morris, K., Hacking, R. and Clifton, S. (2004) *A Provisional Inventory of Parkland and Wood-pasture in the East Midlands*, English Nature Research Report No. 595, English Nature, Peterborough.

Hodder, K.H., Bullock, J.M., Buckland P.C. and Kirby K.J. (2005) *Large herbivores in the Wildwood and Modern Naturalistic Grazing Systems*, English Nature Research Report No. 648, English Nature, Peterborough.

Hodder, K.H. and Bullock, J.M. (2009) 'Really wild? Naturalistic grazing in modern landscapes', *British Wildlife*, 20(5): 37–43.

Hopkins J.J., Allison, H.M., Walmsley, C.A., Gaywood, M. and Thurgate, G. (2007) *Conserving Biodiversity in a Changing Climate: Guidance on Building Capacity to Adapt*, Defra on behalf of the Biodiversity Partnership, Defra, London.

Jonsson, B.G., Kruys, N. and Ranius, T. (2005) 'Ecology of species living on dead wood – lessons for dead wood management', *Silva Fennica*, 39(2): 289–309.

Kirby, K.J. and Drake, C.M. (eds) (1993) *Deadwood Matters: The Ecology and Cconservation of Saproxylic Invertebrates in Britain*, English Nature Science Series No. 7, Proceedings of a British Ecological Society meeting held at Dunham Massey Park, English Nature, Peterborough.

Kirby, K.J., Thomas, R.C., Key, R.S., Mclean, I.F.G. and Hodgetts, N. (1995) 'Pasture woodland and its conservation in Britain', *Biological Journal of the Linnean Society*, 56 (Suppl.): 135–53.

Kirby, K.J., Reid, C.M., Thomas, R.C. and Goldsmith, F.B. (1998) 'Preliminary estimates of fallen dead wood and standing dead trees in managed and unmanaged forests in Britain', *Journal of Applied Ecology*, 35: 148–55.

Kirby, K.J., Perry, S.C. and Brodie-James, T. (2010) 'Possible implications of new tree diseases for nature conservation', *Quarterly Journal of Forestry*, 104(4): 277–84.

Kirby, P. (2001) *Habitat Management for Invertebrates: A Practical Handbook*, RSPB, Sandy, Bedfordshire.

Lovelace, D. (2010) 'Native woodland restoration in Herefordshire project summary report 2006–2009: Ancient Woodlands and Trees of Herefordshire our heritage revealed', January, River Wye Preservation Trust,

Lush, M., Robertson, H.J., Alexander, K.N., Giavarini, V., Hewins, E., Mellings, J., Stevenson, C.R., Storey, M. and Whitehead, P.F. (2009) *Value of Orchards: Biodiversity Studies of Six Traditional Orchards in England*, Natural England Research Report No. 025 Natural England, Sheffield

McLeod, C.R., Yeo, M., Brown, A.E., Burn, A.J., Hopkins, J.J. and Way, S.F. (eds) (2005) *The Habitats Directive: Selection of Special Areas of Conservation in the UK*, 2nd edition, Joint Nature Conservancy Committee, Peterborough.

Mitchell, P.L. (1989) 'Repollarding large neglected pollards: A review of current practice and results', *Arboricultural Journal*, 13: 125–42.

Mitchell, P.L., Morecroft, M.D., Acreman, M., Crick, H.Q.P., Frost, M., Harley, M., Maclean, I.M.D., Mountford, O., Pip, J., Pontier, H., Rehfisch, M.M., Ross, L.C., Smithers, R.J., Stott, A., Walmsley, C., Watts, O. and Wilson, E. (2007) *England Biodiversity Strategy – Towards Adaptation to Climate Change*, Defra, London, http://nora.nerc.ac.uk/915/1/Mitchelletalebs-climate-change.pdf

Montagnini, F. and Nair, P.K.R. (2004) 'Carbon sequestration: An underexploited environmental benefit of agroforestry systems', *Agroforestry Systems*, 61: 281–95.

Natural England (2008) *State of the Natural Environment*, Natural England, Sheffield.

Natural England (2010) *Illustrated Guide to Trees, Woodlands and Scrub*, Technical Information Note No. 078, Natural England, Sheffield.

Naturvårdsverket (2004) *Åtgärdsprogram för skyddsvärda träd i odlingslandskapet. Naturvårdsverket*, (Action Plan for Trees of Nature Conservation Value in the Rural Landscape), Swedish Environmental Protection Agency, Stockholm.

NCC (Nature Conservancy Council) (1989) *Guidelines for the Selection of biological SSSIs; Rationale: Operational Approach and Criteria. Detailed Guidelines for Habitats and Species-groups*, Nature Conservancy Council, Peterborough.

Neroche/GAP (2005) 'Woodland grazing workshop report, 08.11.05', Grazing Animals Project, www.grazinganimalsproject.org.uk

Nunez, V., Hernando, A., Velazquez, J. and Tejera, R. (2012) 'Livestock management in Natura 2000: A case study in a *Quercus pyrenaica* neglected coppice forest', *Journal for Nature Conservation*, 20(1): 1–9.

Peterken, G.F. (1992) 'Conservation of old growth: A European perspective', *Natural Areas Journal*, 12: 10–19.

Phillips, A. (1998) 'The nature of cultural landscapes – a nature conservation perspective', *Landscape Research*, 23: 21–38.

Quelch, P.R. (2001) *Ancient Wood-pasture in Scotland: An illustrated Guide to Ancient Wood Pasture in Scotland*, Forestry Commission Scotland, Edinburgh.

Rackham, O. (1980) *Ancient Woodland: Its History, Vegetation and Uses in England*, Edward Arnold, London.

Rackham, O. (1998) 'Savanna in Europe', in K.J. Kirby and C. Watkins (eds) *The Ecological History of European Forests*, CABI Publishing, Wallingford, Oxon, pp1–24.

Rackham, O. (2008) 'Ancient woodland: modern threats', *New Phytologist*, 180(3): 571–86.

Rackham, O. (2010) *Woodlands*, Collins, London.

Ranius, T. and Jansson N. (2000) 'The influence of forest regrowth, original canopy cover and tree size on saproxylic beetles associated with old oaks', *Biological Conservation*, 95: 85–94.

Rayner, A.D.M. (1993) 'The fundamental importance of fungi in woodlands', *British Wildlife*, 4: 205–15.

Read, H. (ed.) (1996) *Pollarded and Veteran Tree Management II*, incorporating the proceedings of the meeting hosted by the Corporation of London at Epping Forest in 1993, Richmond Publishing/Corporation of London, London.

Read, H. (ed.) (2000) *Veteran Trees: A Guide to Good Management*, Veteran Trees Initiative/English Nature, English Nature, Peterborough.

Shigo, A.L. (1999) *Modern Arboriculture: A Systems Approach to the Care of Trees and their Associates*, Shigo and Trees, USA.

Sibbett, N. (1999) *Ancient Tree Survey of Staverton Park and the Thicks SSSI, Suffolk*, English Nature Research Report No. 334, English Nature, Peterborough.

Smith, K.W. (2007) 'The utilisation of dead wood resources by woodpeckers in Britain', *Ibis*, 149, Suppl. 2: 183–92.

Smith, M.N. (2011) *Great Stag Hunt III: National Stag Beetle Survey 2006–2007*, Peoples' Trust for Endangered Species, London.

Smith, M. and Bunce, R.G.H. (2004) 'Veteran trees in the landscape: A methodology for assessing landscape features with special reference to two ancient landscapes', in *Landscape Ecology of Trees and Forests*, Proceedings of the 12th annual IALE (UK) Conference, pp168–75.

Smithers, R.J., Cowan. C., Harley, M., Hopkins, J.J., Pontier, H. and Watts, O. (2008) *England Biodiversity Strategy: Climate Change Adaptation Principles: Conserving Biodiversity in a Changing Climate*, Defra, London.

Stiven, R. and Holl, K. (2004) *Wood Pasture*, Booklet in the Natural Heritage Management Series, Scottish Natural Heritage, Battleby.

UK Biodiversity Group (1998) *Tranche 2 Action Plans Volume II – Terrestrial and Freshwater Habitats*, English Nature, Peterborough.

UKNEA (UK National Ecosystem Assessment) (2011) *The UK National Ecosystem Assessment Technical Report*, UKNEP-WCMC, Cambridge.

Vandekerkhove, K., De Keersmaeker, L., Walleyn, R., Kohler, F., Crevecoeur, L., Govaere, L., Thomaes, A. and Verheyen, K. (2011) 'Reappearance of old-growth elements in lowland woodland in northern Belgium: Do the associated species follow?', *Silva Fennica*, 45(5): 909–35.

Vera, F.W.M. (2000) *Grazing Ecology and Forest History*, CABI Publishing, Wallingford, Oxon.

Webb, J.R. (2002) *The Invertebrates of Living and Decaying Timber in Britain and Ireland. A Provisional Annotated Checklist*, English Nature Research Report No. 467, English Nature, Peterborough.

Webb, J.R. (2004) *Revision of the Index of Ecological Continuity as Used for Saproxylic Beetles*, English Nature Research Report No. 574, English Nature, Peterborough.

Webb, J.R. and Bowler, J. (2001) *County Surveys of Parkland: The Staffordshire Experience*, English Nature Research Report No. 416, English Nature, Peterborough.

Webb, J.R., Drewitt, A.L. and Measure, G.H. (2010) *Managing for Species: Integrating the Needs of England's Priority Species into Habitat Management. Part 1 Report*, Natural England Research Report No. 024, Natural England, Sheffield.

Woodland Trust (2012) 'Ancient tree hunt website', www.ancient-tree-hunt.org.uk/, accessed May 2012.

Part VI
Summary and conclusions

26 Re-wilding trees for ancients of the future

Jill Butler and Keith Alexander

Introduction

Ancient and veteran trees are rich reservoirs of biodiversity and cultural icons, and in the UK we are lucky that there are so many in our landscape. However, the disappearance of commons, orchards and parkland since the 1850s and the removal of hedgerows have led to significant losses of open-grown trees – both ancient trees and those mature trees that are the next in line to age into the next generation.

Trees are among the most ubiquitous of plants; they have the potential to grow anywhere other than the wettest and highest places. Although they are our constant companions, by being so familiar and 'everyday' they are constantly overlooked. Trees are only recognised in habitats when they are clustered together as woods and then largely for their timber value. In other habitats they and their values seem to 'disappear' – they are just part of the backcloth of our lives until one day they suddenly disappear and too late we realise what we have lost. As individual trees in the wider landscape they are often considered a nuisance rather than respected for all the good work that they do and because they are not rare species are underestimated for their biodiversity value.

Working closely with ancient trees, the Ancient Tree Forum has become increasingly aware of a widespread lack of new generations of trees that will form the ancient trees of the future – mature trees are under threat across the modern human landscape and young trees are not establishing in sufficient quantity. Our current tree populations need to maintain a balance between mortality and recruitment, and if we wish to enhance our 'treescape' then we need to give considerable attention to enabling further recruitment of the right tree in the right place. In particular the establishment of trees that have the potential to form full crowns in open situations so that they can then 'grow downwards' as the crown retrenches into old age.

Furthermore, many of our oldest trees started life in very unplanned ways – perhaps the least influenced by man in Western Europe – but areas where self-sown, 'wild' trees can regenerate naturally in open wood-pasture or parkland are now extremely rare. There are advantages to such natural processes, and efforts should be aimed at promoting them where possible so we can learn more about how to naturally create open-crowned trees in the landscape. Sometimes however, it may also be necessary to plant trees. While there is plenty of existing guidance on planting to create or restock woods, plantations, orchards and hedgerows, and on planting trees in gardens and streets, there are some factors to take into account if those trees are to become the ancients of the future. In most cases, the places where wild trees established are where the landscape has been kept open by grazing animals. These are the places that in the past have been celebrated by owners and landscape designers, artists and writers as some of the most beautiful landscapes in the UK and are the most important today for species associated with ancient and veteran trees in old growth (Figure 26.1).

Figure 26.1 Ancient parkland tree at Windsor Great Park
Source: Jill Butler and Keith Alexander

Wild spaces needed for wild trees

Trees have been reproducing themselves naturally for millennia; however the landscape is now subject to so many competing interests that trees and shrubs are usually only allowed to grow exactly where we want them. In some places, people value the absence of trees so much – in calcareous grasslands and heathlands for example – that even scattered trees and shrubs are no longer seen as desirable and considerable resources are invested in clearing 'scrub'.

We all love trees and woodlands, and invest considerable resources in planting them across new road verges or as new farm woodlands, for example. So why is a planted tree better than a naturally grown one? A planted tree can be placed just where we want it and those that grow 'wild' are often seen as 'neglected' and 'untidy'. If we love large old trees, with full spreading canopies, then why is most tree planting carried out in close blocks where the trees will grow up tall and spindly, overcrowded? Ancient trees can only develop where they have the space to develop, without competition from neighbouring trees. Wood-pastures and parkland where wild regeneration of trees and shrubs can take place are now rare and very special and there is still a great deal to learn about the process.

The challenges for self-sown trees – hitting the spot

The seeds of most temperate trees have specialist mechanisms that allow them to disperse naturally either by wind – the seed cases are light and have wings or other structures to help give them lift so they fly a long way, or by birds or other animals – natural selection has enhanced their attractiveness to animals for example by making them a protein-rich source of food. Where seeds escape being eaten they also benefit from having a rich energy source for early development.

Natural selection

It is clear that natural regeneration in Britain is not truly 'natural' as human activities have considerable influences, and have done for millennia. There are many benefits from enabling natural regeneration and these should ideally be considered in all plans for tree establishment. Unlike planting, which tends to be carried out in uniform blocks of even-aged and regularly spaced trees according to forestry guidelines, wild regeneration results in a variety of density of tree establishment and leads to a structurally rich mosaic of great diversity and therefore value to wildlife.

When wild regeneration occurs, the tree 'selects' its own spot. The seed that lands in an unsuitable place falls by the wayside but any seedling that does take root starts its life without the stresses of being grown in modern nurseries and the shock of relocation.

Rooted to the spot

Seedling development may proceed under a wide variety of situations provided soil conditions are suitable. The root from the seed needs to be able to explore the soil below where it was deposited and develop in response to the local conditions. If the soil is in good condition and contains an undisturbed and undamaged microbiology then the root may immediately be colonised by mycorrhizal fungi, which help with water and mineral gathering, as well as protecting it from pathogens. Immediate attention from mycorrhizal fungi will increase the chances of the seedling's survival and eventual growth of a healthy tree.

Figure 26.2 Trees should be grown from seed in organic soils to benefit their essential mycorrhizal fungi
Source: Jill Butler and Keith Alexander

Light and shade

In open environments such as grasslands, both shade-tolerant and light-demanding trees can establish but as shade-tolerant trees may be more vulnerable to drought they may not be able to survive so well in the open where moisture levels may fluctuate more widely. In woods with continuous canopies, only shade-tolerant trees such as beech will develop and persist under the canopies of other trees to await an opportunity for a gap to occur in the canopy so they can reach maturity. Some more light-demanding trees may hang on in closed-canopy conditions but are unlikely to flower and therefore seed to provide future generations.

Light-demanding seedlings cannot survive under the crown of the parent tree or in shady conditions so the seeds of these trees are dispersed further afield. Scots pine, oak, hawthorn and juniper are examples of very light-demanding species needing open spaces in which to establish and develop.

Table 26.1 Comparison of light requirements of common tree species and the height of the tallest specimen in UK

Tree	Shade tolerance (approx. %)	Height (m)
Scots pine	33	40
Common hawthorn	39	15
Silver birch	41	30
Midland hawthorn	49	15
Pedunculate oak	49	38
Ash	53	30
Alder	54	28
Rowan	55	25
Sessile oak	55	42
Sweet chestnut	63	36
Field maple	64	25
Sycamore	75	38
Holly	77	23
Hornbeam	79	30
Small-leaved lime	84	42
Yew	89	25
Beech	91	40

Note: The ranking by shade tolerance is derived from a desk study (Niinemets and Valladares, 2006). They analysed the rankings provided in a wide body of literature. They applied some statistical rigour to the ranking of the species but this does not mean that that ranking holds true across all geographical and genetics range of variation. In particular, yew in the UK is regarded as more shade tolerant than beech, despite their reversed ranking in the table.

Phoenix regeneration

Many trees are able to regenerate vegetatively by suckering or layering. If collapsed trees are allowed to remain in situ, intact roots may survive and the tree may continue to grow from a horizontal position. In this way many trees appear to have 'walked' across the landscape. Layering branches will act as flying buttresses propping up aging trunks and making them more stable.

Natural scrub tree shelters

The thorn is the mother of the oak.

(Humphry Repton, 1803)

Browsing by livestock affects the establishment, species composition and development of trees and shrubs. However, it is essential to have sufficient browsing by wild animals and domesticated stock to keep landscapes open enough for light-demanding trees and shrubs to thrive – but not too much browsing or from the wrong animals so that seedlings are all eaten and their leading shoots cannot grow on out of reach of animals.

The palatability of leaves to animals influences tree establishment. One reason why most of our deer parks are dominated by oak is that it is much less palatable to deer than ash, for example, and so oak seedlings are more likely to survive. However, many of our cultural landscapes are rich in

Figure 26.3 Browsing animals in wood-pasture
Source: Jill Butler and Keith Alexander

veteran and ancient trees managed as pollards such as ash pollards in the Cotswolds and the Lake District, specifically because their branches were cut for fodder in summer droughts or kept as tree hay for feeding to the animals in winter.

Young trees often survive better if they develop within thorny scrub, for example blackthorn or bramble, which is less affected by browsing – jays tend to bury their acorn hoards into the edge of thorn clumps. At Hatfield Forest in Essex there was a tradition of gathering acorns and tossing them into thorn scrub to promote the development of new wild trees. Wild regeneration will also occur if other protection is available such as fallen branches or even unpalatable plants – either with chemical (e.g. hound's tongue) or structural (e.g. thistle) defences.

Figure 26.4 Ash growing in the protection of bramble
Source: Jill Butler and Keith Alexander

Grazing by too many browsing animals or by those that can break through thorny defences may mean trees are unable to compete. Goats and sheep from southern and eastern Europe are adapted to browsing leaves through the thorny scrub and are specifically chosen as a 'tool' controlling scrub development in herb-rich grasslands and heathland. Exotic muntjac deer from Asia appear to be able to eat plants that are otherwise unpalatable to native wild animals and rare breeds.

Bearing in mind the crown size of mature trees and the need to have a continuous age structure of young trees through to ancient trees, it is preferable to have too little regeneration than too much, especially of very long-lived light-demanding trees. It is difficult to remove trees once they are well established and too many of the same age at one time may lead to closed canopy woodland. It may be easier to think in terms of how many trees need to be planted and how frequently and then compare this with what is happening in the wild situation.

The key elements are open-grown trees, open grass, heath or moorland with wild flowers, scrub and some groups of trees and occasional establishment of trees to provide succession within the area as a whole. There are no recipes. Some suggestions are provided but we would like to hear from people who have successfully developed dynamic systems.

Key factors in 'wild' tree regeneration

Key factors appear to be:

- Identify places where tree establishment can occur through wild processes.
- Grazing with large herbivores in type and quantity that allow sufficient regeneration but not too much.
- Grazing needs to limit regeneration so that some individual trees can grow to maturity and into old age without their crowns touching or being outcompeted by other neighbouring trees.
- Cattle, Exmoor or other rare breed horses and deer are probably the best browsers to use.
- Pigs may be an option in the right location or used during the autumn when there is sufficient mast for them.
- Between 25 and 75 per cent open ground is needed with the rest a mix of single open-crowned trees and a few groups of trees.
- If there is already too much young-tree establishment, young trees may be coppiced or pollarded to reduce canopy cover.
- If there is insufficient scrub, more can be planted or cut thorny scrub with mature fruit, for example bramble, juniper, hawthorn and blackthorn, can be introduced into areas to kick start scrub regeneration so as to provide a nursery for wild seedling protection.
- Wild regeneration needs to be monitored carefully and stocking densities changed where necessary to ensure the right level of regeneration.
- Protect wild saplings that have regenerated if it is difficult to manage browsing stock and there is no regeneration happening.
- Encourage more fallen deadwood to act as a shelter for wild regeneration.
- 'Phoenix' regeneration of trees should be valued and encouraged.

Why plant trees?

Tree planting may be necessary where wild regeneration is too unpredictable and may be too slow. For example where:
- a particular tree species is required for a particular purpose and local natural seed sources are unavailable;
- current age structures are notably poor and new establishments are urgently required;
- important amenity tree features need to be maintained;
- promoting the interest and involvement of people and especially children is important.

The right tree in the right place

Even when planting, the light-demanding or shade-tolerant characteristics of the trees to be planted have to be born in mind. Planting blocks of even aged and regularly spaced trees creates structural uniformity that has its limitations in its value to wildlife and will disadvantage light-demanding species of tree. Shade-tolerant trees will outcompete light-demanding saplings and quickly reduce the variety of trees within the first few decades. When planting more formal groups of trees such as roundels or avenues, the choice of species and spacing of trees is therefore very important.

Trees and shrubs grow best when root growth is good and competition for moisture and nutrients from other plants is low. Fertilisers negatively affect the growth of mycorrizhal fungi and should not be used.

Although native trees may have advantages, some non-native trees, especially of European origin, may be useful in reducing the age gap in tree populations because they grow more quickly. Sweet chestnut for example has heartwood similar to oak.

Too close for comfort

Scattered, open-crowned trees have been shown to be keystone structures in a wide range of landscapes and are objects of great beauty. Trees intended as future replacements must not be planted so close to the older tree that the two trees will be competing with each other for light. The maintenance of a tree-free zone around them is important or it will lead to the older tree being damaged by shading from the more vigorous younger tree as it matures and the shape of the younger tree being affected as it grows.

The minimum amount of space required by a tree or shrub to grow to its full crown potential – especially lower side branches – depends on the species (and genetics) and also on its situation, for example soil characteristics and landscape. Many good tree guides will provide information on the height that trees will achieve. As a minimum, trees should have a space that has a diameter equalling that height.

The temptation to plant too many trees at one time

It is often said that denser plantings allow for the predictable occasional death of a few trees, but such failures are unpredictable and the result generally is an overcrowded stand with a few small gaps. Early thinning to compensate is also often recommended, but rarely actually takes place. It is better to plant at the desired end density and to replace failures as and when they occur. The biggest problem is with the financial drivers of such projects that normally demand large inputs of activity within a single financial year. This is actually an argument for the final-density planting approach as this will produce a closer approximation to the desired outcome than overcrowded plantations and is much less demanding in ongoing maintenance.

Allow for additional planting in decades and centuries to come so that a wide age structure is developed. As oaks can live for 900 years there has to be plenty of space in which future generations can be established.

'Planting' dead trees

Many of the landscape designers of the English romantic garden period in the mid-eighteenth century, such as William Kent, introduced dead trees in their landscapes to help create an immediate 'natural' impact and mood.

Key factors when planting open grown trees

- Trees require sufficient space to grow to full crown potential. With oak, for example, lower-branch growth can extend as much as 15m from the trunk, maybe more in some situations.
- Planting needs to take account of the longevity of trees – more space is needed for trees that may live for many centuries.
- For smaller trees, for example fruit trees, the correct minimum density has evolved from the fruit production industry, enabling optimal flowering and fruiting of trees. In a mature orchard, trees are at about a 10m spacing.
- Use thorny shrubs such as hawthorn to protect young trees and especially to provide long-term trunk protection. This was a common practice used by landscape gardeners such as Humphry Repton and 'Capability' Brown.
- A more rapid crown effect can be achieved by planting a group of trees very close together to form a 'bundle'.
- European non-natives might provide veteran characteristics more quickly than native trees for wildlife benefits where it is necessary to ensure continuity of the ancient tree habitat.
- Dead standing trees may be 'planted' to enhance wildlife habitat and aesthetic impact.

Bibliography and references

Manning, A.D., Fischer, J. and Lindenmayer, D.B. (2006) 'Scattered trees are keystone structures – implications for conservation', *Biological Conservation*, 132: 311–21.

Niinemets, U. and Valladares, F. (2006) 'Tolerance to shade, drought and waterlogging of temperate northern hemisphere trees and shrubs', *Ecological Monographs*, 76(4): 521–47.

Read, H. (ed.) (2000) *Veteran Trees: A Guide to Good Management*, Natural England, Peterborough.

Repton, H. (1803) *Observations on the Theory and Practice of Landscape Gardening*, no publisher given.

Smit, C., Den Ouden, J. and Müller-Schärer, H. (2006) 'Unpalatable plants facilitate tree sapling survival in wooded pastures', *Journal of Applied Ecology*, 43: 305–12.

Vera, F.W.M. (2000) *Forest History and Grazing Ecology*, CABI, Oxon.

Woodland Trust (2009) *Ancient Tree Guide No 4: What Are Ancient, Veteran and Other Trees of Special Interest?*, Woodland Trust, Grantham.

27 Summary and conclusions

Ian D. Rotherham

Introduction

With the publication of this book we have brought the most up-to-date work on European grazed treescapes together in one volume. As is often the case, however, much remains unanswered and the chapters raise ever-more questions. However, we can now consider the roles and impacts of grazing and browsing herbivores in European wooded landscapes, both now and in the past. Chapters written with authority present insights into the history and biology of the animals themselves, the landscape and its carrying capacities, and the ecology and histories of the environments involved. Contributions from Rackham, Yalden and Vera for example, set the scene in terms of the landscape history, the known histories of the mammals involved, and the interpretation of the interactions between landscape and beast over time.

Furthermore, the contributions help us to consider from a position of sound information and insight the issues of new grazing schemes and approaches to the re-wilding of landscapes. Indeed, there is also the question raised as to whether we are really 're-wilding' or creating anew. These areas are cultural landscapes (Agnoletti *et al.*, 2007 and 2008) and modern-day re-wilding may be the latest layer of a long-term palimpsest of management. As such, as conservationists we should be careful not to damage or compromise important archaeology and aspects of historic and cultural interest (Figure 27.1).

Nevertheless, with changing conservation priorities, with the challenges of climate change, and with funding issues raised because of the new austerity, re-wilding of some areas may present unique and valuable opportunities. It is also increasingly clear that grazing was formerly much more widespread and the ending of its cultural utilisation has become a major threat to both landscape and ecology. Yet there are problems in terms of how humans have carved up the landscape resource and the consequent vulnerability of species and of other resources should areas be 're-wilded' but the animals constrained within relatively small areas. In a primeval landscape, species were able to move across wide tracts of connected ecosystems and so local

Figure 27.1 Conservation grazing in Devon with Exmoor ponies
Source: Ian D. Rotherham

extinction might be followed by re-colonisation. In a modern fragmented landscape this is simply not possible, and so local extinctions become permanent losses. This problem becomes even more serious if we consider ecosystem issues of 'wild' herbivores but an absence of large carnivores. The challenges are great but this is an exciting time for both researchers and practitioners attempting to address these issues.

It is also increasingly obvious that large areas of what are in effect woodpastures, grazed wooded landscapes, have survived to the present time. Many of these have been largely overlooked or at least not recognised for what they are. Research into landscape history across Europe is highlighting the importance of the cultural utilisation of many of these landscapes and this has often included grazing or related management such as pollarding, shredding or hay-making. Here, history, ecology and culture meet to provide a uniquely informative insight into how landscapes have been formed and thus how they might be managed in the future. Abandonment and cultural severance are not good for conservation of biodiversity or of landscape quality. An obvious question is why these sites have been overlooked or misunderstood when many have been subject to surveys and assessments, and often over many years.

Summary and conclusions 391

The appliance of science

Della Hooke (this volume) has noted that the reintroduction of large herbivores should only be undertaken with caution because of potential damage to the historic landscape and to archaeological features (Figure 27.2). For wildlife conservation too, these attempts should not be undertaken without due care and attention. These approaches are not cheap and nor are they necessarily a magic bullet to solve nature conservation problems. In particular, they will not by themselves resolve long-term issues of cultural severance. Unfortunately, many academics, practitioners and especially media pundits fail to understand the basic demands of issues such as site carrying capacity and suitability, of animal welfare and of so-called eco-tourism and its economics. Some of the views expressed are misleading and potentially damaging. Vidal (2005) in the *Guardian* newspaper ran a story entitled 'Wild herds may stampede across Britain under plan for huge reserves', which was potentially exciting but in its content sadly misleading. The suggestion that the high Pennine moors might support free-roaming herds of Hecke cattle and even of reindeer is ridiculous beyond belief. Furthermore, suggestions that this is a 'wild and natural' landscape is entirely misleading and that current upland farmers might be replaced in the economy by 'eco-tourism' shows zero understanding of basic tourism economics.

Figure 27.2 Goats grazing and browsing – habitat management, UK
Source: Ian D. Rotherham

Additionally, close observation of conservation sites and their management over a period of 30 years suggests to me that much current grazing management of moors, heaths, grasslands and wooded habitats is potentially very damaging to ecological interests. In North Yorkshire's Farndale there are ancient woodland sites now in severe decline due to both replanting with exotic species and intensive 'under-grazing' with domestic stock. These are not traditionally grazed wooded sites and the current regimes, though grant-aided, are deleterious. Even more worrying has been the imposition of year-round grazing by cattle for example, on species-rich pasture sites (Shire Brook Local Nature Reserve of Sheffield City Council and Woodhouse Washlands Nature Reserve of Yorkshire Wildlife Trust) due to enforced guidance by agencies administering stewardship grants for site management. These same sites were rescued from decline due to cultural severance and abandonment, by carefully targeted programmes of seasonal grazing by cattle. Indeed, over a period of around 20 years, the recovery of species-rich grassland was remarkably successful. The consequences of the recent mismanagement have included the local extinction of the species for which the reserves were originally established. On Blackamoor, an upland local nature reserve owned by Sheffield City Council but managed by Sheffield Wildlife Trust, the site has been damaged by grazing with cattle at inappropriate times of the year. One locally rare grassland plant, in part a reason the site was established as a reserve originally, has been virtually eradicated by summer grazing and the associated neutral grassland including several orchid species has been very badly degraded. The rare peat-bog flora, again a major reason for designation as a SSSI, has also been damaged and driven to localized extinction. Interestingly, the problematic areas of bramble and bracken-dominated encroachment appear in some areas to have worsened; again to the detriment of target conservation communities. These changes are again due to recent management guidance from National Government.

It is clear therefore that while grazing by large herbivores has historic and even prehistoric validity, it comes with a warning label of potential damage (Figure 27.3). Simply placing cattle on an otherwise unmanaged site and expecting everything to be good for conservation is naïve. There is major potential for conservation and habitat creation in targeted, preferably large, production sites. In other areas, the animals may need to be applied in ways that certainly in the 1980s and 1990s were accepted as standard conservation methods i.e. grazing in late summer, autumn and into the winter, and then removal of stock from sensitive sites during spring and summer. The catchall management prescription of year-round gazing that Natural England has been applying to many sites is disastrous for many areas. There is then a danger that the current fashion will backfire and the more creative and positive applications of grazing stock may be seen as problematic. One real problem is that many of the modern grazing schemes are not rooted in any understanding of site history or of the complexities of stock management and controls in traditional sites. Historically, different species were applied to sites seasonally and sequentially, and the whole system was subject to long-term cultural controls. People whose livelihoods and even survival depended on the sustainability of the resource closely monitored impacts of grazing and other utilisation, and if things went wrong, they intervened. The other major difference, and therefore a significant challenge, is that the earlier grazing treed ecosystems were extensive systems in extensive landscapes. Today's sites are mostly small (relatively) and isolated one from another. Over-grazing can and does simply lead to extinction of key species with little hope of re-colonisation. In the original landscapes it is suggested that the extensive grazing and fluid nature of the ecosystems allowed for survival, for patchworks of low-grazing impacts, and for easy movement within and between sites. All this has changed.

Figure 27.3 Sheep grazing on heath
Source: Ian D. Rotherham

The removal of the evidence

In writing about heathlands and their origins (Rotherham, 2009a, b and c), I have drawn attention to the upland–lowland divide with particular reference to the UK. This is a reflection of landscapes and the resistance to cultivation and 'improvement' of the great mass of northern upland landscapes and vegetation communities. As enclosures and drainage swept through the English lowlands in the 1700s and 1800s, much of the uplands remained relatively intact. They were not unchanged, as drainage and intensive use for sheep and grouse had major impacts, but they stayed stolidly resistant to removal. In the lowlands, the once extensive and expansive moors, fens, bogs, marshes, woods, forests and commons spiralled into a terminal decline from which there has been no recovery. A consequence of the parliamentary enclosures was a separation in fates between the upland and lowland situations. With a few examples and exceptions where perhaps forest or parks prevailed, the extensive medieval landscapes and, I suggest, their 'Vera-esque' connections were almost entirely removed (see Vera, 2000). In the uplands, though becoming drier and ecologically simplified by drainage and land 'improvement', the ecological landscape retains more integrity and therefore more connectivity to the past.

In the search for the shadows of the primeval or the medieval, and for our wooded landscapes outside the 'woods', it is here that we must look (Figure 27.4). The evidence in upland heath, moor, bog and sheep walk, and in lowland heath, common and ancient grassland is there to be found; and many of these are or have been grazed 'wooded' landscapes. Evidence of woodland connections and past ecologies are provided by patches of indicator plants in remote areas and often at relatively high altitudes. The same sites are marked out by woodland soils and often by veteran worked

Figure 27.4 Grazed upland shadow wood, Peak District, England
Source: Ian D. Rotherham

trees. Holistic approaches to site assessment are provided by the *Woodland Heritage Manual* (Rotherham *et al.*, 2008) by Rotherham (2011a), and for indicators specifically by Glaves *et al.* (2009a, b and c).

Looking for indicators and identifying woods

Woodland indicators can provide a vital clue in tracing anciently wooded landscapes and even in identifying breaks in long-term management and past episodes of say agricultural use of an otherwise wooded landscape. This then becomes a useful way to link our understanding of history to past site management. Part of our current research has been to investigate the uses of ancient woodland indicator species; something that has previously been done largely to confirm the ancient status of medieval coppice-wood sites. However, I now argue that this approach can be enormously helpful in the understanding of the wider array of wooded and grazed wooded landscapes.

In Britain, 'ancient' woodland (Peterken, 1981 and 1996; Rackham, 1976, 1978, 1980, 1986 and 2004) is that which has existed since at least AD1600, and possibly much longer. Prior to this date, planting of woodland in Britain was very uncommon. In this case, if a wood was present in 1600, it is likely to have been there for some considerable time previously. It has even been argued that it may be a remnant of the original 'wildwood' that once covered most of Britain, though this idea is generally considered untenable (see Beswick and Rotherham, 1993; Pigott, 1993). However, assuming that ancient woods have a significant degree of antiquity, the continuity of woodland cover at a site will have provided refuge for a great variety of plants and animals over the centuries. While this is the case within the wood, there will have been major changes in the surrounding landscapes. Consequently, ancient woods are often very rich in wildlife and have undisturbed soil profiles and natural water features.

Della Hooke (this volume) has argued that the historic landscape and archaeology must be taken into account when considering re-wilding or other related conservation management. Furthermore, Rackham (e.g. 1986) has provided robust evidence of the importance of ancient woods in terms of their unique heritage related to past management. Ancient woodland can also provide a living record of past woodland management practices and the organisation of the landscape. This is through the presence of features such as wood-banks, old pollards and coppice stools, remnant charcoal pits, ore furnaces and kilns (Rotherham *et al.*, 2008). In terms of standard assessment approaches there have been two broad types of ancient woodland identified in Britain: 1) that which has been continuously wooded since AD1600 and is composed of native tree species that have not obviously been planted – this is known as semi-natural ancient woodland; and 2) that which has been continuously wooded since AD1600 but where former tree cover has been replaced with planted trees (often conifers or exotic broadleaved species) – this is known as replanted ancient woodland.

Furthermore, it is only relatively recently that wood-pasture, as opposed to relict coppice-wood, has been considered as 'woodland' or indeed, of high nature conservation value. However, research emphasises very clearly that most evidence for ancient woodlands and the use of botanical indicators in particular relates to coppice-woods of various forms (Glaves *et al.*, 2009a, b and c). The formal recognition of the heritage value of the ancient trees still remains problematic. In Britain, ancient woodland (and particularly that which remains semi-natural) is considered the most important for wildlife conservation and a vital link to past landscapes and heritage. As discussed by Rotherham (2011a), identifying ancient woodland is therefore essential to:

1 promote appropriate management for such woods and their surrounding landscapes;
2 ensure the importance of ancient woodland is recognised where it is affected by planning proposals or other land-use changes;
3 encourage new woodland planting to take account of the needs of ancient woodland;
4 highlight possibilities for restoration (of replanted ancient woodland) to more 'natural' condition.

In this context, it is important to recognise the diversity of woodland types and origins that exists. The main forms of potentially ancient wooded landscape fall into the following categories or variants of these: mediaeval and industrial coppice, park and pasture-wood, ancient forest, wooded common, linear remnant and fragment (Rotherham, 2007a and b). The studies include ongoing research undertaken over a 20-year period and long-term action research with expert stakeholders (Rotherham, 2007c), together with detailed regional audits in the English Midlands and in Cumbria. A two-year project to develop the *Woodland Heritage Manual* (Rotherham *et al.*, 2008) led to a series of expert stakeholder workshops to examine and review issues of woodland indicators, woodland inventories and related landscape issues. The outcomes of the research included major reviews of literature and of evidence bases and approaches (Glaves *et al.*, 2009a, b and c), and a toolkit for practitioners (to be published elsewhere). The latter involves a decision-making tree and an evidence-based ancient woodland status grid (Rotherham, 2011a). The approach has resulted in a revisiting of the use and interpretation of indicator species and of inventory lists, and the advocacy of a more holistic evidence-based evaluation of wooded landscapes and woodland sites.

Given the conservation interest in ancient woods and old-growth forests, we need to consider the range of information sources for antiquity or ancientness in woods and forests. In particular, the recognition of grazed wooded landscapes, and of 'shadow woods', is seen as a priority. Recognition of the wider resource of ancient wooded landscapes provides a further impetus for this work and a challenge to the models. The information used in evaluation may be ecological, pedological, historical or archaeological. Glaves *et al.* (2009a, b and c) and the *Woodland Heritage Manual* (Rotherham *et al.*, 2008) describe in detail the possible approaches to gathering and interrogating information, and the key sources of the data. In undertaking surveys and in listing woods in ancient woodland inventories, botanical indicators of antiquity have played an important role, but this is inadequate for the broader applications.

A part of the story that emerges from this research, and an answer to some of the quandaries, may lie in exactly what we define first as 'woodland', and then as 'ancient'. Unfortunately, this is not always as simple as might be assumed, and most sites recognised through both indicators and historical sources are in fact medieval coppice-woods. This implies that many sites with different timelines and histories may be omitted when ancient status is assessed. The consequences for conservation and for management may be serious. There is a need for robust indicators and for these to be placed in a broader framework for integration and interrogation. Rotherham (2011a) suggests conclusions from these studies:

1   The need for robust and appropriate indicators: the classification of a site as an ancient woodland provides it with extra protection (at least under British planning policy guidance). Species evidence can and has been used to help determine ancient woodland character. However, there are concerns about the robustness of current indicator lists and their misuse. It was concluded from the studies that there is a need for robust and locally appropriate indicators to aid reliable identification of ancient woodland sites. Current lists generally do not take into account other factors that may influence the presence or absence of an indicator, for example internal variation within woodland, woodland size, soil acidity and wetness. Unless these factors are considered and accounted for, there is a tendency to identify base-rich woodlands (with greater diversity) as probable ancient woodlands and to omit more acidic sites of comparable antiquity. Ancient coppice-woods subjected at some point to prolonged and intensive grazing (such as for example, in the uplands of northern England) may also be omitted because they are species-poor. Analysis of local and regional trends and specific sites and datasets is needed to provide a richer context for interpretation.
2   Use of indicator species to confirm ancientness: an important conclusion from the study is that indicator species can be used to confirm historic records of woodland continuity and ancientness of a wood. However, if historic records are lacking for the particular location then woodland indicators can also 'indicate' ancient woodland status and a degree of confidence can be placed on such indication. This is based on an integrated interrogation and, in the absence of other information, does not necessarily confirm status. It was also noted that there

may be no individual species that can be shown to be confined to ancient woodland, and therefore no single perfect ancient woodland indicator. All species and communities of species must be assessed in the wider context of the location.

3   Perceptions and uses of the term 'ancient woodland': important in the context of this book, ancient woodlands are considered more valuable than non-ancient woodland. The 'age' or lineage of a wood is used as a surrogate for nature conservation or historical value. In other words the longer something has been there, the longer features will have accumulated: cultural artefacts, biodiversity etc. However, there is a need for better underpinning science of how strongly particular species are associated with woodland continuity on a site. This includes improved understanding of how the rates at which species can colonise woodland vary, and how quickly key species decline when conditions change. Issues related to this are discussed below. The heritage importance of the woodland archaeology and of worked trees also present problems of evaluation, of comparison and of recognition. Wood-pastures and other grazed woodland sites frequently have a different range of ground flora plant species and also different assemblages of heritage features.

4   Indicators of ancientness or indicators of continuity of environmental conditions and management: as noted, ancientness implies a long continuity of woodland and its associated environmental characteristics (generally including shade, high humidity etc.), though these relate mainly to ancient woods such as medieval coppices and there are notable exceptions. In open wood-pastures these conditions simply do not apply, though there may be pockets of such areas where topography or protection from intensive grazing allow, and Gulliver's chapter (this volume) on refuges is important in this context. It is in part this continuity of environmental conditions that determines the species present. Are we really talking about 'species associated with environmental continuity' or an 'index of ecological continuity' rather than indicators of ancient woodland? In addition, most if not all European woodlands will have been cleared to some extent in the past. Therefore, woodlands that appeared on early maps and again on later maps could have been clear felled and treated as arable land in the intervening period. It is uncertain how woodland indicators respond to such gaps in continuity, but recent work on Ecclesall Woods, Sheffield, is providing some insight into this process (Rotherham, 2011b). How long can a clearing remain unwooded and still be regarded as ancient woodland when the canopy returns. Woodland management has changed over history, and many but not all ancient woodland indicators are associated with dense enclosed high forest. However, in the past, other types of wooded environments including wood-pasture were common and there is a need for indicators for these other types of ancient treed and wooded environments.

Imposed on these ideas relating to ancient woodland are emerging concepts of '*wooded landscapes*' and of '*shadow woods*', and it seems that the roles of large grazing and browsing mammals are critical in both the pasts and the futures of these sites. These further complicate an already multi-faceted paradigm. However, to ignore or overlook these issues potentially devalues the whole system and puts many good sites at risk. A history of grazing can affect a contemporary site assessment based on indicators unless the wider wooded landscape context is better understood. The observations by Stephen Hall at Chillingham Park in Northumberland, England (this volume) are most informative in considering the long-term roles of large grazing herbivores in a specific landscape context. Moving from the great herbivores to wood-boring invertebrates, the importance of deadwood indicators for pasture-woods has been well established and Keith Alexander (this volume) explains the importance of this resource and its management.

The ebb and flow of woodland and grazing

Understanding large-scale and long-term landscape change is essential if we are to use the past to inform present-day management. In such a context the studies of Shaw and Whyte (this volume) provide an important and timely contribution. A Europe-wide context is also enhanced by contributions from Plieninger from the Mediterranean and Rupp from Germany. Indeed, detailed, focused case studies, such as that by Jones from South Yorkshire, England, prove very informative and together these individual researches give a coherent international overview. The overarching observations by Steer enhance this wider context.

One question to be answered is what has happened to the grazed wooded landscapes of Europe? In Britain and across Europe what became of the thousands of medieval parks, large and small, that dotted the landscape, and the truth is that for many of these we simply don't know the answer. It is assumed that most were disparked and converted to other uses, or in many cases they were subsumed into other wooded landscapes. However, the fate of the extensive non-parkland grazed woodlands is even more obscure. For some landscapes there are tantalising glimpses of their fate, and the approach that we have developed provides a fascinating insight into landscape change. Ecclesall Woods in Sheffield is the region's premier conservation woodland today, but detailed studies of deadwood indicators of ancient woodlands, undertaken in the 1980s, identified it as an anomaly. Ecclesall Woods lacks key species of invertebrates that its assumed antiquity would suggest that it should have. Following in-depth studies of field archaeology and archival research, however, the circumstances make eminent sense. Ecclesall Woods was open farmland with small areas of very wet and riverside woodland throughout a long period of the Late Neolithic and through the Bronze Age, Iron Age and Romano-British periods even until the late Saxon. Following the Norman Conquest, the lands changed hands and by AD1317 the area has its origins as a medieval hunting park. In 1317, Robert de Ecclesall was granted a licence to impark, and this is reflected in modern place-names such as Parkhead, Warren Wood, Park Field and Old Park (Hart, 1993). An overview of the issues of interpretation for this particular landscape is given in Rotherham and Ardron (2006). As noted by Hart (1993) there is further evidence of the use of the woods for hunting, with a set of depositions taken on 2 October 1587. These were from George Sixth Earl of Shrewsbury. He stated that he, his father and his grandfather:

> 'used sett and placed Crosbowes for to Kyll the Deare in Ecclesall Afforesaied and to hunte at all tymes when it so pleased them there.' *Thomas Creswick noted that* '{...}ye said Erle George grandfather to ye said now Erle of Shrewsbury hath sett Netts and long bowes to kill deare in Ecclesall and hunted dyvers tymes there and he thinketh that ye said Erle ffrancis father to ye Erle that now is did the lyke.' *Richard Roberts confirmed that* '{...}he hath sene the lord ffrancis hunting in Ecclesall byerlow and that said lords officers sett decoers there at such places as they thought convenyent'.
>
> (Hart, 1993)

In the early 1700s, there were also livestock pastured in the woods with horses, mares, foals, cows, heifers, calves and sterks recorded. Gelly's map of 1725 shows a 'laund' in the centre of the woods and this was planted up in 1752 (Jones and Walker, 1997). In the 1587 deposition (Hart, 1993), it is also clear that wood and underwood are also being taken, and it was this use that was to dominate the former deer park for the next few centuries. It seems despite references to deer hunts from the late 1400s and early 1500s, perhaps its use for hunting was falling from fashion by the late 1500s. Was this the reason for the deposition? Excitingly, in the late 1990s, Paul Ardron, working with the author, located the western boundary bank of the medieval park (Rotherham and Ardron, 2001). Here we have some insight into the evolution of a specific wooded landscape, for which the medieval imparkation was probably the critical moment in it becoming woodland today. However, this 'ancient' woodland is not all it seems, and its ecology and pedology reflect its unique history. From the 1500s onwards, the individual coppices were named 'woods' and being exploited for intensive manufacture of charcoal and whitecoal. By the mid-1800s, the coppice exploitation ended and gradually the woodland was converted to high forest with planted exotic trees. In the twentieth century it was then largely abandoned as 'amenity woodland'. This site is now locked within a sea of urbanisation and separated from its past by the process of 'cultural severance' (Rotherham, 2008 and 2010). However, the key issue is that for long periods of time this site was mostly unwooded and included large areas of arable land, and for much of the rest of its history it was grazed parkland.

Today, culturally severed from its working past and managed as an urban amenity space, it is rapidly becoming 'parkified' and aside from occasional deer and rabbits, there is no grazing. However, red deer are now re-colonising (Rotherham and Derbyshire, 2012) and it remains to be seen how either people or ecology respond to this uncontrolled 're-wilding'.

We see similar processes at across the English landscape, as by the late nineteenth and early twentieth centuries, many great houses, parks and gardens were subject to neglect or became financial liabilities. In the 1950s, even famous and now highly valued locations for wood-pastures, such as Chatsworth Park in Derbyshire, were seriously considered for demolition. Many smaller houses and their parks have long since gone. Other imposed parks on farming landscapes, such as Oakes Park at Norton (formerly north Derbyshire), are now among the richest ecological sites in their region. However, despite the well-documented conservation value, they lie uncared for and neglected, social misfits in landscapes of urban sprawl. The losses and severance of the landscape lineage is beyond calculation, and the more so for genuinely medieval parks. The loss of Ongar Great Park, Essex, a pre-Conquest survival, was possibly the worst loss of a visible Anglo-Saxon antiquity in the twentieth century (Rackham, 1986). So what have we left? The nineteenth-century clergyman and diarist, the Rev. Francis Kilvert gives some idea of the best sites, describing the ancient oaks of Moccas Park, Herefordshire: 'grey, gnarled, low-browed, knock-kneed, bowed, bent, huge, strange, long-armed, deformed, hunchbacked, misshapen, oakmen with both feet in the grave yet tiring down and seeing out generation after generation' (Plomer, 1938).

Conclusions

It is important when we read Kilvert's evocative description that we recognise that these trees only survived because the grazed woodlands, whether parks or other grazed wooded areas such as commons or heaths, were working lands. When their economically or socially driven functions changed, the sites and their trees were under threat. Parks and great trees may '*survive*' in new landscapes, housing or agriculture, but most are erased from land and memory. Even if the trees survive, there is no means to replace them as time and nature run their course. Therefore, while the remaining sites are conservation icons, they are often isolated in time and space. These sites and their trees may possess unique ecological resources of lichens, bryophytes, insects, spiders and more. Enmeshed with a cultural lineage from the great forests of north-western Europe the ancient parks are absolutely irreplaceable conservation resources (Figure 27.5). Not only are these wonderful ecological features, but they are also remarkable monuments of nature and to past human communities too. The trees and their landscapes, and indeed the trees 'in' their landscapes, are both dramatic and iconic, but also of increasing economic value as tourism attractants too.

Figure 27.5 Ancient oak
Source: Ian D. Rotherham

The great trees of Sherwood or Chatsworth contribute enormously to the history and the myths that draw tourists from around the world.

However, we are now discovering that beyond the park pale is a further resource but one largely overlooked and unrecognised. Research across Europe by Read (1999), by Butler and by Green (this volume) has highlighted the previously overlooked legacy of mostly retired working trees and just a few areas of still working veterans. But even more than this, we are now searching for shadows and ghosts of grazed woodlands past. Similar to the shadow woods I have already described, Quelch and colleagues have found extensive areas of upland wood-pastures in northern England and in Scotland. Furthermore, the importance of the Scottish pinewoods has long been understood. All these are landscapes driven, at least in the past, by grazing.

Looking forwards, conservation policy makers and land managers are seeking to conserve sites that are known and to create new sites to replace in part what we are bound to lose through continuing processes of abandonment, destruction and just natural decay. Chapters by Perry and by Butler and Alexander (this volume) give examples of strategies to take this work forwards. Other contributions present case studies of either site assessment or equally importantly, of ongoing implementation projects (such as by Gareth Browning and John Gorst from Cumbria). Work by Samojlik and Kuijper, by Hall (e.g. 2007) by Buttenschøn and Buttenschøn and by Newton *et al.* (this volume) are assisting managers in understanding how forests and grazed landscape have functioned in the past and how such processes can be harnessed today. Similarly, Boström *et al.* give insights into relationships between grazing management and fire risk in fire-prone forests. The contribution by Kirby and Baker to this book provides an important link between academic research and practical implementation, and Iris Glimmerveen also considers future potential of grazed woodlands. Finally, an important part of the debate on the re-wilding of large-scale landscapes and the reversion to more naturalistic management systems often implies a hands-off approach to allow large mammal to range freely to drive the ecology. The ongoing studies of Goulding (reported in this volume) present challenges in terms of a large mammal now establishing in wooded landscapes across Britain. Should the wild boar be managed or might we allow them to roam free? Moreover, of course, who will decide? We can demonstrate from examples and studies across Europe that large grazing mammals were important in many wooded landscapes. It remains to be seen how significant they might be in the future.

Bibliography and references

Agnoletti, M., Anderson, S., Johann, E., Kulvik, M., Saratsi, E., Kushlin, A., Mayer, P., Montiel, C., Parrotta, J. and Rotherham, I.D. (2007) *Guidelines for the Implementation of Social and Cultural Values in Sustainable Forest Management: A Scientific Contribution* to the Implementation of MCPFE – Vienna Resolution 3, IUFRO Occasional Paper No. 19, ISSN 1024-414X, IUFRO Headquarters, Vienna, Austria.

Agnoletti, M., Anderson, S., Johann, E., Kulvik, M., Saratsi, E., Kushlin, A., Mayer, P., Montiel, C., Parrotta, J. and Rotherham, I.D. (2008) 'The introduction of historical and cultural values in the sustainable management of European forests', *Global Environment*, 2: 172–93.

Beswick, P. and Rotherham, I.D. (eds) (1993) 'Ancient Woodlands – their archaeology and ecology – a coincidence of interest', *Landscape Archaeology and Ecology*, 1.

Glaves, P., Rotherham, I.D., Wright, B., Handley, C. and Birbeck, J. (2009a) *A Report to the Woodland Trust. The Identification of Ancient Woodland: Demonstrating Antiquity and Continuity – Issues and Approaches*, Hallam Environmental Consultants Ltd., Biodiversity and Landscape History Research Institute, and Geography, Tourism and Environment Change Research Unit, Sheffield Hallam University, Sheffield.

Glaves, P., Rotherham, I.D., Wright, B., Handley, C. and Birbeck, J. (2009b) *A Report to the Woodland Trust. A Survey of the Coverage, Use and Application of Ancient Woodland Indicator Lists in the UK*, Hallam Environmental Consultants Ltd., Biodiversity and Landscape History Research Institute, and Geography, Tourism and Environment Change Research Unit, Sheffield Hallam University, Sheffield.

Glaves, P., Rotherham, I.D., Wright, B., Handley, C. and Birbeck, J. (2009c) *A Report to the Woodland Trust. Field Surveys for Ancient Woodlands: Issues and Approaches*, Hallam Environmental Consultants Ltd., Biodiversity and Landscape History Research Institute, and Geography, Tourism and Environment Change Research Unit, Sheffield Hallam University, Sheffield.

Hall, S. (2007) 'Chillingham Wild Cattle Park, Northumberland', *Landscape Archaeology and Ecology*, 6: 53–7.

Hart, C.R. (1993) 'The ancient woodland of Ecclesall Woods, Sheffield', in Beswick, P. and Rotherham, I.D. (eds) 'Proceedings of the National Conference on Ancient Woodlands: Their Archaeology and Ecology – A Coincidence of Interest, Sheffield 1992', *Landscape Archaeology and Ecology*, 1: 49–66.

Jones, M. and Walker, P. (1997) 'From coppice-with-standards to high forest: the management of Ecclesall Woods 1715–1901', in Rotherham, I.D. and Jones, M. (eds) 'The Natural History of Ecclesall Woods, Pt 1', *Peak District Journal of Natural History and Archaeology Special Publication*, 1: 11–20.

Peterken, G.F. (1981) *Woodland Conservation and Management*, Chapman and Hall, London

Peterken, G.F. (1996) *Natural Woodland: Ecology and Conservation in Northern Temperate Regions*, Cambridge University Press, Cambridge.

Pigott, C.D. (1993) 'The history and ecology of ancient woodlands', *Landscape Archaeology and Ecology*, 1: 1–11.

Plomer, W. (ed.) (1938) *Kilvert's Diary, 1870–1879*, Jonathan Cape, London.

Rackham, O. (1976) *Trees and Woodland in the British Landscape*, J.M. Dent and Sons Ltd, London.

Rackham, O. (1978) 'Archaeology and land-use history', in Corke, D. (ed.) 'Epping Forest – the Natural Aspect?', *Essex Nat., N.S.* 2: 16–57.

Rackham, O. (1980) *Ancient Woodland; its History, Vegetation and Uses in England*, Arnold, London.

Rackham, O. (1986) *The History of the Countryside*, J.M. Dent and Sons Ltd, London.

Rackham, O. (2004) 'Pre-existing trees and woods in country-house parks', *Landscapes*, 5(2): 1–16.

Read, H. (1999) *Veteran Trees: A Guide to Good Management*, English Nature, Peterborough.
Rotherham, I.D. (2007a) 'The historical ecology of medieval deer parks and the implications for conservation', in Liddiard, R. (ed.) *The Medieval Deer Park: New Perspectives*, Windgather Press, Macclesfield, pp79–96.
Rotherham, I.D (2007b) 'The ecology and economics of medieval deer parks', *Landscape Archaeology and Ecology*, 6: 86–102.
Rotherham, I.D. (2007c) 'The implications of perceptions and cultural knowledge loss for the management of wooded landscapes: A UK case-study', *Forest Ecology and Management*, 249: 100–15.
Rotherham, I.D. (2008) 'The importance of cultural severance in landscape ecology research', in Dupont, A. and Jacobs, H. (eds) (2008) *Landscape Ecology Research Trends*. Nova Science Publishers Inc., New York, pp71–87.
Rotherham, I.D. (2009a) 'Hanging by a thread: A brief overview of the heaths and commons of the north-east midlands of England', in Rotherham, I.D. and Bradley, J. (eds) *Lowland Heaths: Ecology, History, Restoration and Management*, Wildtrack Publishing, Sheffield, pp30–47.
Rotherham, I.D. (2009b) 'Habitat fragmentation and isolation in relict urban heaths: The ecological consequences and future potential', in Rotherham, I.D. and Bradley, J. (eds) (2009) *Lowland Heaths: Ecology, History, Restoration and Management*, Wildtrack Publishing, Sheffield, pp106–15.
Rotherham, I.D. (2009c) 'Cultural severance in landscapes and the causes and consequences for lowland heaths', in Rotherham, I.D. and Bradley, J. (eds) (2009) *Lowland Heaths: Ecology, History, Restoration and Management*, Wildtrack Publishing, Sheffield, pp130–43.
Rotherham, I.D. (2010) 'Cultural severance and the end of tradition', *Landscape Archaeology and Ecology*, 8: 178–99.
Rotherham, I.D. (2011a) 'A landscape history approach to the assessment of ancient woodlands', in Wallace, E.B. (ed.) *Woodlands: Ecology, Management and Conservation*, Nova Science Publishers Inc., USA, pp161–84.
Rotherham, I.D. (2011b) 'Animals, man and treescapes – perceptions of the past in the present', in Rotherham, I.D. and Handley, C. (eds) (2011) *Animals, Man and Treescapes: The Interactions between Grazing Animals, People and Wooded Landscapes*, Wildtrack Publishing, Sheffield, pp1–32.
Rotherham, I.D. and Ardron, P.A. (eds) (2001) *Ecclesall Woods Millenium Archaeology Project*, Sheffield Hallam University, Sheffield.
Rotherham, I.D. and Ardron, P.A. (2006) 'The archaeology of woodland landscapes: Issues for managers based on the case-study of Sheffield, England and four thousand years of human impact', *Arboricultural Journal*, 29(4): 229–43.
Rotherham, I.D. and Derbyshire, M.J. (2012) 'Deer in and around the Peak District and its urban fringe', *British Wildlife*, 23(4): 256–64.
Rotherham, I.D. and Handley, C. (eds) (2011) *Animals, Man and Treescapes: The Interactions between Grazing Animals, People and Wooded Landscapes*, Wildtrack Publishing, Sheffield.
Rotherham, I.D., Jones, M., Smith, L. and Handley, C. (eds) (2008) *The Woodland Heritage Manual: A Guide to Investigating Wooded Landscapes*, Wildtrack Publishing, Sheffield.
Vera, F.W.M. (2000) *Grazing Ecology and Forest History*, CABI Publishing, Oxon, UK.
Vidal, J. (2005) 'Wild herds may stampede across Britain under plan for huge

reserves', *Guardian*, 27 October.

Yalden, D.W. (1999) *The History of British Mammals*, T. & A.D. Poyser, London.

Yalden, D.W. (2001) 'Mammals as climatic indicators', in Brothwell, D.R. and Pollard, A.M. (eds) *Handbook of Archaeological Sciences*, Wiley, Chichester, pp147–54.

Yalden, D.W. (2003) 'Mammals in Britain: A historical perspective', *British Wildlife*, 14: 243–51.

Index

Act of Commons 5, 82
agricultural abandonment 218, 239
agri-environment scheme 92, 94, 250, 297, 368
agroforestry 97, 180–8, 222, 251, 253, 368, 370, 373
Ancient Tree Forum ix, 128, 135, 138, 358, 365, 370, 371, 378
ancient wood-pasture 16, 44, 341, 374
ancient wood 13, 34, 74, 78, 82, 122, 395, 396, 398
ancient woodland xi, 3, 6, 9, 13, 19, 22, 34, 49, 60, 68, 74, 75, 78, 80, 84–86, 123, 128, 141, 165, 178, 225, 236, 258, 272, 300, 317, 318, 325, 352, 373–4, 392, 395–400, 404–5
animal welfare legislation 94
archival research 55, 57, 399
arolla pine 140
ash xi–xiii, xvi, xvii, 11, 21, 42, 106–7, 112, 120, 129, 245, 247, 282, 296–7, 342, 345, 348, 349, 382–4
Ashtead Common 16
Asian wild ass ix, 108
aspen xiv, xviii, 68–9, 194–196, 198–9, 201–2, 204–207, 300, 319–21, 323, 349
aurochs 36, 64–69, 89, 96, 98–100, 103, 107–8, 143–4, 147–8, 154, 159, 162–3

beaver 36, 67–8, 163
beech xiii, 42, 51, 59, 60, 88–9, 106–7, 112, 122–3, 125, 127, 129, 165, 167, 170, 175, 246–7, 257, 266, 293, 319, 323, 328, 331, 336, 347, 360, 365, 370, 371, 382
beech-mast 42, 51
beechnuts 113
beetle xiii, 10, 40, 80, 88–9, 92, 98, 116, 125, 127, 129–30, 179, 330–4, 336–8, 360, 374–5
belted Galloways 47
Berberis vulgaris xv, 289–90
bird cherry 112, 196–198, 201
Białowieża Primeval Forest vi, xiv, 95, 143–4, 146, 149, 158–60, 266
bison xiv, 63, 65, 67, 90, 99, 103, 119, 142, 147–8, 153, 155, 157, 159–60, 163
Bison priscus 63
Black Death 27, 78, 140
Blackamoor 392
blackthorn xiii, 6, 16, 80, 104, 108–10, 113, 120, 145, 159, 164, 169–71, 175–6, 297, 384–5
blossom 309, 331
bluebell 6, 47, 80, 258, 262
bracken xi, 12, 20, 46, 245, 276, 279, 289, 366, 392
Bradgate Park 16, 77
bramble xvii, 6, 46, 62, 80, 169, 170–1, 349, 362, 384, 385, 392
Bronze Age 36, 39, 90, 269, 399
browsing herbivores 8, 283, 389
Burnham Beeches 360, 365, 370

Caledonian forest 228
Cannock Chase 43
Carpathian Basin xii, 52, 53
charcoal 31, 44, 47, 49, 135, 152, 209, 302, 346, 395, 400
Chatsworth Park 82, 401
Chillingham Park vii, 80, 242, 247, 251–3, 399

408 *Index*

Chillingham Wild Cattle Association 243, 253
climate change 6, 54, 57, 59, 93, 97, 175, 180, 225, 251, 270–1, 341, 373–4, 389
climax 102–3, 208, 223–4, 226–7, 302
conservation grazing xvii, xviii, 343–5, 350, 390
conservation strategies 5
Corylus avellana 122, 124, 168, 206
Cotswolds 384
crab apple 169–71, 322
Crane-fly 331
Crataegus laevigata 322
Crataegus monogyna xv, 108, 145, 167–9, 215, 282, 289, 321
cultural ecology 3, 72, 226
cultural landscape 87, 92, 94–6, 100–1, 103, 119, 167–9, 215, 225, 232, 236, 238, 240, 248, 330, 333–5, 337, 356, 374, 383, 389

Danish Forest Act 317
deer park xii, 8, 9, 24, 27–34, 73, 78, 80, 83, 85–6, 293, 355, 372, 383, 400, 405
Denmark ix, xiii, 59, 64–6, 68, 105, 121–2, 145, 177, 317, 329
dog's mercury 6, 80
Dark Peak 25
deadwood vii, xiii, 82–3, 91, 129–30, 314, 330–5, 337, 339, 356, 360, 362, 372–3, 385, 399
dendroecology 55
Domesday Book 3, 16, 43, 49, 51, 53, 85
Dorset 38, 96, 256
downy currant 196–7, 199, 201, 205
dung 80, 89, 114, 166–7, 217–9, 295, 314, 316–7, 322, 324–5, 328, 329, 344, 350

Ecclesall Woods 31, 33, 398, 399, 404, 405
ectomycorrhiza 80
ecosystem services 185, 209, 220, 232–3, 240, 358, 372
Ennerdale Valley vii, 46, 269, 270
Epping Forest 8, 16, 47, 85, 90, 96, 98, 331, 334, 338, 365, 370
Equus przewalski 99, 108,163
EU Habitats Directive 221, 249, 361

eucalypts 19
Euonymus europeus 112
grazed landscapes i, xviii, 3, 5, 75–6, 145, 180, 220, 224, 284
European Strategy for the Conservation of Invertebrates 332, 337
extinction rate 96, 332

fagot 134–5
fallow deer xii, 43, 77, 165, 258–9
Farndale 392
felling xi, 11, 13, 16, 34, 46, 66–7, 134, 243, 245, 293, 363
fire vi, 7, 16–7, 19, 44, 49, 51, 57–8, 72, 87, 90, 96, 144, 152, 161, 165, 173–4, 180–8, 193, 202, 208–10, 289, 346, 347, 366, 392, 403
FireSmart 180–1, 183–5, 187–8
fodder bundle 134
Forest of Wyre xii, 47
Fraxinus excelsior 106, 167–8, 282, 313
Fraxinus ornus 215
fringe and mantle 342, 350
funding issues 389
fungus gnat 331

Galloway cattle vii, 269, 350
Germany vii, x, xiii, xiv, xvi, xvii, 63, 66, 102, 104, 106, 109, 126, 145–6, 157, 252–4, 262, 267, 301–3, 307, 310, 331, 337, 353–4
glacial refuge 87, 89, 118
Glen Affric vii, xiv, 223, 226–31, 234, 237, 238, 240–1,
goat xvii, 42–3, 66, 87, 144, 192, 194, 217
golden oriole 92
gorse xv, 28, 39, 67, 165, 169–72, 289, 297
grazed upland shadow wood xii, xvii, 79, 394
Grazing Advice Partnership 365
Grazing Animals Project 365, 374
grazing mammals 62, 63, 66, 76, 80, 128
Grazing Refuge Habitats vi, xvii, 190–5, 197–203, 205, 207
great yellow gentian 109
Green Monuments Campaign 139
grove vi, xiii, 36, 68, 109, 110, 113–6, 139, 140, 164, 355
guelder rose 112, 196, 199, 201, 349

Hatfield Forest xi, 15, 17, 112, 370, 384
hawthorn xii, xiii, xv, xvi, 16, 29–30, 79, 82, 104, 108, 110, 111, 113, 145, 167, 169, 171, 282, 285, 289, 296–7, 322, 331, 347–9, 362, 382, 385, 387
hay 3, 8, 39, 43, 58, 79, 82, 88, 98, 101, 134–5, 152, 233, 234, 243–4, 246, 248, 249, 285, 294, 296–9, 342, 346, 347, 355, 384, 390
hay-making 390
heath xi, xvii, 5, 7, 8–9, 12, 16, 19–21, 28, 38, 72–4, 76–8, 80, 82–3, 85–6, 91–2, 100, 103–5, 120, 128, 142, 177, 178, 227, 240, 272, 329, 339, 356, 385, 393,–4, 401, 405
heather 62, 165, 171–2, 246, 248, 289
heathland 47, 163, 165–6, 169, 170, 172–5, 178, 223, 229, 236, 238, 245, 257, 267, 278–9, 317, 336, 356, 379, 384
Hedgerow Campaign 139
high forest hypothesis 143
Higher Level Stewardship Scheme xviii, 249, 369, 370
Highland cattle xvi, 46, 275, 351
hogweed 331
honeysuckle 320
hornbeam xiv, 106, 112, 127, 135, 149–51, 158, 167, 266
Humberhead Levels 25
Hungary x, xii, 52–6, 59, 61
identifying woods 395

Ilex 67, 77, 167–9, 172, 174, 196, 206, 314, 315, 360
inbreeding 244, 252
indicators 5, 9, 44, 69, 83, 89, 221, 252, 253, 395–9, 406
indicators of ancientness 398
indicators of continuity 398
Industrial Revolution 232, 302
Institute of Terrestrial Ecology 8, 84, 96, 121, 205, 238, 332, 337, 372
Irish elk 63, 64
IUCN European Red List of Saproxylic Beetles 330
ivy 67, 129, 135

jay 110–3, 120, 127, 139–40, 384
juniper 17, 65, 195–7, 200, 201–3, 205, 207, 215, 246, 248, 313–4, 321, 325, 349, 382, 385
Juniperus 17, 65, 195–6, 207, 215, 313, 321, 325

Kashmir 282
Kilvert, Rev Francis 368, 401, 404
Knepp Castle 94, 365, 370

Lake District xv, xvi, 197, 203, 269, 274, 282–3, 287, 289, 294, 296, 297, 384
Lake District National Park 269
land improvement 40, 83, 232, 393
Late Glacial 63, 64–5, 67
launds 28–30, 77
leaf hay 342, 346–7, 355
light demanding xiii, 105–6, 112–3, 115, 118–9, 127–8, 156, 158, 176, 291–3, 321, 331, 382–3, 385–6
lime 20, 88, 106, 112, 150, 158, 293, 294, 349, 382
Lincoln Red cattle 250
longhorn cattle 47, 250
Lonicera periclymenum 321, 325
lowland heath 9, 38, 78, 85–6, 128, 142, 174, 177–8, 394, 405

Malus sylvestris xvi, 115, 167, 168, 319, 321–2, 325, 329
mantle and fringe xiii, 104, 109, 111–2,116
mast xii, 42, 51–2, 54–6, 59–61, 110, 125, 165, 210, 257, 265, 385
medieval parks 9, 27–8, 44, 73–8, 84, 142, 336–7, 356, 399, 401
Mesolithic 65, 67, 68, 87, 96, 201
Moccas Park xii, 45, 365, 401
moorland 25, 38–9, 78, 97, 193, 202, 238, 240, 246, 385
moose xiv, 36, 147–8, 155, 157, 161
Morocco 282
mountain hare 62, 64, 68
muntjac deer 384
muskox 63
mycorrhiza xvii, 45, 72, 80, 136, 142, 291, 380–1
myxomatosis 127, 139, 291

National Ecosystem Assessment 268, 374
National Park Dalby Söderkog xiii, 107
National Tree Week 139
Natura 2000 Network 334–5
Natural England x, 91–2, 95, 97, 166, 232, 253, 270, 355–6, 359, 363–5, 370–1, 373, 375, 388, 392
Natural Environment and Rural Communities Act 358, 363
Natural Environment White Paper 367, 372
natural landscape xii, 45, 80, 87.91, 92, 94, 99, 101, 174, 232, 236
naturalistic grazing 8, 49, 93–6, 164, 177, 300, 372
Nature Conservancy Council 97, 332, 359, 371–3
Nature Conservation Review 97, 331, 337
Neolithic vi, 19, 60, 66–8, 87, 89, 90–7, 100, 239, 294, 399
nettle 109, 291
New Forest vi, xiii, xvii, 67–8, 90, 96, 115, 121, 123, 125, 134, 143–4, 152, 161–8, 170, 173–9, 249, 254, 316, 331, 360, 369
New Forest National Park 165
noble chafer beetle 92
nutcracker 140

old-growth forest 146, 158, 396
olive 187, 212
Ongar Great Park 401
Oostvaardersplassen 93, 95, 98, 143, 164, 179, 241
open-crowned tree 379, 385, 386
open-grown tree xviii, 5, 49, 113, 116, 127–9, 131, 246, 332, 336, 337, 345, 364, 385
Open Woodlands Through Pasture vii, 301, 303, 305, 307, 309, 311, 313, 315
Ovibos moschatus 63

Pakistan 282
palaeoecology 84, 95, 224–7, 235
palynology 224, 238
parliamentary enclosure xvi, 26, 83, 297, 394
pasture-woodland 3, 8, 74, 75, 84, 96, 119, 121, 177, 253, 337, 372

pats 322, 324, 325
pedunculate oak vi, xii, 7, 106–7, 115, 195, 201, 203–4, 317, 382
phoenix regeneration 383, 385
plains xi, , 15, 28, 30, 77, 360
pollard xi, xii, xiii, xvi, 7, 15, 16, 27, 29, 31, 44–5, 69, 74, 83, 132–5, 245, 296–7, 336, 339, 346, 349, 364, 368, 372, 374, 384–5, 395
pollarding 12, 82, 131, 134, 141, 342, 347–8, 366, 370, 373, 390
pony 171, 249
Populus alba 168
Populus tremula xiv, 68, 194,–6, 205, 207, 319, 321, 325
post-glacial v, 62, 63, 65, 67, 69, 87–8, 90, 143, 163, 224, 225–7, 229, 237
Pre-Neolithic vi, 19, 87, 89–7, 239
Przewalski horse 108
Primula elatior 20
privet 112
Prunus padus 112, 196–7, 205
Prunus spinosa 108, 120, 145, 159, 164, 167–9, 297, 314, 321, 322
public safety 260

Quercus cerris 215
Quercus fusiformis 17
Quercus grisea xii, 18
Quercus petraea 106, 122, 195, 282
Quercus pubescens 215
Quercus robur vii, 7, 52, 81, 119–20, 124, 126, 145, 158, 167–9, 172, 174, 195, 317, 319, 321, 323, 325, 360

rainforest 19
Ratcliffe, Derek 91–2, 97, 331, 337
refuges 62, 64, 87, 89, 122, 191–3, 197, 202, 203, 205, 248, 398
refugia 63, 69, 118
regrowth
reindeer 63–5, 67, 391
replanted ancient woodland 395–6
re-wilding v, viii, 35–9, 41, 43, 45, 47–9, 93, 94, 96, 164, 178, 179, 228, 231, 239, 335, 378–9, 381, 385, 387, 389, 395, 401, 403
Ribblesdale vii, xv, 223, 226, 231–5, 237, 240
Richmond Park 367
rinderpest 16, 139

Road Traffic Accidents 261
rock whitebeam 197, 200, 201
roe deer xiv, 62, 64–7, 69, 103–4, 147, 148, 155, 157, 159, 163, 259, 318, 319
Romania 53, 344, 346
rook 140
Rosa canina 215, 322
Rosa dumalis 322, 323
Rosa rubiginosa 322, 323
rose 112, 169–71, 196, 199, 201, 322, 323, 349
Rothschild, Charles 92
Royal Forest 27, 74, 77, 152, 165, 356

saproxylic invertebrate 330, 332–7, 359, 368, 371–3
savanna xi, xii, 12, 16–22, 374
savannah 3–5, 7, 72, 73, 78, 84, 91, 108, 116, 127, 131, 140, 141, 253
Scots pine 72–3, 78, 84, 91, 108, 116, 127, 131, 140–1, 253
Scottish Rural Development Programme 344
scrub xv, xvi, 30, 38, 47, 64, 64, 97, 110–4, 128, 131, 169, 172, 175, 196, 204, 232, 233, 276, 278–280, 289, 296, 317–8, 328–9, 339, 340, 362, 366, 373, 379, 383–5
semi-natural ancient woodland 395
shade tolerant xiii, 89, 106–7, 112–3, 115, 118, 127–9, 288, 292–3, 323, 328, 331, 382, 386
shadow xiii, 89, 106–7, 112–3, 115, 118, 127–9, 288, 292, 293, 323, 328, 331, 382, 386
shadow wood vi, xii, xvii, 72, 75, 78–9, 82, 294, 394
sheep xiv, xv, xvii, 14, 20–1, 28, 29, 31–3, 36–7, 39–42, 46–7, 66, 73, 77, 87, 91, 93, 103–5, 120, 136, 144, 155, 159, 191–2, 194, 202, 213–4, 217, 231–2, 244–5, 252, 258, 263, 271–3, 286, 294, 296, 298, 314, 329, 341, 350, 384, 393–4
Sheffield i, x, xii, 2, 5, 9, 20, 25–7, 30–4, 85–6, 97, 124, 128, 131, 141–2, 188, 205–6, 234, 281, 329, 337, 371, 373, 375, 392
Sherwood Forest xv, 7, 288, 331, 359, 367, 371–2

shifting baseline syndrome 99, 100–1, 107, 123, 124, 239, 240
Shire Brook Local Nature Reserve 392
shredding 390
shreds 132, 134
silva 25, 25, 34
silva glandinosa 52
silva minuta 24
silva modica 24
silva pastilis 24, 25
silvopastoral system 209, 252
Slovakia 53
Slovenia xv, 209, 252, 282, 284, 289–90
Sorbus aria 168
South Yorkshire v, xii, 24–9, 31, 34, 40, 85, 399
spindle tree 112
Statute of Merton 5, 76, 82
Staverton xi, 12, 17, 44, 286, 359, 360, 374
steppe 62, 64, 98–9, 108, 143, 157
Strict Forest Reserves 334–5
Succisa pratensis 19

Tankersley Park xii, 29, 33
tarpan 66, 99, 100, 103, 107, 123, 143–4, 147–8, 154, 163
traditional management 56, 84, 178, 225–6, 232–6, 240, 366, 370
tree archaeology 132
Tree Council 139
Trees for Life 228, 231, 241
treescape iii, vii, x, xv, 2, 48, 77, 86, 128, 137–8, 142, 205, 234, 236, 277, 279–82, 29, 290, 292–3, 329, 331, 378, 389, 405
tundra 62–5, 99, 143

UK Biodiversity Action Plan 92, 197, 356–7
Ukraine 53
Ulex xv, 67, 168–9, 172, 289, 297
Ulmus crassifolia 17
upland-lowland divide 393

Vera Hypothesis 2, 225
vesicular-arbuscular mycorrhiza 72, 80
Veteran Tree Initiative 138, 370, 372
veteran tree xvii, xviii, 3, 9, 19–21, 44, 85, 91, 97, 116, 119, 138, 300,

333–4, 336, 339, 356–67, 369–72, 374, 378–9, 388
veteran worked tree   394
Viburnum opulus   112, 168, 205
von Cotta, Heinrich   101–2, 120

waste   27, 76–8, 80, 82, 83
Weald   42, 44, 48
whitebeam   115, 196–7, 200–1, 206
wild apple   110, 115
wild cherry   115
Wild Ennerdale   ix, 270–3, 275–7, 281
wild pear   115
wildfire   180–2, 188, 208, 210
wildwood   8, 10, 19–20, 24, 96, 98, 177, 179, 235, 300, 356, 372, 395
Windsor Forest   41, 360, 371
Windsor Park   131, 142, 367
wood decay invertebrate   vii, 330
wood-boring beetle   129
wooded landscape   vi, 2–5, 8, 9, 67, 72–9, 81–6, 88, 97, 100, 120, 160, 172, 174, 176–7, 236, 239, 251, 280, 300, 332, 389–90, 394–6, 399–400, 403, 405
Woodhouse Washlands Nature Reserve   392
Woodland Grazing Toolbox   350
Woodland Trust   ix, 352, 355, 359, 365, 375, 388, 404
wood-spurge   xi, 13
woolly mammoth   63–4, 69
worked trees   7, 394, 398
wood anemone   6, 47, 80, 262, 266
wood mouse   110, 112
woodland herbs   14
Woodland Heritage Manual   4, 9, 84, 86, 396–7, 405
worked trees   7, 394, 398

yellow archangel   6, 258
Yellowstone National Park   69, 90, 97, 283, 300
Younger Dryas   64–5

zebra   108

Printed in Great Britain
by Amazon